Multi-Carrier and Spread Spectrum Systems

to

my parents, my wife Miriam,
my daughters Sarah, Sophia, and Susanna
(K.F.)

my wife Susanna,
my sons Lukas and Philipp and my daughter Anna
(S.K.)

Contents

Foreword   xi

Preface   xiii

Acknowledgments   xv

   Introduction   1

1 Fundamentals   15
   1.1 Radio Channel Characteristics   15
      1.1.1 Understanding Radio Channels   15
      1.1.2 Channel Modeling   16
      1.1.3 Channel Fade Statistics   18
      1.1.4 Inter-Symbol (ISI) and Inter-Channel Interference (ICI)   19
      1.1.5 Examples of Discrete Multipath Channel Models   20
      1.1.6 Multi-Carrier Channel Modeling   21
      1.1.7 Diversity   22
   1.2 Multi-Carrier Transmission   24
      1.2.1 Orthogonal Frequency Division Multiplexing (OFDM)   25
      1.2.2 Advantages and Drawbacks of OFDM   30
      1.2.3 Applications and Standards   30
   1.3 Spread Spectrum Techniques   30
      1.3.1 Direct Sequence Code Division Multiple Access   34
      1.3.2 Advantages and Drawbacks of DS-CDMA   37
      1.3.3 Applications of Spread Spectrum   37
   1.4 Multi-Carrier Spread Spectrum   41
      1.4.1 Principle of Various Schemes   41
      1.4.2 Advantages and Drawbacks   43
      1.4.3 Examples of Future Application Areas   44
   1.5 References   45

2 MC-CDMA and MC-DS-CDMA   49
   2.1 MC-CDMA   49
      2.1.1 Signal Structure   49
      2.1.2 Downlink Signal   50
      2.1.3 Uplink Signal   51
      2.1.4 Spreading Techniques   51

	2.1.5 Detection Techniques	57
	2.1.6 Pre-Equalization	65
	2.1.7 Soft Channel Decoding	67
	2.1.8 Flexibility in System Design	72
	2.1.9 Performance Analysis	74
2.2	MC-DS-CDMA	83
	2.2.1 Signal Structure	83
	2.2.2 Downlink Signal	86
	2.2.3 Uplink Signal	86
	2.2.4 Spreading	86
	2.2.5 Detection Techniques	87
	2.2.6 Performance Analysis	87
2.3	References	90

3 Hybrid Multiple Access Schemes — 93

3.1	Introduction	93
3.2	Multi-Carrier FDMA	94
	3.2.1 Orthogonal Frequency Division Multiple Access (OFDMA)	95
	3.2.2 OFDMA with Code Division Multiplexing: SS-MC-MA	100
	3.2.3 Interleaved FDMA (IFDMA)	104
3.3	Multi-Carrier TDMA	105
3.4	Ultra Wide Band Systems	107
	3.4.1 Pseudo-Random PPM UWB Signal Generation	107
	3.4.2 UWB Transmission Schemes	109
3.5	Comparison of Hybrid Multiple Access Schemes	110
3.6	References	112

4 Implementation Issues — 115

4.1	Multi-Carrier Modulation and Demodulation	116
	4.1.1 Pulse Shaping in OFDM	119
	4.1.2 Digital Implementation of OFDM	119
	4.1.3 Virtual Sub-Carriers and DC Sub-Carrier	120
	4.1.4 D/A and A/D Conversion, I/Q Generation	120
4.2	Synchronization	123
	4.2.1 General	125
	4.2.2 Effects of Synchronization Errors	126
	4.2.3 Maximum Likelihood Parameter Estimation	129
	4.2.4 Time Synchronization	132
	4.2.5 Frequency Synchronization	136
	4.2.6 Automatic Gain Control (AGC)	139
4.3	Channel Estimation	139
	4.3.1 Two-Dimensional Channel Estimation	140
	4.3.2 One-Dimensional Channel Estimation	143
	4.3.3 Filter Design	144
	4.3.4 Implementation Issues	145
	4.3.5 Performance Analysis	147
	4.3.6 Time Domain Channel Estimation	151
	4.3.7 Decision Directed Channel Estimation	152

	4.3.8 Blind and Semi-Blind Channel Estimation	153
	4.3.9 Channel Estimation in MC-SS Systems	154
	4.3.10 Channel Estimation in MIMO-OFDM Systems	158
4.4	Channel Coding and Decoding	158
	4.4.1 Punctured Convolutional Coding	159
	4.4.2 Concatenated Convolutional and Reed–Solomon Coding	159
	4.4.3 Turbo Coding	162
	4.4.4 OFDM with Code Division Multiplexing: OFDM-CDM	166
4.5	Signal Constellation, Mapping, Demapping, and Equalization	167
	4.5.1 Signal Constellation and Mapping	167
	4.5.2 Equalization and Demapping	169
4.6	Adaptive Techniques in Multi-Carrier Transmission	170
	4.6.1 Nulling of Weak Sub-Carriers	171
	4.6.2 Adaptive Channel Coding and Modulation	171
	4.6.3 Adaptive Power Control	172
4.7	RF Issues	172
	4.7.1 Phase Noise	173
	4.7.2 Non-Linearities	177
	4.7.3 Narrowband Interference Rejection in MC-CDMA	185
	4.7.4 Link Budget Evaluation	188
4.8	References	189

5 Applications — 195

5.1	Introduction	195
5.2	Cellular Mobile Communications Beyond 3G	198
	5.2.1 Objectives	198
	5.2.2 Network Topology and Basic Concept	199
	5.2.3 System Parameters	200
5.3	Wireless Local Area Networks	203
	5.3.1 Network Topology	205
	5.3.2 Channel Characteristics	206
	5.3.3 IEEE 802.11a, HIPERLAN/2, and MMAC	206
	5.3.4 Transmission Performance	208
5.4	Fixed Wireless Access below 10 GHz	210
	5.4.1 Network Topology	211
	5.4.2 Channel Characteristics	212
	5.4.3 Multi-Carrier Transmission Schemes	212
	5.4.4 Transmission Performance	220
5.5	Interaction Channel for DVB-T: DVB-RCT	220
	5.5.1 Network Topology	221
	5.5.2 Channel Characteristics	223
	5.5.3 Multi-Carrier Uplink Transmission	223
	5.5.4 Transmission Performance	229
5.6	References	230

6 Additional Techniques for Capacity and Flexibility Enhancement — 233

6.1	Introduction	233
6.2	General Principle of Multiple Antenna Diversity	234

	6.2.1 BLAST Architecture	235
	6.2.2 Space–Time Coding	236
	6.2.3 Achievable Capacity	239
6.3	Diversity Techniques for Multi-Carrier Transmission	240
	6.3.1 Transmit Diversity	240
	6.3.2 Receive Diversity	244
	6.3.3 Performance Analysis	245
	6.3.4 OFDM and MC-CDMA with Space–Frequency Coding	248
6.4	Examples of Applications of Diversity Techniques	253
	6.4.1 UMTS-WCDMA	253
	6.4.2 FWA Multi-Carrier Systems	254
6.5	Software-Defined Radio	255
	6.5.1 General	255
	6.5.2 Basic Concept	257
	6.5.3 MC-CDMA-Based Software-Defined Radio	258
	References	260

Definitions, Abbreviations, and Symbols — **263**
    Definitions — 263
    Abbreviations — 265
    Symbols — 270

Index — **275**

Foreword

This book discusses multi-carrier modulation and spread spectrum techniques, recognized as the most promising candidate modulation methods for the 4th generation (4G) of mobile communications systems. The authors of this book were the first to propose MC-CDMA for the next generation of mobile communications, and are still continuing their contribution towards beyond 3G. Considering the requirements of 4G systems, multi-carrier and spread spectrum systems appear to be the most suitable as they provide higher flexibility, higher transmission rates and frequency usage efficiency. This is the first book on these methods, providing the reader with the fundamentals of the technologies involved and the related applications.

The book deals with the principles through definitions of basic technologies and the multipath channel over which the signals are transmitted. It defines MC-CDMA as a frequency PN pattern and MC-DS-CDMA as a straight extension of DS-CDMA; and argues that these twin asymmetric technologies are most suitable for 4G since MC-CDMA is suitable for the downlink and MC-DS-CDMA is suitable for the uplink in the cellular systems. Although MC-CDMA performs better than MC-DS-CDMA, it needs chip synchronization between users, and is therefore difficult to deploy in the uplink. Thus, for this asymmetric structure it is very important to understand the multi-carrier spread spectrum methods. Hybrid multiple access schemes like Multi-Carrier FDMA, Multi-Carrier TDMA, and Ultra Wide Band systems are discussed as more extended systems. Implementation issues, including synchronization, channel estimation, and RF issues, are also discussed in depth. Wireless local area networks, broadcasting transmission, and cellular mobile radio are shown to realize seamless networking for 4G. Although cellular systems have not yet been combined with other wireless networks, different wireless systems should be seamlessly combined. The last part of this book discusses capacity and flexibility enhancement technologies like diversity techniques, space–time/frequency coding, and SDR (Software Defined Radio).

This book greatly assists not only theoretical researchers, but also practicing engineers of the next generation of mobile communications systems.

March 2003
Prof. Masao Nakagawa
Department of Information and Computer Science
Keio University, Japan

Preface

Nowadays, multi-carrier transmission is considered to be an old concept. Its basic idea goes back to the mid-1960s. Nevertheless, behind any old technique there are always many simple and exciting ideas, the terrain for further developments of new efficient schemes.

Our first experience with the simple and exciting idea of OFDM started in early 1991 with *digital audio broadcasting (DAB)*. From 1992, our active participation in several research programmes on *digital terrestrial TV broadcasting (DVB-T)* gave us further opportunities to look at several aspects of the OFDM technique with its new advanced digital implementation possibilities. The experience gained from the joined specification of several OFDM-based demonstrators within the German HDTV-T and the EU-RACE dTTb research projects served as a basis for our commitment in 1995 to the final specifications of the DVB-T standard, relaying on the multi-carrier transmission technique.

Parallel to the HDTV-T and the dTTb projects, our further involvement from 1993 in the EU-RACE CODIT project, with the scope of building a first European 3G testbed, following the DS-CDMA scheme, inspired our interest in another old technique, *spread spectrum*, being as impressive as multi-carrier transmission. Although the final choice of the specification of the CODIT testbed was based on wideband CDMA, an alternative multiple-access scheme exploiting the new idea of combining OFDM with spread spectrum, i.e., *multi-carrier spread spectrum* (MC-SS), was considered as a potential candidate and discussed widely during the definition phase of the first testbed.

Our strong belief in the efficiency and flexibility of multi-carrier spread spectrum compared to W-CDMA for applications such as beyond 3G motivated us, from the introduction of this new multiple access scheme at PIMRC '93, to further contribute to it, and to investigate different corresponding system level aspects.

Due to the recognition of the merits of this combination by well-known international experts, since the PIMRC '93 conference, MC-SS has rapidly become one of the most widespread independent research topics in the field of mobile radio communications. The growing success of our organized series of international workshops on MC-SS since 1997, the large number of technical sessions devoted in international conferences to multi-carrier transmission, and the several special editions of the *European Transactions on Telecommunications (ETT)* on MC-SS highlight the importance of this combination for future wireless communications.

Several MC-CDMA demonstrators, e.g., one of the first built within DLR and its live demonstration during the 3rd international MC-SS workshop, a multitude of recent international research programmes like the research collaboration between DoCoMo-Eurolabs and DLR on the design of a future broadband air interface or the EU-IST MATRICE, 4MORE and WINNER projects, and especially the NTT-DoCoMo research initiative to build a demonstrator for beyond 3G systems based on the multi-carrier spread spectrum technique, emphasize the commitment of the international research community to this new topic.

Our experience gained during the above-mentioned research programmes, our current involvement in the ETSI-BRAN project, our yearly seminars organized within Carl Granz Gesellschaft (CCG) on digital TV broadcasting and on WLAN/WLL have given us sufficient background knowledge and material to take this initiative to collect in this book most important aspects on multi-carrier, spread spectrum and multi-carrier spread spectrum systems.

We hope that this book will contribute to a better understanding of the principles of multi-carrier and spread spectrum and may motivate further investigation into and development of this new technology.

K. Fazel, S. Kasier

Acknowledgements

The authors would like to express their sincere thanks to Prof. M. Nakagawa from Keio University, Japan, for writing the foreword. Many thanks go to Dr. H. Attarachi, Dr. N. Maeda, Dr. S. Abeta, and Dr. M. Sawahashi from NTT-DoCoMo for providing us with material regarding their multi-carrier spread spectrum activities. Many thanks also for the support of Dr. E. Auer from Marconi Communications and for helpful technical discussions with members of the Mobile Radio Transmission Group from DLR. Further thanks also go to I. Cosovic from DLR who provided us with results for the uplink, especially with pre-equalization.

K. Fazel, S. Kasier

Introduction

The common feature of the next generation wireless technologies will be the convergence of multimedia services such as speech, audio, video, image, and data. This implies that a future wireless terminal, by guaranteeing high-speed data, will be able to connect to different networks in order to support various services: switched traffic, IP data packets and broadband streaming services such as video. The development of wireless terminals with generic protocols and multiple-physical layers or software-defined radio interfaces is expected to allow users to seamlessly switch access between existing and future standards.

The rapid increase in the number of wireless mobile terminal subscribers, which currently exceeds 1 billion users, highlights the importance of wireless communications in this new millennium. This revolution in the information society has been happening, especially in Europe, through a continuous evolution of emerging standards and products by keeping a seamless strategy for the choice of solutions and parameters. The adaptation of wireless technologies to the user's rapidly changing demands has been one of the main drivers of this revolution. Therefore, the worldwide wireless access system is and will continue to be characterized by a heterogeneous multitude of standards and systems. This plethora of wireless communication systems is not limited to cellular mobile telecommunication systems such as GSM, IS-95, D-AMPS, PDC, UMTS or cdma2000, but also includes wireless local area networks (WLANs), e.g., HIPERLAN/2, IEEE 802.11a/b and Bluetooth, and wireless local loops (WLL), e.g., HIPERMAN, HIPERACCESS, and IEEE 802.16 as well as broadcast systems such as digital audio broadcasting (DAB) and digital video broadcasting (DVB).

These trends have accelerated since the beginning of the 1990s with the replacement of the first generation analog mobile networks by the current 2nd generation (2G) systems (GSM, IS-95, D-AMPS and PDC), which opened the door for a fully digitized network. This evolution is still continuing today with the introduction of the deployment of the 3rd generation (3G) systems (UMTS, IMT-2000 and cdma2000). In the meantime, the research community is focusing its activity towards the next generation beyond 3G, i.e. *fourth generation (4G)* systems, with more ambitious technological challenges.

The primary goal of next-generation wireless systems (4G) will not only be the introduction of new technologies to cover the need for higher data rates and new services, but also the *integration* of existing technologies in a common platform. Hence, the selection of a *generic* air-interface for future generation wireless systems will be of great importance. Although the exact requirements for 4G have not yet been commonly defined, its new air interface shall fulfill at least the following requirements:

— *generic architecture*, enabling the integration of existing technologies,
— *high spectral efficiency*, offering higher data rates in a given scarce spectrum,
— *high scalability*, designing different cell configurations (hot spot, ad hoc), hence better coverage,
— *high adaptability and reconfigurability*, supporting different standards and technologies,
— *low cost*, enabling a rapid market introduction, and
— *future proof*, opening the door for new technologies.

From Second- to Third-Generation Multiple Access Schemes

2G wireless systems are mainly characterized by the transition of analog towards a fully digitized technology and comprise the GSM, IS-95, PDC and D-AMPS standards.

Work on the pan-European digital cellular standard *Global System for Mobile* communications (GSM) started in 1982 [14][37], where now it accounts for about two-thirds of the world mobile market. In 1989, the technical specifications of GSM were approved by the European Telecommunication Standard Institute (ETSI), where its commercial success began in 1993. Although GSM is optimized for circuit-switched services such as voice, it offers low-rate data services up to 14.4 kbit/s. High speed data services up to 115.2 kbit/s are possible with the enhancement of the GSM standard towards the *General Packet Radio Service* (GPRS) by using a higher number of time slots. GPRS uses the same modulation, frequency band and frame structure as GSM. However, the *Enhanced Data rate for Global Evolution* (EDGE) [3] system which further improves the data rate up to 384 kbit/s introduces a new modulation scheme. The final evolution from GSM is the transition from EDGE to 3G.

Parallel to GSM, the American IS-95 standard [43] (recently renamed *cdmaOne*) was approved by the Telecommunication Industry Association (TIA) in 1993, where its first commercial application started in 1995. Like GSM, the first version of this standard (IS-95A) offers data services up to 14.4 kbit/s. In its second version, IS-95B, up to 64 kbit/s data services are possible.

Meanwhile, two other 2G mobile radio systems have been introduced: *Digital Advanced Mobile Phone Services* (D-AMPS/IS-136), called TDMA in the USA and the *Personal Digital Cellular* (PDC) in Japan [28]. Currently PDC hosts the most convincing example of high-speed internet services to mobile, called *i-mode*. The high amount of congestion in the PDC system will urge the Japanese towards 3G and even 4G systems.

Trends towards more capacity for mobile receivers, new multimedia services, new frequencies and new technologies have motivated the idea of 3G systems. A unique international standard was targeted: *Universal/International Mobile Telecommunication System* (UMTS/IMT-2000) with realization of a new generation of mobile communications technology for a world in which personal communication services will dominate. The objectives of the third generation standards, namely UMTS [17] and cdma2000 [44] went far beyond the second-generation systems, especially with respect to:

— the wide range of multimedia services (speech, audio, image, video, data) and bit rates (up to 2 Mbit/s for indoor and hot spot applications),

- the high quality of service requirements (better speech/image quality, lower bit error rate (BER), higher number of active users),
- operation in mixed cell scenarios (macro, micro, pico),
- operation in different environments (indoor/outdoor, business/domestic, cellular/cordless),
- and finally flexibility in frequency (variable bandwidth), data rate (variable) and radio resource management (variable power/channel allocation).

The commonly used multiple access schemes for second and third generation wireless mobile communication systems are based on either *Time Division Multiple Access* (TDMA), *Code Division Multiple Access* (CDMA) or the combined access schemes in conjunction with an additional *Frequency Division Multiple Access* (FDMA) component:

- The GSM standard, employed in the 900 MHz and 1800 MHz bands, first divides the allocated bandwidth into 200 kHz FDMA sub-channels. Then, in each sub-channel, up to 8 users share the 8 time slots in a TDMA manner [37].
- In the IS-95 standard up to 64 users share the 1.25 MHz channel by CDMA [43]. The system is used in the 850 MHz and 1900 MHz bands.
- The aim of D-AMPS (TDMA IS-136) is to coexist with the analog AMPS, where the 30 kHz channel of AMPS is divided into three channels, allowing three users to share a single radio channel by allocating unique time slots to each user [27].
- The recent ITU adopted standards for 3G (UMTS and cdma2000) are both based on CDMA [17][44]. For UMTS, the CDMA-FDD mode, which is known as wideband CDMA, employs separate 5 MHz channels for both the uplink and downlink directions. Within the 5 MHz bandwidth, each user is separated by a specific code, resulting in an end-user data rate of up to 2 Mbit/s per carrier.

Table 1 summarizes the key characteristics of 2G and 3G mobile communication systems.

Beside tremendous developments in mobile communication systems, in public and private environments, operators are offering wireless services using WLANs in selected

Table 1 Main parameters of 2G and 3G mobile radio systems

Parameter	2G systems		3G systems
	GSM	IS-95	IMT-2000/UMTS (WARC'92 [39])
Carrier frequencies	900 MHz 1800 MHz	850 MHz 1900 MHz	1900–1980 MHz 2010–2025 MHz 2110–2170 MHz
Peak data rate	64 kbit/s	64 kbit/s	2 Mbit/s
Multiple access	TDMA	CDMA	CDMA
Services	Voice, low rate data	Voice, low rate data	Voice, data, video

Table 2 Main parameters of WLAN communication systems

Parameter	Bluetooth	IEEE 802.11b	IEEE 802.11a	HIPERLAN/2
Carrier frequency	2.4 GHz (ISM)	2.4 GHz (ISM)	5 GHz	5 GHz
Peak data rate	1 Mbit/s	5.5 Mbit/s	54 Mbit/s	54 Mbit/s
Multiple access	FH-CDMA	DS-CDMA with carrier sensing	TDMA	TDMA
Services	Ethernet	Ethernet	Ethernet	Ethernet, ATM

spots such as hotels, train stations, airports and conference rooms. As Table 2 shows, there is a similar objective to go higher in data rates with WLANs, where multiple access schemes TDMA or CDMA are employed [15][30].

FDMA, TDMA and CDMA are obtained if the transmission bandwidth, the transmission time or the spreading code are related to the different users, respectively [2].

FDMA is a multiple access technology widely used in satellite, cable and terrestrial radio networks. FDMA subdivides the total bandwidth into N_c narrowband sub-channels which are available during the whole transmission time (see Figure 1). This requires bandpass filters with sufficient stop band attenuation. Furthermore, a sufficient guard band is left between two adjacent spectra in order to cope with frequency deviations of local oscillators and to minimize interference from adjacent channels. The main advantages of FDMA are in its low required transmit power and in channel equalization that is either not needed or much simpler than with other multiple access techniques. However, its drawback in a cellular system might be the implementation of N_c modulators and demodulators at the base station (BS).

TDMA is a popular multiple access technique, which is used in several international standards. In a TDMA system all users employ the same band and are separated by allocating short and distinct time slots, one or several assigned to a user (see Figure 2).

In TDMA, neglecting the overhead due to framing and burst formatting, the multiplexed signal bandwidth will be approximately N_c times higher than in an FDMA system, hence,

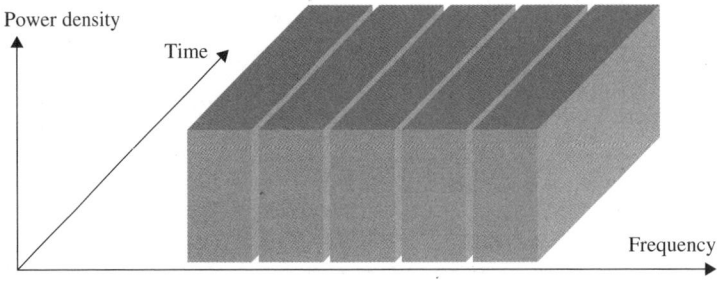

Figure 1 Principle of FDMA (with $N_c = 5$ sub-channels)

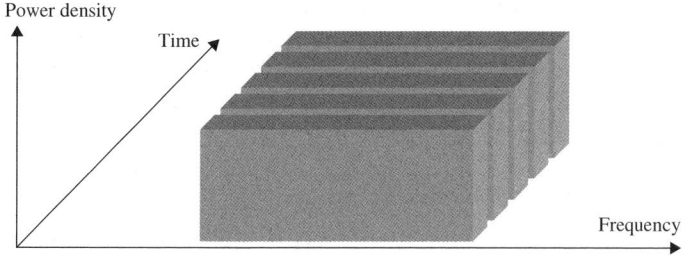

Figure 2 Principle of TDMA (with 5 time slots)

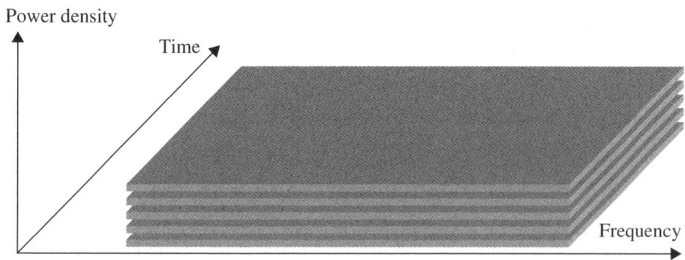

Figure 3 Principle of CDMA (with 5 spreading codes)

Table 3 Advantages and drawbacks of different multiple access schemes

Multiple access scheme	Advantages	Drawbacks
FDMA	– Low transmit power – Robust to multipath – Easy frequency planning – Low delay	– Low peak data rate – Loss due to guard bands – Sensitive to narrow band interference
TDMA	– High peak data rate – High multiplexing gain in case of bursty traffic	– High transmit power – Sensitive to multipath – Difficult frequency planning
CDMA	– Low transmit power – Robust to multipath – Easy frequency planning – High scalability – Low delay	– Low peak data rate – Limited capacity per sector due to multiple access interference

leading to quite complex equalization, especially for high-data rate applications. The channel separation of TDMA and FDMA is based on the orthogonality of signals. Therefore, in a cellular system, the co-channel interference is only present from the reuse of frequency.

On the contrary, in CDMA systems all users transmit at the same time on the same carrier using a wider bandwidth than in a TDMA system (see Figure 3). The signals of users are distinguished by assigning different spreading codes with low cross-correlation properties. Advantages of the spread spectrum technique are immunity against multipath distortion, simple frequency planning, high flexibility, variable rate transmission and resistance to interference.

In Table 3, the main advantages and drawbacks of FDMA, TDMA and CDMA are summarized.

From Third- to Fourth-Generation Multiple Access Schemes

Besides offering new services and applications, the success of the next generation of wireless systems (4G) will strongly depend on the choice of the concept and technology innovations in architecture, spectrum allocation, spectrum utilization and exploitation [38][39]. Therefore, new high-performance physical layer and multiple access technologies are needed to provide high speed data rates with flexible bandwidth allocation. A low-cost *generic radio interface*, being operational in mixed-cell and in different environments with scalable bandwidth and data rates, is expected to have better acceptance.

The technique of *spread spectrum* may allow the above requirements to be at least partially fulfilled. As explained earlier, a multiple access scheme based on direct sequence code division multiple access (DS-CDMA) relies on spreading the data stream using an assigned spreading code for each user in the time domain [40][45][47][48]. The capability of minimizing multiple access interference (MAI) is given by the cross-correlation properties of the spreading codes. In the case of severe multipath propagation in mobile communications, the capability of distinguishing one component from others in the composite received signal is offered by the autocorrelation properties of the spreading codes [45]. The so-called rake receiver should contain multiple correlators, each matched to a different resolvable path in the received composite signal [40]. Therefore, the performance of a DS-CDMA system will strongly depend on the number of active users, the channel characteristics, and the number of arms employed in the rake. Hence, the system capacity is limited by self-interference and MAI, which results from the imperfect auto- and cross-correlation properties of spreading codes. Therefore, it will be difficult for a DS-CDMA receiver to make full use of the received signal energy scattered in the time domain and hence to handle full load conditions [40].

The technique of *multi-carrier transmission* has recently been receiving wide interest, especially for high data-rate broadcast applications. The history of orthogonal multi-carrier transmission dates back to the mid-1960s, when Chang published his paper on the synthesis of band-limited signals for multichannel transmission [5][6]. He introduced the basic principle of transmitting data simultaneously through a band-limited channel without interference between sub-channels (without *inter-channel interference*, ICI) and without interference between consecutive transmitted symbols (without *inter-symbol interference*, ISI) in time domain. Later, Saltzberg performed further analyses [41]. However, a major contribution to multi-carrier transmission was presented in 1971 by Weinstein and Ebert [49] who used Fourier transform for base-band processing instead of a bank of sub-carrier oscillators. To combat ICI and ISI, they introduced the well-known *guard time* between the transmitted symbols with raised cosine windowing.

The main advantages of multi-carrier transmission are its robustness in frequency selective fading channels and, in particular, the reduced signal processing complexity by equalization in the frequency domain.

The basic principle of multi-carrier modulation relies on the transmission of data by dividing a high-rate data stream into several low-rate sub-streams. These sub-streams are modulated on different sub-carriers [1][4][9]. By using a large number of sub-carriers, a high immunity against multipath dispersion can be provided since the useful symbol duration T_s on each sub-stream will be much larger than the channel time dispersion. Hence, the effects of ISI will be minimized. Since the amount of filters and oscillators necessary is considerable for a large number of sub-carriers, an efficient digital implementation of a special form of multi-carrier modulation, called orthogonal frequency division multiplexing (OFDM), with rectangular pulse-shaping and guard time was proposed in [1]. OFDM can be easily realized by using the discrete Fourier transform (DFT). OFDM, having densely spaced sub-carriers with overlapping spectra of the modulated signals, abandons the use of steep band-pass filters to detect each sub-carrier as it is used in FDMA schemes. Therefore, it offers a high spectral efficiency.

Today, progress in digital technology has enabled the realization of a DFT also for large numbers of sub-carriers (up to several thousand), through which OFDM has gained much importance. The breakthrough of OFDM came in the 1990s as it was the modulation chosen for ADSL in the USA [8], and it was selected for the European DAB standard [11]. This success continued with the choice of OFDM for the European DVB-T standard [13] in 1995 and later for the WLAN standards HIPERLAN/2 and IEEE802.11a [15][30] and recently in the interactive terrestrial return channel (DVB-RCT) [12]. It is also a potential candidate for the future fixed wireless access standards HIPERMAN and IEEE802.16a [16][31]. Table 4 summarizes the main characteristics of several standards employing OFDM.

The advantages of multi-carrier modulation on one hand and the flexibility offered by the spread spectrum technique on the other hand have motivated many researchers to investigate the combination of both techniques, known as *Multi-Carrier Spread Spectrum* (MC-SS). This combination, published in 1993 by several authors independently [7][10][18][25][35][46][50], has introduced new multiple access schemes called MC-CDMA and MC-DS-CDMA. It allows one to benefit from several advantages of both multi-carrier modulation and spread spectrum systems by offering, for instance, high flexibility, high

Table 4 Examples of wireless transmission systems employing OFDM

Parameter	DAB	DVB-T	IEEE 802.11a	HIPERLAN/2
Carrier frequency	VHF	VHF and UHF	5 GHz	5 GHz
Bandwidth	1.54 MHz	8 MHz (7 MHz)	20 MHz	20 MHz
Max. data rate	1.7 Mbit/s	31.7 Mbit/s	54 Mbit/s	54 Mbit/s
Number of sub-carriers (FFT size)	192 up to 1536 (256 up to 2048)	1705 and 6817 (2048 and 8196)	52 (64)	52 (64)

spectral efficiency, simple and robust detection techniques and narrow band interference rejection capability.

Multi-carrier modulation and multi-carrier spread spectrum are today considered potential candidates to fulfill the requirements of next generation (4G) high-speed wireless multimedia communications systems, where *spectral efficiency* and *flexibility* will be considered the most important criteria for the choice of the air interface.

Multi-Carrier Spread Spectrum

Since 1993, various combinations of multi-carrier modulation with the spread spectrum technique as multiple access schemes have been introduced. It has been shown that multi-carrier spread spectrum (MC-SS) offers high spectral efficiency, robustness and flexibility [29].

Two different philosophies exist, namely MC-CDMA (or OFDM-CDMA) and MC-DS-CDMA (see Figure 4 and Table 5).

MC-CDMA is based on a serial concatenation of direct sequence (DS) spreading with multi-carrier modulation [7][18][25][50]. The high-rate DS spread data stream of processing gain P_G is multi-carrier modulated in the way that the chips of a spread data symbol are transmitted in parallel and the assigned data symbol is simultaneously transmitted on each sub-carrier (see Figure 4). As for DS-CDMA, a user may occupy the total bandwidth for the transmission of a single data symbol. Separation of the user's signal is performed in the code domain. Each data symbol is copied on the sub-streams before multiplying it with a chip of the spreading code assigned to the specific user. This reflects that an MC-CDMA system performs the spreading in frequency direction and, thus, has an additional degree of freedom compared to a DS-CDMA system. Mapping of the chips

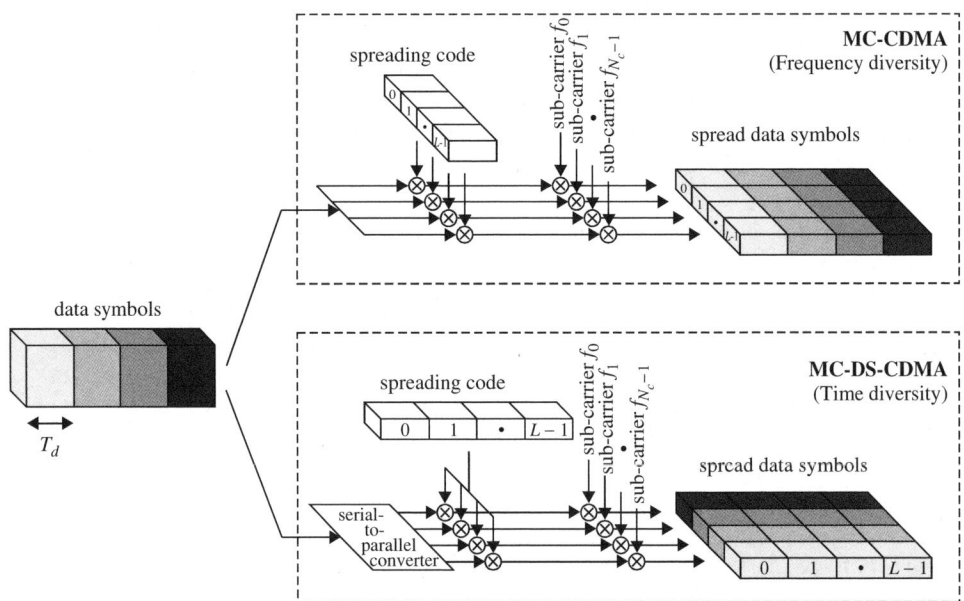

Figure 4 General principle of MC-CDMA and MC-DS-CDMA systems

Table 5 Main characteristics of different MC-SS concepts

Parameter	MC-CDMA	MC-DS-CDMA
Spreading	Frequency direction	Time direction
Sub-carrier spacing	$F_S = \frac{P_G}{N_c T_d}$	$F_S \geqslant \frac{P_G}{N_c T_d}$
Detection algorithm	MRC, EGC, ZF, MMSE equalization, IC, MLD	Correlation detector (coherent rake)
Specific characteristics	Very efficient for the synchronous downlink by using orthogonal codes	Designed especially for an asynchronous uplink
Applications	Synchronous uplink and downlink	Asynchronous uplink and downlink

in the frequency direction allows for simple methods of signal detection. This concept was proposed with OFDM for optimum use of the available bandwidth. The realization of this concept implies a guard time between adjacent OFDM symbols to prevent ISI or to assume that the symbol duration is significantly larger than the time dispersion of the channel. The number of sub-carriers N_c has to be chosen sufficiently large to guarantee frequency nonselective fading on each sub-channel. The application of orthogonal codes, such as Walsh–Hadamard codes for a synchronous system, e.g., the downlink of a cellular system, guarantees the absence of MAI in an ideal channel and a minimum MAI in a real channel. For signal detection, single-user detection techniques such as maximum ratio combining (MRC), equal gain combining (EGC), zero forcing (ZF) or minimum mean square error (MMSE) equalization, as well as multiuser detection techniques like interference cancellation (IC) or maximum likelihood detection (MLD), can be applied.

As depicted in Figure 4, MC-DS-CDMA modulates sub-streams on sub-carriers with a carrier spacing proportional to the inverse of the chip rate. This will guarantee orthogonality between the spectra of the sub-streams [42]. If the spreading code length is smaller or equal to the number of sub-carriers N_c, a single data symbol is not spread in the frequency direction, instead it is spread in the time direction. Spread spectrum is obtained by modulating N_c time spread data symbols on parallel sub-carriers. By using high numbers of sub-carriers, this concept benefits from time diversity. However, due to the frequency nonselective fading per sub-channel, frequency diversity can only be exploited if channel coding with interleaving or sub-carrier hopping is employed or if the same information is transmitted on several sub-carriers in parallel. Furthermore, higher frequency diversity could be achieved if the sub-carrier spacing is chosen larger than the chip rate. This concept was investigated for an asynchronous uplink scenario. For data detection, N_c coherent receivers can be used.

It can be noted that both schemes have a generic architecture. In the case where the number of sub-carriers $N_c = 1$, the classical DS-CDMA transmission scheme is obtained, whereas without spreading ($P_G = 1$) it results in a pure OFDM system.

By using a variable spreading factor in frequency and/or time and a variable sub-carrier allocation, the system can easily be adapted to different environments such as *multicell* and *single cell* topologies, each with different coverage areas.

Today, the field of multi-carrier spread spectrum communications is considered to be an independent and important research topic; see [19] to [23], [26], [36]. Several deep system analysis and comparisons of MC-CDMA and MC-DS-CDMA with DS-CDMA have been performed that show the superiority of MC-SS [24][29][32][33][34]. In addition, new application fields have been proposed such as high-rate cellular mobile (4G), high-rate wireless indoor and fixed wireless access (FWA). In addition to system-level analysis, a multitude of research activities have been addressed to develop appropriate strategies for detection, interference cancellation, channel coding, modulation, synchronization (especially uplink) and low-cost implementation design.

The Aim of this Book

The interest in multi-carrier transmission, especially in multi-carrier spread spectrum, is still growing. Many researchers and system designers are involved in system aspects and the implementation of these new techniques. However, a comprehensive collection of their work is still missing.

The aim of this book is first to describe and analyze the basic concepts of the combination of multi-carrier transmission with spread spectrum, where the different architectures and the different detection strategies are detailed. Concrete examples of its applications for future cellular mobile communications systems are given. Then, we examine other derivatives of MC-SS (e.g., OFDMA, SS-MC-MA and interleaved FDMA) and other variants of the combination of OFDM with TDMA, which are today part of WLAN, WLL and DVB-RCT standards. Basic OFDM implementation issues, valid for most of these combinations, such as channel coding, modulation, digital I/Q-generation, synchronization, channel estimation, and effects of phase noise and nonlinearity are further analyzed.

Chapter 1 covers the fundamentals of today's wireless communications. First a detailed analysis of the radio channel (outdoor and indoor) and its modeling are presented. Then the principle of OFDM multi-carrier transmission is introduced. In addition, a general overview of the spread spectrum technique, especially of DS-CDMA, is given. Examples of applications of OFDM and DS-CDMA for broadcast, WLAN, and cellular systems (IS-95, UMTS) are briefly presented.

Chapter 2 describes the combinations of multi-carrier transmission with the spread spectrum technique, namely MC-CDMA and MC-DS-CDMA. It includes a detailed description of the different detection strategies (single-user and multiuser) and presents their performance in terms of bit error rate (BER), spectral efficiency and complexity. Here a cellular system with a point- to multi-point topology is considered. Both downlink and uplink architectures are examined.

Hybrid multiple access schemes based on MC-SS, OFDM or spread spectrum are analyzed in Chapter 3. This chapter covers OFDMA, being a derivative of MC-CDMA, OFDM-TDMA, SS-MC-MA, interleaved FDMA and ultra wide band (UWB) schemes. All these multiple access schemes have recently received wide interest. Their concrete application fields are detailed in Chapter 5.

The issues of digital implementation of multi-carrier transmission systems, essential especially for system- and hardware designers, are addressed in Chapter 4. Here, the different functions such as digital I/Q generation, analog/digital conversion, digital multi-carrier modulation/demodulation, synchronization (time, frequency), channel estimation, coding/decoding and other related RF issues such as nonlinearities, phase noise and narrow band interference rejection are analyzed.

In Chapter 5, concrete application fields of MC-SS, OFDMA and OFDM-TDMA for cellular mobile (4G), wireless indoor (WLAN), fixed wireless access (FWA/WLL) and interactive multimedia communication (DVB-T return channel) are outlined, where for each of these systems, the multi-carrier architecture and their main parameters are described. The capacity advantages of using adaptive channel coding and modulation, adaptive spreading and scalable bandwidth allocation are discussed.

Finally, Chapter 6 covers further techniques that can be used to enhance system capacity or offer more flexibility for the implementation and deployment of the transmission systems examined in Chapter 5. Here, diversity techniques such as space time/frequency coding and Tx/Rx antenna diversity in MIMO concepts and software-defined radio (SDR) are introduced.

References

[1] Alard M. and Lassalle R., "Principles of modulation and channel coding for digital broadcasting for mobile receivers," *European Broadcast Union Review*, no. 224, pp. 47–69, Aug. 1987.

[2] Bhargava V.K., Haccoun D., Matyas R. and Nuspl P.P., *Digital Communications by Satellite*. New York: John Wiley & Sons, 1981.

[3] Bi Q., Zysman G.I. and Menkes H, "Wireless mobile communications at the start of the 21st century," *IEEE Communications Magazine*, vol. 39, pp. 110–116, Jan. 2001.

[4] Bingham J.A.C., "Multicarrier modulation for data transmission: An idea whose time has come," *IEEE Communications Magazine*, vol. 28, pp. 5–14, May 1990.

[5] Chang R.W., "Synthesis of band-limited orthogonal signals for multi-channel data transmission," *Bell Labs Technical Journal*, no. 45, pp. 1775–1796, Dec. 1966.

[6] Chang R.W. and Gibby R.A., "A theoretical study of performance of an orthogonal multiplexing data transmission scheme," *IEEE Transactions on Communication Technology*, vol. 16, pp. 529–540, Aug. 1968.

[7] Chouly A., Brajal A. and Jourdan S., "Orthogonal multicarrier techniques applied to direct sequence spread spectrum CDMA systems," in *Proc. IEEE Global Telecommunications Conference (GLOBECOM'93)*, Houston, USA, pp. 1723–1728, Nov./Dec. 1993.

[8] Chow J.S., Tu J.-C., and Cioffi J.M., "A discrete multitone transceiver system for HDSL applications," *IEEE Journal on Selected Areas in Communications (JSAC)*, vol. 9, pp. 895–908, Aug. 1991.

[9] Cimini L.J., "Analysis and simulation of a digital mobile channel using orthogonal frequency division multiplexing," *IEEE Transactions on Communications*, vol. 33, pp. 665–675, July 1985.

[10] DaSilva V. and Sousa E.S., "Performance of orthogonal CDMA codes for quasi-synchronous communication systems," in *Proc. IEEE International Conference on Universal Personal Communications (ICUPC'93)*, Ottawa, Canada, pp. 995–999, Oct. 1993.

[11] ETSI DAB (EN 300 401 V1.3.1), "Radio broadcasting systems; digital audio broadcasting (DAB) to mobile, portable and fixed receivers," Sophia Antipolis, France, April 2000.

[12] ETSI DVB RCT (EN 301 958 V1.1.1), "Interaction channel for digital terrestrial television (RCT) incorporating multiple access OFDM," Sophia Antipolis, France, March 2001.

[13] ETSI DVB-T (EN 300 744 V1.2.1), "Digital video broadcasting (DVB); framing structure, channel coding and modulation for digital terrestrial television," Sophia Antipolis, France, July 1999.

[14] ETSI GSM Recommendations, 05 series, Sophia Antipolis, France.

[15] ETSI HIPERLAN/2 (TS 101 475), "Broadband radio access networks HIPERLAN Type 2 functional specification – Part 1: Physical layer," Sophia Antipolis, France, Sept. 1999.

[16] ETSI HIPERMAN (TR 101 856), "High performance metropolitan area network, requirements MAC and physical layer below 11 GHz band," Sophia Antipolis, France, Dec. 2002.

[17] ETSI UMTS (TR-101 112), V 3.2.0, Sophia Antipolis, France, April 1998.
[18] Fazel K., "Performance of CDMA/OFDM for mobile communications system," in *Proc. IEEE International Conference on Universal Personal Communications (ICUPC'93)*, Ottawa, Canada, pp. 975–979, Oct. 93.
[19] Fazel K. and Fettweis G. (eds), *Multi-Carrier Spread-Spectrum*. Boston: Kluwer Academic Publishers, 1997, *Proceedings of the 1st International Workshop on Multi-Carrier Spread-Spectrum (MC-SS'97)*.
[20] Fazel K. and Kaiser S. (eds), *Multi-Carrier Spread-Spectrum & Related Topics*. Boston: Kluwer Academic Publishers, 2000, *Proceedings of the 2nd International Workshop on Multi-Carrier Spread-Spectrum & Related Topics (MC-SS'99)*.
[21] Fazel K. and Kaiser S. (eds), *Multi-Carrier Spread-Spectrum & Related Topics*. Boston: Kluwer Academic Publishers, 2002, *Proceedings of the 3rd International Workshop on Multi-Carrier Spread-Spectrum & Related Topics (MC-SS'01)*.
[22] Fazel K. and Kaiser S. (eds), *Special Issue on Multi-Carrier Spread Spectrum and Related Topics, European Transactions on Telecommunications (ETT)*, vol. 11, no. 6, Nov./Dec. 2000.
[23] Fazel K. and Kaiser S. (eds), *Special Issue on Multi-Carrier Spread Spectrum and Related Topics, European Transactions on Telecommunications (ETT)*, vol. 13, no. 5, Sept. 2002.
[24] Fazel K., Kaiser S. and Schnell M., "A flexible and high performance cellular mobile communications system based on orthogonal multi-carrier SSMA," *Wireless Personal Communications*, vol. 2, nos. 1&2, pp. 121–144, 1995.
[25] Fazel K. and Papke L., "On the performance of convolutionally-coded CDMA/OFDM for mobile communications system," in *Proc. IEEE International Symposium on Personal, Indoor and Mobile Radio Communications (PIMRC'93)*, Yokohama, Japan, pp. 468–472, Sept. 1993.
[26] Fazel K. and Prasad R. (eds), *Special Issue on Multi-Carrier Spread Spectrum, European Transactions on Telecommunications (ETT)*, vol. 10, no. 4, July/Aug. 1999.
[27] Goodman D.J., "Second generation wireless information network," *IEEE Transactions on Vehicular Technology*, vol. 40, no. 2, pp. 366–374, May 1991.
[28] Goodman D.J., "Trends in cellular and cordless communications," *IEEE Communications Magazine*, vol. 29, pp. 31–40, June 1991.
[29] Hara S. and Prasad R., "Overview of multicarrier CDMA," *IEEE Communications Magazine*, vol. 35, pp. 126–133, Dec. 1997.
[30] IEEE-802.11 (P802.11a/D6.0), "LAN/MAN specific requirements – Part 2: Wireless MAC and PHY specifications – high speed physical layer in the 5 GHz band," IEEE 802.11, May 1999.
[31] IEEE 802.16ab-01/01, Draft, "Air interface for fixed broadband wireless access systems – Part A: Systems between 2 and 11 GHz," IEEE 802.16, June 2000.
[32] Kaiser S., "OFDM-CDMA versus DS-CDMA: Performance evaluation for fading channels," in *Proc. IEEE International Conference on Communications (ICC'95)*, Seattle, USA, pp. 1722–1726, June 1995.
[33] Kaiser S., "On the performance of different detection techniques for OFDM-CDMA in fading channels," in *Proc. IEEE Global Telecommunications Conference (GLOBECOM'95)*, Singapore, pp. 2059–2063, Nov. 1995.
[34] Kaiser S., *Multi-Carrier CDMA Mobile Radio Systems – Analysis and Optimization of Detection, Decoding, and Channel Estimation*. Düsseldorf: VDI-Verlag, Fortschrittberichte VDI, series 10, no. 531, 1998, PhD Thesis.
[35] Kondo S. and Milstein L.B., "On the use of multicarrier direct sequence spread spectrum systems," in *Proc. IEEE Military Communications Conference (MILCOM'93)*, Boston, USA, pp. 52–56, Oct. 1993.
[36] Linnartz J.P. and Hara S. (eds), *Special Issue on Multi-Carrier Communications, Wireless Personal Communications*, Kluwer Academic Publishers, vol. 2, nos. 1 & 2, 1995.
[37] Mouly M. and Paulet M.-B., *The GSM System for Mobile Communications* Palaiseau: published by authors, France 1992.
[38] Pereira J.M., "Beyond third generation," in *Proc. International Symposium on Wireless Personal Multimedia Communications (WPMC'99)*, Amsterdam, The Netherlands, Sept. 1999.
[39] Pereira J.M., "Fourth Generation: Now it is personal!," in *Proc. IEEE International Symposium on Personal, Indoor and Mobile Radio Communications (PIMRC 2000)*, London, UK, pp. 1009–1016, Sept. 2000.
[40] Pickholtz R.L., Schilling D.L. and Milstein L.B., "Theory of spread-spectrum communications–a tutorial," *IEEE Transactions on Communication Technology*, vol. 30, pp. 855–884, May 1982.

References

[41] Saltzberg, B.R., "Performance of an efficient parallel data transmission system," *IEEE Transactions on Communication Technology*, vol. 15, pp. 805–811, Dec. 1967.

[42] Sourour E.A. and Nakagawa M., "Performance of orthogonal multi-carrier CDMA in a multipath fading channel," *IEEE Transactions on Communication*, vol. 44, pp. 356–367, March 1996.

[43] TIA/EIA/IS-95, "Mobile station-base station compatibility standard for dual mode wideband spectrum cellular system," July 1993.

[44] TIA/EIA/IS-cdma2000, "Physical layer standard for cdma2000 spread spectrum systems," Aug. 1999.

[45] Turin G.L., "Introduction to spread-spectrum anti-multi-path techniques and their application to urban digital radio," *Proceedings of the IEEE*, vol. 68, pp. 328–353, March 1980.

[46] Vandendorpe L., "Multitone direct sequence CDMA system in an indoor wireless environment," in *Proc. IEEE First Symposium of Communications and Vehicular Technology*, Delft, The Netherlands, pp. 4.1.1–4.1.8, Oct. 1993.

[47] Viterbi A.J., "Spread-spectrum communications–myths and realities," *IEEE, Communications Magazine*, vol. 17, pp. 11–18, May 1979.

[48] Viterbi A.J., *CDMA: Principles of Spread Spectrum Communication*. Reading: Addison-Wesley, 1995.

[49] Weinstein S.B. and Ebert P.M., "Data transmission by frequency-division multiplexing using the discrete Fourier transform," *IEEE Transactions on Communication Technology*, vol. 19, pp. 628–634, Oct. 1971.

[50] Yee N., Linnartz J.-P. and Fettweis G., "Multi-carrier CDMA for indoor wireless radio networks," in *Proc. International Symposium on Personal, Indoor and Mobile Radio Communications (PIMRC'93)*, Yokohama, Japan, pp. 109–113, Sept. 1993.

1

Fundamentals

This chapter describes the fundamentals of today's wireless communications. First a detailed description of the radio channel and its modeling are presented, followed by the introduction of the principle of OFDM multi-carrier transmission. In addition, a general overview of the spread spectrum technique, especially DS-CDMA, is given and examples of potential applications for OFDM and DS-CDMA are analyzed. This introduction is essential for a better understanding of the idea behind the combination of OFDM with the spread spectrum technique, which is briefly introduced in the last part of this chapter.

1.1 Radio Channel Characteristics

Understanding the characteristics of the communications medium is crucial for the appropriate selection of transmission system architecture, dimensioning of its components, and optimizing system parameters, especially since mobile radio channels are considered to be the most difficult channels, since they suffer from many imperfections like multipath fading, interference, Doppler shift, and shadowing. The choice of system components is totally different if, for instance, multipath propagation with long echoes dominates the radio propagation.

Therefore, an accurate channel model describing the behavior of radio wave propagation in different environments such as mobile/fixed and indoor/outdoor is needed. This may allow one, through simulations, to estimate and validate the performance of a given transmission scheme in its several design phases.

1.1.1 Understanding Radio Channels

In mobile radio channels (see Figure 1-1), the transmitted signal suffers from different effects, which are characterized as follows:

Multipath propagation occurs as a consequence of reflections, scattering, and diffraction of the transmitted electromagnetic wave at natural and man-made objects. Thus, at the receiver antenna, a multitude of waves arrives from many different directions with different delays, attenuations, and phases. The superposition of these waves results in amplitude and phase variations of the composite received signal.

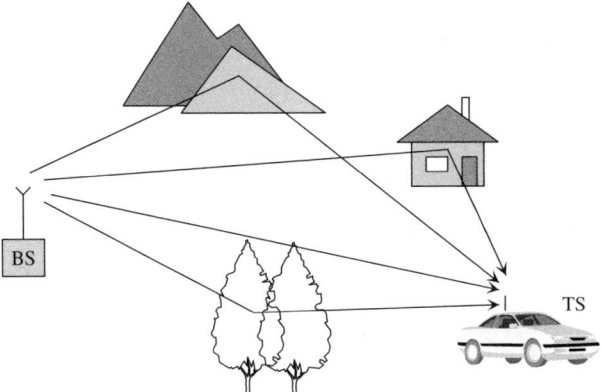

Figure 1-1 Time-variant multipath propagation

Doppler spread is caused by moving objects in the mobile radio channel. Changes in the phases and amplitudes of the arriving waves occur which lead to time-variant multipath propagation. Even small movements on the order of the wavelength may result in a totally different wave superposition. The varying signal strength due to time-variant multipath propagation is referred to as fast fading.

Shadowing is caused by obstruction of the transmitted waves by, e.g., hills, buildings, walls, and trees, which results in more or less strong attenuation of the signal strength. Compared to fast fading, longer distances have to be covered to significantly change the shadowing constellation. The varying signal strength due to shadowing is called slow fading and can be described by a log-normal distribution [36].

Path loss indicates how the mean signal power decays with distance between transmitter and receiver. In free space, the mean signal power decreases with the square of the distance between base station (BS) and terminal station (TS). In a mobile radio channel, where often no line of sight (LOS) path exists, signal power decreases with a power higher than two and is typically in the order of three to five.

Variations of the received power due to shadowing and path loss can be efficiently counteracted by power control. In the following, the mobile radio channel is described with respect to its fast fading characteristic.

1.1.2 Channel Modeling

The mobile radio channel can be characterized by the time-variant channel impulse response $h(\tau, t)$ or by the time-variant channel transfer function $H(f, t)$, which is the Fourier transform of $h(\tau, t)$. The channel impulse response represents the response of the channel at time t due to an impulse applied at time $t - \tau$. The mobile radio channel is assumed to be a wide-sense stationary random process, i.e., the channel has a fading statistic that remains constant over short periods of time or small spatial distances. In environments with multipath propagation, the channel impulse response is composed of a large number of scattered impulses received over N_p different paths,

$$h(\tau, t) = \sum_{p=0}^{N_p-1} a_p e^{j(2\pi f_{D,p} t + \varphi_p)} \delta(\tau - \tau_p), \tag{1.1}$$

where
$$\delta(\tau - \tau_p) = \begin{cases} 1 & \text{if } \tau = \tau_p \\ 0 & \text{otherwise} \end{cases} \quad (1.2)$$

and a_p, $f_{D,p}$, φ_p, and τ_p are the amplitude, the Doppler frequency, the phase, and the propagation delay, respectively, associated with path p, $p = 0, \ldots, N_p - 1$. The assigned channel transfer function is

$$H(f, t) = \sum_{p=0}^{N_p-1} a_p e^{j(2\pi(f_{D,p}t - f\tau_p) + \varphi_p)}. \quad (1.3)$$

The delays are measured relative to the first detectable path at the receiver. The Doppler frequency

$$f_{D,p} = \frac{v f_c \cos(\alpha_p)}{c} \quad (1.4)$$

depends on the velocity v of the terminal station, the speed of light c, the carrier frequency f_c, and the angle of incidence α_p of a wave assigned to path p. A channel impulse response with corresponding channel transfer function is illustrated in Figure 1-2.

The delay power density spectrum $\rho(\tau)$ that characterizes the frequency selectivity of the mobile radio channel gives the average power of the channel output as a function of the delay τ. The mean delay $\bar{\tau}$, the root mean square (RMS) delay spread τ_{RMS} and the maximum delay $\tau_{\max}$ are characteristic parameters of the delay power density spectrum. The mean delay is

$$\bar{\tau} = \frac{\sum_{p=0}^{N_p-1} \tau_p \Omega_p}{\sum_{p=0}^{N_p-1} \Omega_p}, \quad (1.5)$$

where
$$\Omega_p = |a_p|^2 \quad (1.6)$$

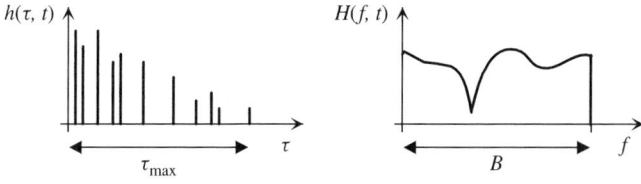

Figure 1-2 Time-variant channel impulse response and channel transfer function with frequency-selective fading

is the power of path p. The RMS delay spread is defined as

$$\tau_{RMS} = \sqrt{\frac{\sum_{p=0}^{N_p-1} \tau_p^2 \Omega_p}{\sum_{p=0}^{N_p-1} \Omega_p} - \bar{\tau}^2}. \quad (1.7)$$

Similarly, the Doppler power density spectrum $S(f_D)$ can be defined that characterizes the time variance of the mobile radio channel and gives the average power of the channel output as a function of the Doppler frequency f_D. The frequency dispersive properties of multipath channels are most commonly quantified by the maximum occurring Doppler frequency $f_{D\max}$ and the Doppler spread $f_{D\text{spread}}$. The Doppler spread is the bandwidth of the Doppler power density spectrum and can take on values up to two times $|f_{D\max}|$, i.e.,

$$f_{D\text{spread}} \leqslant 2|f_{D\max}|. \quad (1.8)$$

1.1.3 Channel Fade Statistics

The statistics of the fading process characterize the channel and are of importance for channel model parameter specifications. A simple and often used approach is obtained from the assumption that there is a large number of scatterers in the channel that contribute to the signal at the receiver side. The application of the central limit theorem leads to a complex-valued Gaussian process for the channel impulse response. In the absence of line of sight (LOS) or a dominant component, the process is zero-mean. The magnitude of the corresponding channel transfer function

$$a = a(f, t) = |H(f, t)| \quad (1.9)$$

is a random variable, for brevity denoted by a, with a Rayleigh distribution given by

$$p(a) = \frac{2a}{\Omega} e^{-a^2/\Omega}, \quad (1.10)$$

where

$$\Omega = E\{a^2\} \quad (1.11)$$

is the average power. The phase is uniformly distributed in the interval $[0, 2\pi]$.

In the case that the multipath channel contains a LOS or dominant component in addition to the randomly moving scatterers, the channel impulse response can no longer be modeled as zero-mean. Under the assumption of a complex-valued Gaussian process for the channel impulse response, the magnitude a of the channel transfer function has a Rice distribution given by

$$p(a) = \frac{2a}{\Omega} e^{-(a^2/\Omega + K_{Rice})} I_0\left(2a\sqrt{\frac{K_{Rice}}{\Omega}}\right). \quad (1.12)$$

The Rice factor K_{Rice} is determined by the ratio of the power of the dominant path to the power of the scattered paths. I_0 is the zero-order modified Bessel function of first kind. The phase is uniformly distributed in the interval $[0, 2\pi]$.

1.1.4 Inter-Symbol (ISI) and Inter-Channel Interference (ICI)

The delay spread can cause inter-symbol interference (ISI) when adjacent data symbols overlap and interfere with each other due to different delays on different propagation paths. The number of interfering symbols in a single-carrier modulated system is given by

$$N_{\text{ISI,single carrier}} = \left\lceil \frac{\tau_{\max}}{T_d} \right\rceil. \quad (1.13)$$

For high data rate applications with very short symbol duration $T_d < \tau_{\max}$, the effect of ISI and, with that, the receiver complexity can increase significantly. The effect of ISI can be counteracted by different measures such as time or frequency domain equalization. In spread spectrum systems, rake receivers with several arms are used to reduce the effect of ISI by exploiting the multipath diversity such that individual arms are adapted to different propagation paths.

If the duration of the transmitted symbol is significantly larger than the maximum delay $T_d \gg \tau_{\max}$, the channel produces a negligible amount of ISI. This effect is exploited with multi-carrier transmission where the duration per transmitted symbol increases with the number of sub-carriers N_c and, hence, the amount of ISI decreases. The number of interfering symbols in a multi-carrier modulated system is given by

$$N_{\text{ISI,multi carrier}} = \left\lceil \frac{\tau_{\max}}{N_c T_d} \right\rceil. \quad (1.14)$$

Residual ISI can be eliminated by the use of a guard interval (see Section 1.2).

The maximum Doppler spread in mobile radio applications using single-carrier modulation is typically much less than the distance between adjacent channels, such that the effect of interference on adjacent channels due to Doppler spread is not a problem for single-carrier modulated systems. For multi-carrier modulated systems, the sub-channel spacing F_s can become quite small, such that Doppler effects can cause significant ICI. As long as all sub-carriers are affected by a common Doppler shift f_D, this Doppler shift can be compensated for in the receiver and ICI can be avoided. However, if Doppler spread in the order of several percent of the sub-carrier spacing occurs, ICI may degrade the system performance significantly. To avoid performance degradations due to ICI or more complex receivers with ICI equalization, the sub-carrier spacing F_s should be chosen as

$$F_s \gg f_{D\max}, \quad (1.15)$$

such that the effects due to Doppler spread can be neglected (see Chapter 4). This approach corresponds with the philosophy of OFDM described in Section 1.2 and is followed in current OFDM-based wireless standards.

Nevertheless, if a multi-carrier system design is chosen such that the Doppler spread is in the order of the sub-carrier spacing or higher, a rake receiver in the frequency domain can be used [22]. With the frequency domain rake receiver each branch of the rake resolves a different Doppler frequency.

1.1.5 Examples of Discrete Multipath Channel Models

Various discrete multipath channel models for indoor and outdoor cellular systems with different cell sizes have been specified. These channel models define the statistics of the discrete propagation paths. An overview of widely used discrete multipath channel models is given in the following.

COST 207 [8]: The COST 207 channel models specify four outdoor macro cell propagation scenarios by continuous, exponentially decreasing delay power density spectra. Implementations of these power density spectra by discrete taps are given by using up to 12 taps. Examples for settings with 6 taps are listed in Table 1-1. In this table for several propagation environments the corresponding path delay and power profiles are given. Hilly terrain causes the longest echoes.

The classical Doppler spectrum with uniformly distributed angles of arrival of the paths can be used for all taps for simplicity. Optionally, different Doppler spectra are defined for the individual taps in [8]. The COST 207 channel models are based on channel measurements with a bandwidth of 8–10 MHz in the 900-MHz band used for 2G systems such as GSM.

COST 231 [9] and COST 259 [10]: These COST actions which are the continuation of COST 207 extend the channel characterization to DCS 1800, DECT, HIPERLAN and UMTS channels, taking into account macro, micro, and pico cell scenarios. Channel models with spatial resolution have been defined in COST 259. The spatial component is introduced by the definition of several clusters with local scatterers, which are located in a circle around the base station. Three types of channel models are defined. The macro cell type has cell sizes from 500 m up to 5000 m and a carrier frequency of 900 MHz or 1.8 GHz. The micro cell type is defined for cell sizes of about 300 m and a carrier frequency of 1.2 GHz or 5 GHz. The pico cell type represents an indoor channel model with cell sizes smaller than 100 m in industrial buildings and in the order of 10 m in an office. The carrier frequency is 2.5 GHz or 24 GHz.

Table 1-1 Settings for the COST 207 channel models with 6 taps [8]

Path #	Rural area (RA)		Typical urban (TU)		Bad urban (BU)		Hilly terrain (HT)	
	delay in μs	power in dB	delay in μs	power in dB	delay in μs	power in dB	delay in μs	power in dB
1	0	0	0	−3	0	−2.5	0	0
2	0.1	−4	0.2	0	0.3	0	0.1	−1.5
3	0.2	−8	0.5	−2	1.0	−3	0.3	−4.5
4	0.3	−12	1.6	−6	1.6	−5	0.5	−7.5
5	0.4	−16	2.3	−8	5.0	−2	15.0	−8.0
6	0.5	−20	5.0	−10	6.6	−4	17.2	−17.7

COST 273: The COST 273 action additionally takes multi-antenna channel models into account, which are not covered by the previous COST actions.

CODIT [7]: These channel models define typical outdoor and indoor propagation scenarios for macro, micro, and pico cells. The fading characteristics of the various propagation environments are specified by the parameters of the Nakagami-m distribution. Every environment is defined in terms of a number of scatterers which can take on values up to 20. Some channel models consider also the angular distribution of the scatterers. They have been developed for the investigation of 3G system proposals. Macro cell channel type models have been developed for carrier frequencies around 900 MHz with 7 MHz bandwidth. The micro and pico cell channel type models have been developed for carrier frequencies between 1.8 GHz and 2 GHz. The bandwidths of the measurements are in the range of 10–100 MHz for macro cells and around 100 MHz for pico cells.

JTC [28]: The JTC channel models define indoor and outdoor scenarios by specifying 3 to 10 discrete taps per scenario. The channel models are designed to be applicable for wideband digital mobile radio systems anticipated as candidates for the PCS (Personal Communications Systems) common air interface at carrier frequencies of about 2 GHz.

UMTS/UTRA [18][44]: Test propagation scenarios have been defined for UMTS and UTRA system proposals which are developed for frequencies around 2 GHz. The modeling of the multipath propagation corresponds to that used by the COST 207 channel models.

HIPERLAN/2 [33]: Five typical indoor propagation scenarios for wireless LANs in the 5 GHz frequency band have been defined. Each scenario is described by 18 discrete taps of the delay power density spectrum. The time variance of the channel (Doppler spread) is modeled by a classical Jake's spectrum with a maximum terminal speed of 3 m/h.

Further channel models exist which are, for instance, given in [16].

1.1.6 Multi-Carrier Channel Modeling

Multi-carrier systems can either be simulated in the time domain or, more computationally efficient, in the frequency domain. Preconditions for the frequency domain implementation are the absence of ISI and ICI, the frequency nonselective fading per sub-carrier, and the time-invariance during one OFDM symbol. A proper system design approximately fulfills these preconditions. The discrete channel transfer function adapted to multi-carrier signals results in

$$H_{n,i} = H(nF_s, iT'_s)$$
$$= \sum_{p=0}^{N_p-1} a_p e^{j(2\pi(f_{D,p}iT'_s - nF_s\tau_p) + \varphi_p)} \quad (1.16)$$
$$= a_{n,i} e^{j\varphi_{n,i}}$$

where the continuous channel transfer function $H(f,t)$ is sampled in time at OFDM symbol rate $1/T'_s$ and in frequency at sub-carrier spacing F_s. The duration T'_s is the total OFDM symbol duration including the guard interval. Finally, a symbol transmitted on sub-channel n of the OFDM symbol i is multiplied by the resulting fading amplitude $a_{n,i}$ and rotated by a random phase $\varphi_{n,i}$.

The advantage of the frequency domain channel model is that the IFFT and FFT operation for OFDM and inverse OFDM can be avoided and the fading operation results in one complex-valued multiplication per sub-carrier. The discrete multipath channel models introduced in Section 1.1.5 can directly be applied to (1.16). A further simplification of the channel modeling for multi-carrier systems is given by using the so-called uncorrelated fading channel models.

1.1.6.1 Uncorrelated Fading Channel Models for Multi-Carrier Systems

These channel models are based on the assumption that the fading on adjacent data symbols after inverse OFDM and de-interleaving can be considered as uncorrelated [29]. This assumption holds when, e.g., a frequency and time interleaver with sufficient interleaving depth is applied. The fading amplitude $a_{n,i}$ is chosen from a distribution $p(a)$ according to the considered cell type and the random phase $\varphi_{n,I}$ is uniformly distributed in the interval $[0, 2\pi]$. The resulting complex-valued channel fading coefficient is thus generated independently for each sub-carrier and OFDM symbol. For a propagation scenario in a macro cell without LOS, the fading amplitude $a_{n,i}$ is generated by a Rayleigh distribution and the channel model is referred to as an uncorrelated Rayleigh fading channel. For smaller cells where often a dominant propagation component occurs, the fading amplitude is chosen from a Rice distribution. The advantages of the uncorrelated fading channel models for multi-carrier systems are their simple implementation in the frequency domain and the simple reproducibility of the simulation results.

1.1.7 Diversity

The coherence bandwidth $(\Delta f)_c$ of a mobile radio channel is the bandwidth over which the signal propagation characteristics are correlated and it can be approximated by

$$(\Delta f)_c \approx \frac{1}{\tau_{\max}}. \tag{1.17}$$

The channel is frequency-selective if the signal bandwidth B is larger than the coherence bandwidth $(\Delta f)_c$. On the other hand, if B is smaller than $(\Delta f)_c$, the channel is frequency nonselective or flat. The coherence bandwidth of the channel is of importance for evaluating the performance of spreading and frequency interleaving techniques that try to exploit the inherent frequency diversity D_f of the mobile radio channel. In the case of multi-carrier transmission, frequency diversity is exploited if the separation of sub-carriers transmitting the same information exceeds the coherence bandwidth. The maximum achievable frequency diversity D_f is given by the ratio between the signal bandwidth B and the coherence bandwidth,

$$D_f = \frac{B}{(\Delta f)_c}. \tag{1.18}$$

The coherence time of the channel $(\Delta t)_c$ is the duration over which the channel characteristics can be considered as time-invariant and can be approximated by

$$(\Delta t)_c \approx \frac{1}{2 f_{D\max}}. \qquad (1.19)$$

If the duration of the transmitted symbol is larger than the coherence time, the channel is time-selective. On the other hand, if the symbol duration is smaller than $(\Delta t)_c$, the channel is time nonselective during one symbol duration. The coherence time of the channel is of importance for evaluating the performance of coding and interleaving techniques that try to exploit the inherent time diversity D_O of the mobile radio channel. Time diversity can be exploited if the separation between time slots carrying the same information exceeds the coherence time. A number of N_s successive time slots create a time frame of duration T_{fr}. The maximum time diversity D_t achievable in one time frame is given by the ratio between the duration of a time frame and the coherence time,

$$D_t = \frac{T_{fr}}{(\Delta t)_c}. \qquad (1.20)$$

A system exploiting frequency and time diversity can achieve the overall diversity

$$D_O = D_f D_t. \qquad (1.21)$$

The system design should allow one to optimally exploit the available diversity D_O. For instance, in systems with multi-carrier transmission the same information should be transmitted on different sub-carriers and in different time slots, achieving uncorrelated faded replicas of the information in both dimensions.

Uncoded multi-carrier systems with flat fading per sub-channel and time-invariance during one symbol cannot exploit diversity and have a poor performance in time and frequency selective fading channels. Additional methods have to be applied to exploit diversity. One approach is the use of data spreading where each data symbol is spread by a spreading code of length L. This, in combination with interleaving, can achieve performance results which are given for $D_O \geqslant L$ by the closed-form solution for the BER for diversity reception in Rayleigh fading channels according to [40]

$$P_b = \left[\frac{1-\gamma}{2}\right]^L \sum_{l=0}^{L-1} \binom{L-1+l}{l} \left[\frac{1+\gamma}{2}\right]^l, \qquad (1.22)$$

where $\binom{n}{k}$ represents the combinatory function,

$$\gamma = \sqrt{\frac{1}{1+\sigma^2}}, \qquad (1.23)$$

and σ^2 is the variance of the noise. As soon as the interleaving is not perfect or the diversity offered by the channel is smaller than the spreading code length L, or MC-CDMA with multiple access interference is applied, (1.22) is a lower bound. For $L = 1$, the performance of an OFDM system without forward error correction (FEC) is obtained,

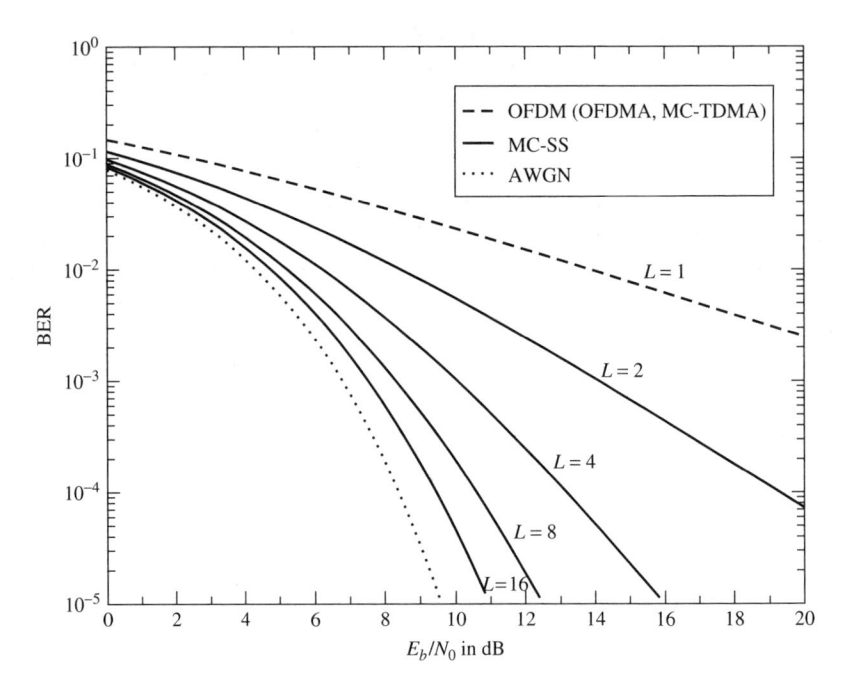

Figure 1-3 Diversity in OFDM and MC-SS systems in a Rayleigh fading channel

which cannot exploit any diversity. The BER according to (1.22) of an OFDM (OFDMA, MC-TDMA) system and a multi-carrier spread spectrum (MC-SS) system with different spreading code lengths L is shown in Figure 1-3. No other diversity techniques are applied. QPSK modulation is used for symbol mapping. The mobile radio channel is modeled as uncorrelated Rayleigh fading channel (see Section 1.1.6). As these curves show, for large values of L, the performance of MC-SS systems approaches that of an AWGN channel.

Another form of achieving diversity in OFDM systems is channel coding by FEC, where the information of each data bit is spread over several code bits. Additional to the diversity gain in fading channels, a coding gain can be obtained due to the selection of appropriate coding and decoding algorithms.

1.2 Multi-Carrier Transmission

The principle of multi-carrier transmission is to convert a serial high-rate data stream onto multiple parallel low-rate sub-streams. Each sub-stream is modulated on another sub-carrier. Since the symbol rate on each sub-carrier is much less than the initial serial data symbol rate, the effects of delay spread, i.e., ISI, significantly decrease, reducing the complexity of the equalizer. OFDM is a low-complex technique to efficiently modulate multiple sub-carriers by using digital signal processing [5][14][26][46][49].

An example of multi-carrier modulation with four sub-channels $N_c = 4$ is depicted in Figure 1-4. Note that the three-dimensional time/frequency/power density representation is used to illustrate the principle of various multi-carrier and multi-carrier spread spectrum

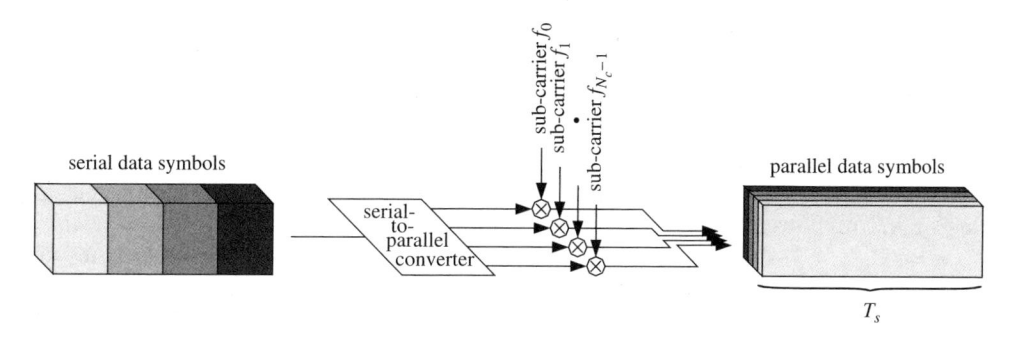

Figure 1-4 Multi-carrier modulation with $N_c = 4$ sub-channels

systems. A cuboid indicates the three-dimensional time/frequency/power density range of the signal, in which most of the signal energy is located and does not make any statement about the pulse or spectrum shaping.

An important design goal for a multi-carrier transmission scheme based on OFDM in a mobile radio channel is that the channel can be considered as time-invariant during one OFDM symbol and that fading per sub-channel can be considered as flat. Thus, the OFDM symbol duration should be smaller than the coherence time $(\Delta t)_c$ of the channel and the sub-carrier spacing should be smaller than the coherence bandwidth $(\Delta f)_c$ of the channel. By fulfilling these conditions, the realization of low-complex receivers is possible.

1.2.1 Orthogonal Frequency Division Multiplexing (OFDM)

A communication system with multi-carrier modulation transmits N_c complex-valued source symbols[1] $S_n, n = 0, \ldots, N_c - 1$, in parallel on N_c sub-carriers. The source symbols may, for instance, be obtained after source and channel coding, interleaving, and symbol mapping. The source symbol duration T_d of the serial data symbols results after serial-to-parallel conversion in the OFDM symbol duration

$$T_s = N_c T_d. \qquad (1.24)$$

The principle of OFDM is to modulate the N_c sub-streams on sub-carriers with a spacing of

$$F_s = \frac{1}{T_s} \qquad (1.25)$$

in order to achieve orthogonality between the signals on the N_c sub-carriers, presuming a rectangular pulse shaping. The N_c parallel modulated source symbols $S_n, n = 0, \ldots, N_c - 1$, are referred to as an OFDM symbol. The complex envelope of an OFDM symbol with rectangular pulse shaping has the form

$$x(t) = \frac{1}{N_c} \sum_{n=0}^{N_c-1} S_n e^{j2\pi f_n t}, \quad 0 \leqslant t < T_s. \qquad (1.26)$$

---

[1] Variables which can be interpreted as values in the frequency domain like the source symbols S_n, each modulating another sub-carrier frequency, are written with capital letters.

The N_c sub-carrier frequencies are located at

$$f_n = \frac{n}{T_s}, \quad n = 0, \ldots, N_c - 1. \qquad (1.27)$$

The normalized power density spectrum of an OFDM symbol with 16 sub-carriers versus the normalized frequency fT_d is depicted as a solid curve in Figure 1-5. Note that in this figure the power density spectrum is shifted to the center frequency. The symbols S_n, $n = 0, \ldots, N_c - 1$, are transmitted with equal power. The dotted curve illustrates the power density spectrum of the first modulated sub-carrier and indicates the construction of the overall power density spectrum as the sum of N_c individual power density spectra, each shifted by F_s. For large values of N_c, the power density spectrum becomes flatter in the normalized frequency range of $-0.5 \leqslant fT_d \leqslant 0.5$ containing the N_c sub-channels.

Only sub-channels near the band edges contribute to the out-of-band power emission. Therefore, as N_c becomes large, the power density spectrum approaches that of single-carrier modulation with ideal Nyquist filtering.

A key advantage of using OFDM is that multi-carrier modulation can be implemented in the discrete domain by using an IDFT, or a more computationally efficient IFFT. When sampling the complex envelope $x(t)$ of an OFDM symbol with rate $1/T_d$ the samples are

$$x_v = \frac{1}{N_c} \sum_{n=0}^{N_c-1} S_n e^{j2\pi nv/N_c}, \quad v = 0, \ldots, N_c - 1. \qquad (1.28)$$

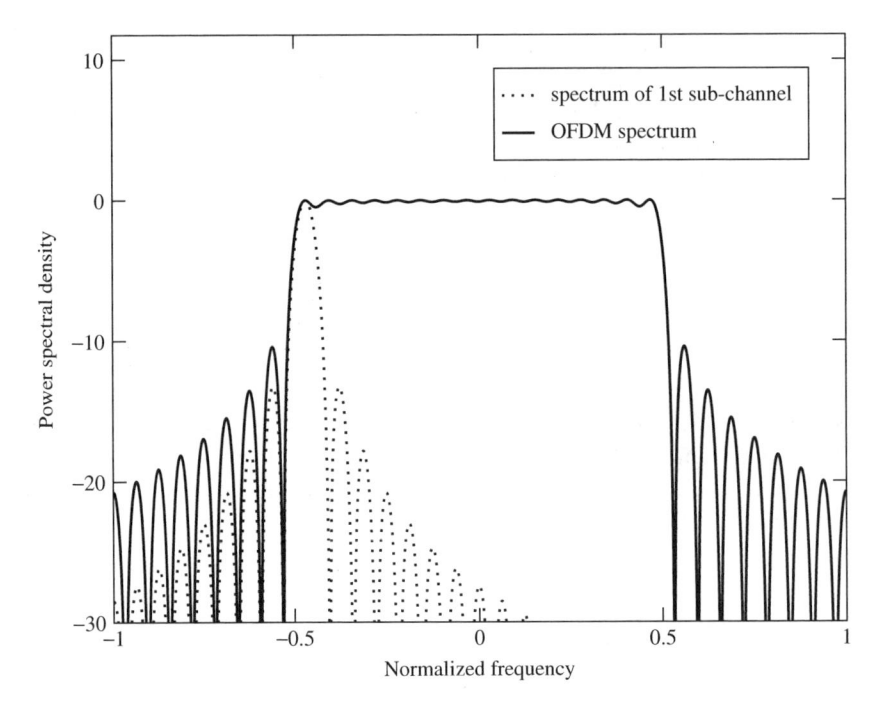

Figure 1-5 OFDM spectrum with 16 sub-carriers

Multi-Carrier Transmission

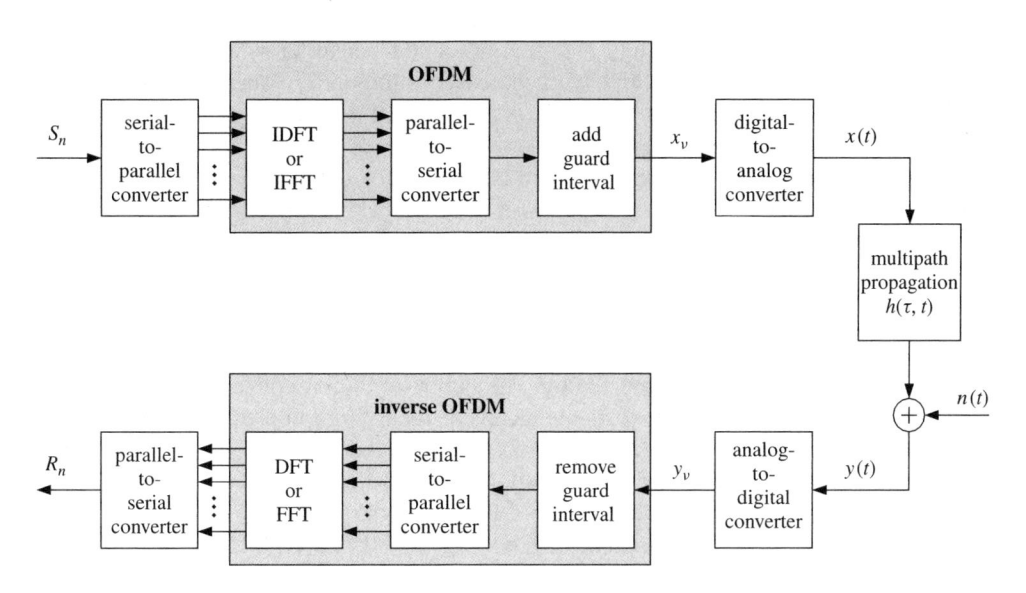

Figure 1-6 Digital multi-carrier transmission system applying OFDM

The sampled sequence x_v, $v = 0, \ldots, N_c - 1$, is the IDFT of the source symbol sequence S_n, $n = 0, \ldots, N_c - 1$. The block diagram of a multi-carrier modulator employing OFDM based on an IDFT and a multi-carrier demodulator employing inverse OFDM based on a DFT is illustrated in Figure 1-6.

When the number of sub-carriers increases, the OFDM symbol duration T_s becomes large compared to the duration of the impulse response $\tau_{\max}$ of the channel, and the amount of ISI reduces.

However, to completely avoid the effects of ISI and, thus, to maintain the orthogonality between the signals on the sub-carriers, i.e., to also avoid ICI, a guard interval of duration

$$T_g \geqslant \tau_{\max} \tag{1.29}$$

has to be inserted between adjacent OFDM symbols. The guard interval is a cyclic extension of each OFDM symbol which is obtained by extending the duration of an OFDM symbol to

$$T'_s = T_g + T_s. \tag{1.30}$$

The discrete length of the guard interval has to be

$$L_g \geqslant \left\lceil \frac{\tau_{\max} N_c}{T_s} \right\rceil \tag{1.31}$$

samples in order to prevent ISI. The sampled sequence with cyclic extended guard interval results in

$$x_v = \frac{1}{N_c} \sum_{n=0}^{N_c-1} S_n e^{j2\pi nv/N_c}, \quad v = -L_g, \ldots, N_c - 1. \tag{1.32}$$

This sequence is passed through a digital-to-analog converter whose output ideally would be the signal waveform $x(t)$ with increased duration T'_s. The signal is up converted and the RF signal is transmitted to the channel (see Chapter 4 regarding RF up/down conversion).

The output of the channel, after RF down conversion, is the received signal waveform $y(t)$ obtained from convolution of $x(t)$ with the channel impulse response $h(\tau,t)$ and addition of a noise signal $n(t)$, i.e.,

$$y(t) = \int_{-\infty}^{\infty} x(t-\tau) h(\tau,t) \, d\tau + n(t). \qquad (1.33)$$

The received signal $y(t)$ is passed through an analog-to-digital converter, whose output sequence y_v, $v = -L_g, \ldots, N_c - 1$, is the received signal $y(t)$ sampled at rate $1/T_d$. Since ISI is only present in the first L_g samples of the received sequence, these L_g samples are removed before multi-carrier demodulation. The ISI-free part $v = 0, \ldots, N_c - 1$, of y_v is multi-carrier demodulated by inverse OFDM exploiting a DFT. The output of the DFT is the multi-carrier demodulated sequence R_n, $n = 0, \ldots, N_c - 1$, consisting of N_c complex-valued symbols

$$R_n = \sum_{v=0}^{N_c - 1} y_v e^{-j 2\pi n v / N_c}, \quad n = 0, \ldots, N_c - 1. \qquad (1.34)$$

Since ICI can be avoided due to the guard interval, each sub-channel can be considered separately. Furthermore, when assuming that the fading on each sub-channel is flat and ISI is removed, a received symbol R_n is obtained from the frequency domain representation according to

$$R_n = H_n S_n + N_n, \quad n = 0, \ldots, N_c - 1, \qquad (1.35)$$

where H_n is the flat fading factor and N_n represents the noise of the nth sub-channel. The flat fading factor H_n is the sample of the channel transfer function $H_{n,i}$ according to (1.16) where the time index i is omitted for simplicity. The variance of the noise is given by

$$\sigma^2 = E\{|N_n|^2\}. \qquad (1.36)$$

When ISI and ICI can be neglected, the multi-carrier transmission system shown in Figure 1-6 can be viewed as a discrete time and frequency transmission system with a set of N_c parallel Gaussian channels with different complex-valued attenuations H_n (see Figure 1-7).

A time/frequency representation of an OFDM symbol is shown in Figure 1-8(a). A block of subsequent OFDM symbols, where the information transmitted within these OFDM symbols belongs together, e.g., due to coding and/or spreading in time and frequency direction, is referred to as an OFDM frame. An OFDM frame consisting of N_s OFDM symbols with frame duration

$$T_{fr} = N_s T'_s \qquad (1.37)$$

is illustrated in Figure 1-8(b).

Multi-Carrier Transmission

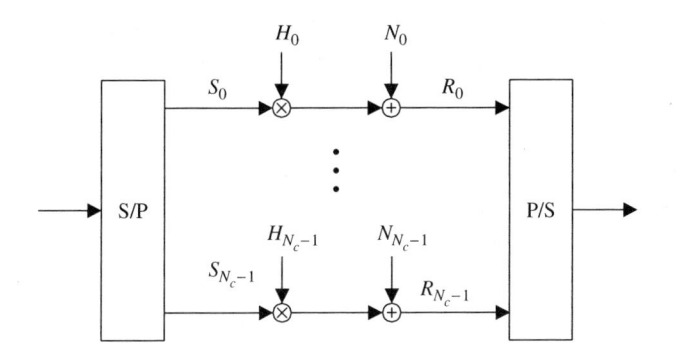

Figure 1-7 Simplified multi-carrier transmission system using OFDM

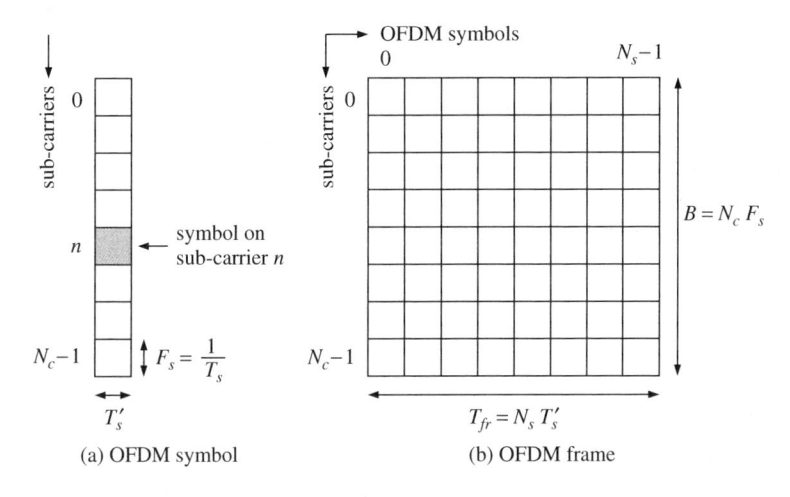

Figure 1-8 Time/frequency representation of an OFDM symbol and an OFDM frame

The following matrix-vector notation is introduced to concisely describe multi-carrier systems. Vectors are represented by boldface small letters and matrices by boldface capital letters. The symbol $(\cdot)^T$ denotes the transposition of a vector or a matrix. The complex-valued source symbols S_n, $n = 0, \ldots, N_c - 1$, transmitted in parallel in one OFDM symbol, are represented by the vector

$$\mathbf{s} = (S_0, S_1, \ldots, S_{N_c-1})^T. \tag{1.38}$$

The $N_c \times N_c$ channel matrix

$$\mathbf{H} = \begin{pmatrix} H_{0,0} & 0 & \cdots & 0 \\ 0 & H_{1,1} & & 0 \\ \vdots & & \ddots & \vdots \\ 0 & 0 & \cdots & H_{N_c-1,N_c-1} \end{pmatrix} \tag{1.39}$$

is of diagonal type in the absence of ISI and ICI. The diagonal components of **H** are the complex-valued flat fading coefficients assigned to the N_c sub-channels. The vector

$$\mathbf{n} = (N_0, N_1, \ldots, N_{N_c-1})^T \quad (1.40)$$

represents the additive noise. The received symbols obtained after inverse OFDM are given by the vector

$$\mathbf{r} = (R_0, R_1, \ldots, R_{N_c-1})^T \quad (1.41)$$

and are obtained by

$$\mathbf{r} = \mathbf{Hs} + \mathbf{n}. \quad (1.42)$$

1.2.2 Advantages and Drawbacks of OFDM

This section summarizes the strengths and weaknesses of multi-carrier modulation based on OFDM.

Advantages:
— High spectral efficiency due to nearly rectangular frequency spectrum for high numbers of sub-carriers.
— Simple digital realization by using the FFT operation.
— Low complex receivers due to the avoidance of ISI and ICI with a sufficiently long guard interval.
— Flexible spectrum adaptation can be realized, e.g., notch filtering.
— Different modulation schemes can be used on individual sub-carriers which are adapted to the transmission conditions on each sub-carrier, e.g., water filling.

Disadvantages:
— Multi-carrier signals with high peak-to-average power ratio (PAPR) require high linear amplifiers. Otherwise, performance degradations occur and the out-of-band power will be enhanced.
— Loss in spectral efficiency due to the guard interval.
— More sensitive to Doppler spreads then single-carrier modulated systems.
— Phase noise caused by the imperfections of the transmitter and receiver oscillators influence the system performance.
— Accurate frequency and time synchronization is required.

1.2.3 Applications and Standards

The key parameters of various multi-carrier-based communications standards for broadcasting, WLAN and WLL, are summarized in Tables 1-2 to 1-4.

1.3 Spread Spectrum Techniques

Spread spectrum systems have been developed since the mid-1950s. The initial applications have been military antijamming tactical communications, guidance systems, and experimental anti-multipath systems [39][43].

Table 1-2 Broadcasting standards DAB and DVB-T

Parameter	DAB			DVB-T	
Bandwidth	1.5 MHz			8 MHz	
Number of sub-carriers N_c	192 (256 FFT)	384 (512 FFT)	1536 (2k FFT)	1705 (2k FFT)	6817 (8k FFT)
Symbol duration T_s	125 µs	250 µs	1 ms	224 µs	896 µs
Carrier spacing F_s	8 kHz	4 kHz	1 kHz	4.464 kHz	1.116 kHz
Guard time T_g	31 µs	62 µs	246 µs	$T_s/32, T_s/16, T_s/8, T_s/4$	
Modulation	D-QPSK			QPSK, 16-QAM, 64-QAM	
FEC coding	Convolutional with code rate 1/3 up to 3/4			Reed Solomon + convolutional with code rate 1/2 up to 7/8	
Max. data rate	1.7 Mbit/s			31.7 Mbit/s	

Table 1-3 Wireless local area network (WLAN) standards

Parameter	IEEE 802.11a, HIPERLAN/2
Bandwidth	20 MHz
Number of sub-carriers N_c	52 (64 FFT)
Symbol duration T_s	4 µs
Carrier spacing F_s	312.5 kHz
Guard time T_g	0.8 µs
Modulation	BPSK, QPSK, 16-QAM, and 64-QAM
FEC coding	Convolutional with code rate 1/2 up to 3/4
Max. data rate	54 Mbit/s

Literally, a spread spectrum system is one in which the transmitted signal is spread over a wide frequency band, much wider than the minimum bandwidth required to transmit the information being sent (see Figure 1-9). Band spreading is accomplished by means of a code which is independent of the data. A reception synchronized to the code is used to despread and recover the data at the receiver [47][48].

Table 1-4 Wireless local loop (WLL) standards

Parameter	Draft IEEE 802.16a, HIPERMAN	
Bandwidth	from 1.5 to 28 MHz	
Number of sub-carriers N_c	256 (OFDM mode)	2048 (OFDMA mode)
Symbol duration T_s	from 8 to 125 µs (depending on bandwidth)	from 64 to 1024 µs (depending on bandwidth)
Guard time T_g	from 1/32 up to 1/4 of T_s	
Modulation	QPSK, 16-QAM, and 64-QAM	
FEC coding	Reed Solomon + convolutional with code rate 1/2 up to 5/6	
Max. data rate (in a 7 MHz channel)	up to 26 Mbit/s	

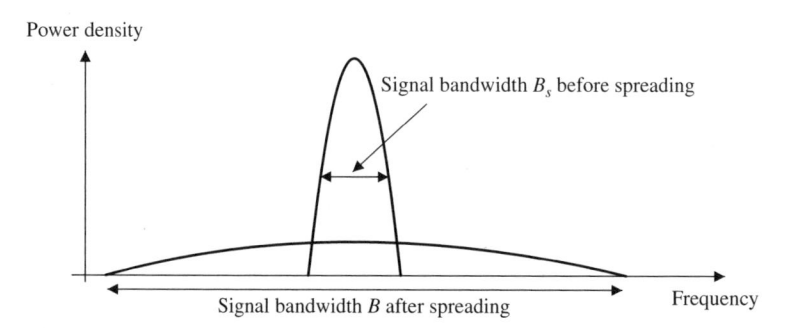

Figure 1-9 Power spectral density after direct sequence spreading

There are many application fields for spreading the spectrum [13]:

— Antijamming,
— Interference rejection,
— Low probability of intercept,
— Multiple access,
— Multipath reception,
— Diversity reception,
— High resolution ranging,
— Accurate universal timing.

There are two primary spread spectrum concepts for multiple access: direct sequence code division multiple access (DS-CDMA) and frequency hopping code division multiple access (FH-CDMA).

The general principle behind DS-CDMA is that the information signal with bandwidth B_s is spread over a bandwidth B, where $B \gg B_s$. The processing gain is specified as

$$P_G = \frac{B}{B_s}. \qquad (1.43)$$

The higher the processing gain, the lower the power density one needs to transmit the information. If the bandwidth is very large, the signal can be transmitted such that it appears like a noise. Here, for instance ultra wide band (UWB) systems (see Chapter 3) can be mentioned as a example [37].

One basic design problem with DS-CDMA is that, when multiple users access the same spectrum, it is possible that a single user could mask all other users at the receiver side if its power level is too high. Hence, accurate power control is an inherent part of any DS-CDMA system [39].

For signal spreading, pseudorandom noise (PN) codes with good cross- and autocorrelation properties are used [38]. A PN code is made up from a number of *chips* for mixing the data with the code (see Figure 1-10). In order to recover the received signal, the code which the signal was spread with in the transmitter is reproduced in the receiver and mixed with the spread signal. If the incoming signal and the locally generated PN code are synchronized, the original signal after correlation can be recovered. In a multiuser environment, the user signals are distinguished by different PN codes and the receiver needs only knowledge of the user's PN code and has to synchronize with it. This principle of user separation is referred to as DS-CDMA. The longer the PN code is, the more noise-like signals appear. The drawback is that synchronization becomes more difficult unless synchronization information such as pilot signals is sent to aid acquisition.

Frequency hopping (FH) is similar to direct sequence spreading where a code is used to spread the signal over a much larger bandwidth than that required to transmit the signal. However, instead of spreading the signal over a continuous bandwidth by mixing the signal with a code, the signal bandwidth is unchanged and is hopped over a number of channels, each having the same bandwidth as the transmitted signal. Although at any instant the transmit power level in any narrowband region may be higher than with DS-CDMA, the signal may be present in a particular channel for a very small time period.

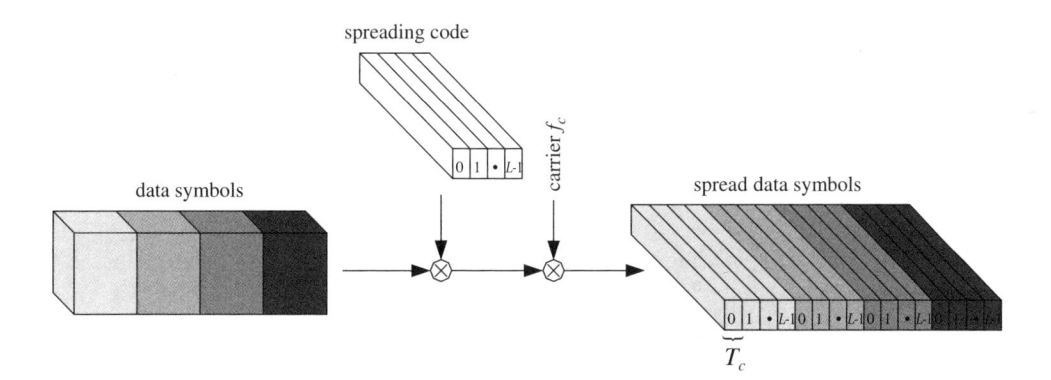

Figure 1-10 Principle of DS-CDMA

For detection, the receiver must know in advance the hopping pattern, unless it will be very difficult to detect the signal. It is the function of the PN code to ensure that all frequencies in the total available bandwidth are optimally used.

There are two kinds of frequency hopping [13]: slow frequency hopping (SFH) and fast frequency hopping (FFH). With SFH many symbols are transmitted per hop. FFH means that there are many hops per symbol. FFH is more resistant to jamming but it is more complex to implement since fast frequency synthesizers are required.

In order to reduce complexity, a hybrid DS/FH scheme can be considered. Here, the signal is first spread over a bandwidth as in DS-CDMA and then hopped over a number of channels, each with bandwidth equal to the bandwidth of the DS spread signal. This allows one to use a much larger bandwidth than with conventional DS spreading by using low cost available components. For instance, if we have a 1 GHz spectrum available, a PN code generator producing 10^9 chips/s or hopping achieving 10^9 hops/s might not be practicable. Alternatively, we could use two code generators: one for spreading the signal and the other for producing the hopping pattern. Both codes could be generated using low cost components.

1.3.1 Direct Sequence Code Division Multiple Access

The principle of DS-CDMA is to spread a data symbol with a spreading sequence $c^{(k)}(t)$ of length L,

$$c^{(k)}(t) = \sum_{l=0}^{L-1} c_l^{(k)} p_{T_c}(t - lT_c), \tag{1.44}$$

assigned to user k, $k = 0, \ldots, K-1$, where K is the total number of active users. The rectangular pulse $p_{T_c}(t)$ is equal to 1 for $0 \leqslant t < T_c$ and zero otherwise. T_c is the chip duration and $c_l^{(k)}$ are the chips of the user specific spreading sequence $c^{(k)}(t)$. After spreading, the signal $x^{(k)}(t)$ of user k is given by

$$x^{(k)}(t) = d^{(k)} \sum_{l=0}^{L-1} c_l^{(k)} p_{T_c}(t - lT_c), \quad 0 \leqslant t < T_d, \tag{1.45}$$

for one data symbol duration $T_d = LT_c$, where $d^{(k)}$ is the transmitted data symbol of user k. The multiplication of the information sequence with the spreading sequence is done bit-synchronously and the overall transmitted signal $x(t)$ of all K synchronous users (case downlink of a cellular system) results in

$$x(t) = \sum_{k=0}^{K-1} x^{(k)}(t). \tag{1.46}$$

The proper choice of spreading sequences is a crucial problem in DS-CDMA, since the multiple access interference strongly depends on the cross-correlation function (CCF) of the used spreading sequences. To minimize the multiple access interference, the CCF values should be as small as possible [41]. In order to guarantee equal interference among all transmitting users, the cross-correlation properties between different pairs of spreading sequences should be similar. Moreover, the autocorrelation function (ACF) of the spreading sequences should have low out-of-phase peak magnitudes in order to achieve a reliable synchronization.

The received signal $y(t)$ obtained at the output of the radio channel with impulse response $h(t)$ can be expressed as

$$y(t) = x(t) \otimes h(t) + n(t) = r(t) + n(t)$$

$$= \sum_{k=0}^{K-1} r^{(k)}(t) + n(t) \qquad (1.47)$$

where $r^{(k)}(t) = x^{(k)}(t) \otimes h(t)$ is the noise-free received signal of user k, $n(t)$ is the additive white Gaussian noise (AWGN), and $\otimes$ denotes the convolution operation. The impulse response of the matched filter (MF) $h_{MF}^{(k)}(t)$ in the receiver of user k is adapted to both the transmitted waveform including the spreading sequence $c^{(k)}(t)$ and to the channel impulse response $h(t)$,

$$h_{MF}^{(k)}(t) = c^{(k)*}(-t) \otimes h^*(-t). \qquad (1.48)$$

The notation x^* denotes the conjugate of the complex value x. The signal $z^{(k)}(t)$ after the matched filter of user k can be written as

$$z^{(k)}(t) = y(t) \otimes h_{MF}^{(k)}(t)$$

$$= r^{(k)}(t) \otimes h_{MF}^{(k)}(t) + \sum_{\substack{g=0 \\ g \neq k}}^{K-1} r^{(g)}(t) \otimes h_{MF}^{(g)}(t) + n(t) \otimes h_{MF}^{(k)}(t). \qquad (1.49)$$

After sampling at the time-instant $t = 0$, the decision variable $\rho^{(k)}$ for user k results in

$$\rho^{(k)} = z^{(k)}(0)$$

$$= \int_0^{T_d + \tau_{max}} r^{(k)}(\tau) h_{MF}^{(k)}(\tau) \, d\tau + \sum_{\substack{g=0 \\ g \neq k}}^{K-1} \int_0^{T_d + \tau_{max}} r^{(g)}(\tau) h_{MF}^{(g)}(\tau) \, d\tau$$

$$+ \int_0^{T_d + \tau_{max}} n(\tau) h_{MF}^{(k)}(\tau) \, d\tau, \qquad (1.50)$$

where τ_{max} is the maximum delay of the radio channel.

Finally, a threshold detection on $\rho^{(k)}$ is performed to obtain the estimated information symbol $\hat{d}^{(k)}$. The first term in the above equation is the desired signal part of user k, whereas the second term corresponds to the multiple access interference and the third term is the additive noise. It should be noted that due to the multiple access interference the estimate of the information bit might be wrong with a certain probability even at high SNRs, leading to the well-known error-floor in the BER curves of DS-CDMA systems.

Ideally, the matched filter receiver resolves all multipath propagation in the channel. In practice a good approximation of a matched filter receiver is a rake receiver [40][43] (see Section 1.3.1.2). A rake receiver has D arms to resolve D echoes where D might be limited by the implementation complexity. In each arm d, $d = 0, \ldots, D-1$, the received signal $y(t)$ is delayed and despread with the code $c^{(k)}(t)$ assigned to user k and weighted with the conjugate instantaneous value $h_d^*, d = 0, \ldots, D-1$, of the time-varying complex channel attenuation of the assigned echo. Finally, the rake receiver combines the results obtained from each arm and makes a final decision.

1.3.1.1 DS-CDMA Transmitter

Figure 1-11 shows a direct sequence spread spectrum transmitter [40]. It consists of a forward error correction (FEC) encoder, mapping, spreader, pulse shaper, and analog front-end (IF/RF part). Channel coding is required to protect the transmitted data against channel errors. The encoded and mapped data are spread with the code $c^{(k)}(t)$ over a much wider bandwidth than the bandwidth of the information signal. As the power of the output signal is distributed over a wide bandwidth, the power density of the output signal is much lower than that of the input signal. Note that the multiplication process is done with a spreading sequence with no DC component.

The chip rate directly influences the bandwidth and with that the processing gain. The wider the bandwidth, the better the resolution in multipath detection. Since the total transmission bandwidth is limited, a pulse shaping filtering is employed (e.g., a root Nyquist filter) so that the frequency spectrum is used efficiently.

1.3.1.2 DS-CDMA Receiver

In Figure 1-12, the receiver block-diagram of a DS-CDMA signal is plotted [40]. The received signal is first filtered and then digitally converted with a sampling rate of $1/T_c$. It is followed by a rake receiver. The rake receiver is necessary to combat multipath, i.e., to combine the power of each received echo path. The echo paths are detected with a resolution of T_c. Therefore, each received signal of each path is delayed by lT_c and

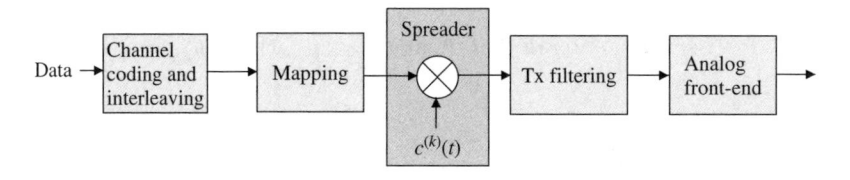

Figure 1-11 DS spread spectrum transmitter block diagram

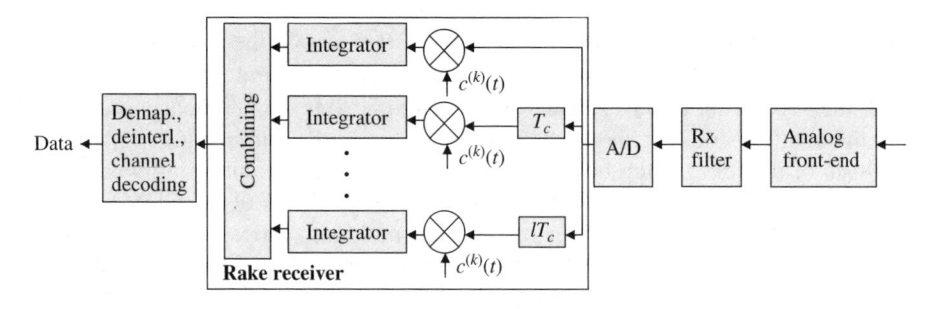

Figure 1-12 DS-CDMA rake receiver block diagram

correlated with the assigned code sequence. The total number of resolution paths depends on the processing gain. Usually in practice 3–4 arms are used. After correlation, the power of all detected paths are combined and, finally, the demapping and FEC decoding are performed to assure the data integrity.

1.3.2 Advantages and Drawbacks of DS-CDMA

Conventional DS-CDMA systems offer several advantages in cellular environments including easy frequency planning, high immunity against interference if a high processing gain is used, and flexible data rate adaptation.

Besides these advantages, DS-CDMA suffers from several problems in multiuser wireless communications systems with limited available bandwidth [25]:

— *Multiple access interference (MAI)*
  As the number of simultaneously active users increases, the performance of the DS-CDMA system decreases rapidly, since the capacity of a DS-CDMA system with moderate processing gain (limited spread bandwidth) is limited by MAI.
— *Complexity*
  In order to exploit all multipath diversity it is necessary to apply a matched filter receiver approximated by a rake receiver with sufficient number of arms, where the required number of arms is $D = \tau_{max}/T_c + 1$ [40]. In addition, the receiver has to be matched to the time-variant channel impulse response. Thus, proper channel estimation is necessary. This leads to additional receiver complexity with adaptive receiver filters and a considerable signaling overhead.
— *Single-/Multitone interference*
  In the case of single-tone or multitone interference the conventional DS-CDMA receiver spreads the interference signal over the whole transmission bandwidth B whereas the desired signal part is despread. If this interference suppression is not sufficient, additional operations have to be done at the receiver, such as Notch filtering in the time domain (based on the least mean square algorithm) or in the frequency domain (based on fast Fourier transform) to partly decrease the amount of interference [30][34]. Hence, this extra processing leads to additional receiver complexity.

1.3.3 Applications of Spread Spectrum

To illustrate the importance of the spread spectrum technique in today's wireless communications we will briefly introduce two examples of its deployment in cellular mobile communications systems. Here we will describe the main features of the IS-95 standard and the third-generation standard UMTS.

1.3.3.1 IS-95

The first commercial cellular mobile radio communication system based on the spread spectrum was the IS-95 standard [42]. This standard was developed in the USA just after the introduction of GSM in Europe. IS-95 is based on frequency division duplex (FDD). The available bandwidth is divided into channels with 1.25 MHz (nominal 1.23 MHz) bandwidth.

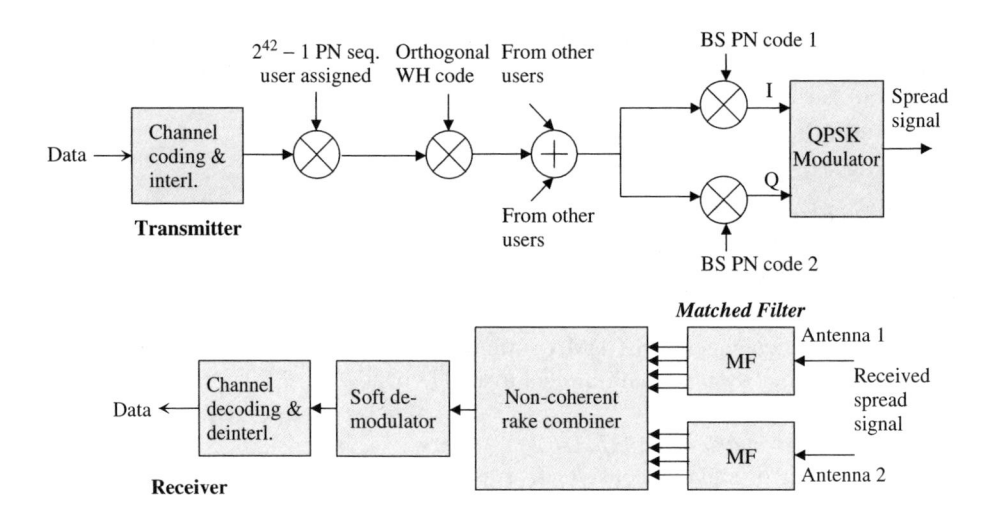

Figure 1-13 Simplified block diagram of the base station IS-95 transceiver

As shown in Figure 1-13, in the downlink, binary PN codes are used to distinguish signals received at the terminal station from different base stations. All CDMA signals share a quadrature pair of PN codes. Signals from different cells and sectors are distinguished by the time offset from the basic code. The PN codes used are generated by linear shift registers that produce a code with a period of 32768 chips. Two codes are generated, one for each quadrature carrier (I and Q) of QPSK type of modulation.

As mentioned earlier, signals (traffic or control) transmitted from a single antenna (e.g., a base station sector antenna) in a particular CDMA radio channel share a common PN code phase. The traffic and control signals are distinguished at the terminal station receiver by using a binary Walsh–Hadamard (WH) orthogonal code with spreading factor of 64.

The transmitted downlink information (e.g., voice of rate 9.6 kbit/s) is first convolutionally encoded with rate 1/2 and memory 9 (see Figure 1-13). To provide communication privacy, each user's signal is scrambled with a user-addressed long PN code sequence. Each data symbol is spread using orthogonal WH codes of length 64. After superposition of the spread data of all active users, the resulting signal is transmitted to the in-phase and to the quadrature components, i.e., QPSK modulated by a pair of PN codes with an assigned offset. Furthermore, in the downlink a pilot signal is transmitted by each cell site and is used as a coherent carrier reference for demodulation by all mobile receivers. The pilot channel signal is the *zero* WH code sequence.

The transmitted uplink information is concatenated encoded (see Figure 1-14). The outer code is a convolutional code of rate 1/3 and memory 9. The encoded information is grouped into 6 symbol groups which are used to select one of the different WH inner code words of length 64 (rate 6/64). The signal from each terminal station is distinguished by the use of a very long ($2^{42} - 1$) PN code (privacy code) with a user address-determined time offset. Finally, the same information is transmitted in the in-phase (I) and quadrature (Q) component of an offset QPSK type modulator, where the I and Q components are multiplied by long PN codes.

Spread Spectrum Techniques

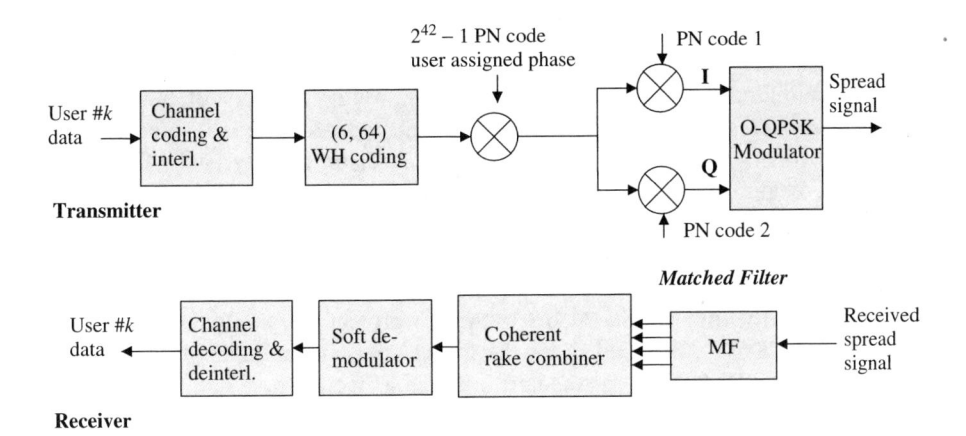

Figure 1-14 Simplified block diagram of the IS-95 terminal station transceiver

In Table 1-5 important parameters of the IS-95 standard are summarized. Note that in IS-95 the WH code in the uplink is used for FEC, which together with convolutional coding results in a very low code rate, hence, guaranteeing very good protection. This is different from the downlink, where the WH code is used for signal spreading. Furthermore, the use of WH codes in the uplink allows one to perform noncoherent detection at the base station. It saves the transmission of pilot symbols from terminal stations.

Table 1-5 Radio link parameters of IS-95

Parameter	IS-95
Bandwidth	1.25 MHz
Duplex scheme	Frequency division duplex (FDD)
Spreading code short/long	Walsh–Hadamard orthogonal code/PN code
Modulation	Coherent QPSK for the downlink Noncoherence offset QPSK for the uplink
Channel coding	DL: Convolutional $R = 1/2$, memory 9 UL: Convolutional $R = 1/3$, memory 9 with WH(6,64)
Processing gain	19.3 dB
Max. data rate	14.4 kbit/s for data and 9.6 kbit/s for voice
Diversity	Rake + antenna
Power control	Fast power control based on signal-to-interference ratio (SIR) measurement

1.3.3.2 UMTS

The major services of the 2nd generation mobile communication systems are limited to voice, facsimile and low-rate data transmission. With a variety of new high-speed multimedia services such as high-speed internet and video/high quality image transmission the need for higher data rates increases. The research activity on UMTS started in Europe at the beginning of the 1990s. Several EU-RACE projects such as CODIT and A-TDMA were dealing deeply with the study of the third-generation mobile communications systems. Within the CODIT project a wideband CDMA testbed was built, showing the feasibility of a flexible CDMA system [3]. Further detailed parameters for the 3G system were specified within the EU-ACTS FRAMES project [4]. In 1998, ETSI decided to adopt wideband CDMA (W-CDMA) for the frequency division duplex bands [18]. Later on, ARIB approved W-CDMA as standard in Japan as well, where both ETSI and ARIB use the same W-CDMA concept.

The third-generation mobile communication systems, called International Mobile Telecommunications-2000 (IMT-2000) or Universal Mobile Telecommunications System (UMTS), are designed to support wideband services with data rates up to 2 Mbit/s. The carrier frequency allocated for UMTS is about 2 GHz. In case of FDD, the allocated total bandwidth is 2×60 MHz: the uplink carrier frequency is 1920–1980 MHz and the downlink carrier frequency is 2110–2170 MHz.

In Table 1-6 key parameters of UMTS are outlined. In Figures 1-15 and 1-16, simplified block diagrams of a base station and a terminal station are illustrated. In contrast to IS-95, the UMTS standard applies variable length orthogonal spreading codes and coherent QPSK detection for both uplink and downlink directions [1]. The generation of the orthogonal variable spreading code [12] is illustrated in Figure 1-17. Note that for scrambling and spreading, complex codes are employed.

Table 1-6 Radio link parameters of UMTS

Parameter	UMTS
Bandwidth (MHz)	1.25/5/10/20
Duplex scheme	FDD and TDD
Spreading code short/long	Tree-structured orthogonal variable spreading factor (VSF)/PN codes
Modulation	Coherent QPSK (DL/UL)
Channel coding	Voice: Convolutional $R = 1/3$, memory 9 Data: Concatenated Reed Solomon (RS) + convolutional High rate high quality services: Convolutional Turbo codes
Diversity	Rake + antenna
Power control	Fast power control based on SIR measurement

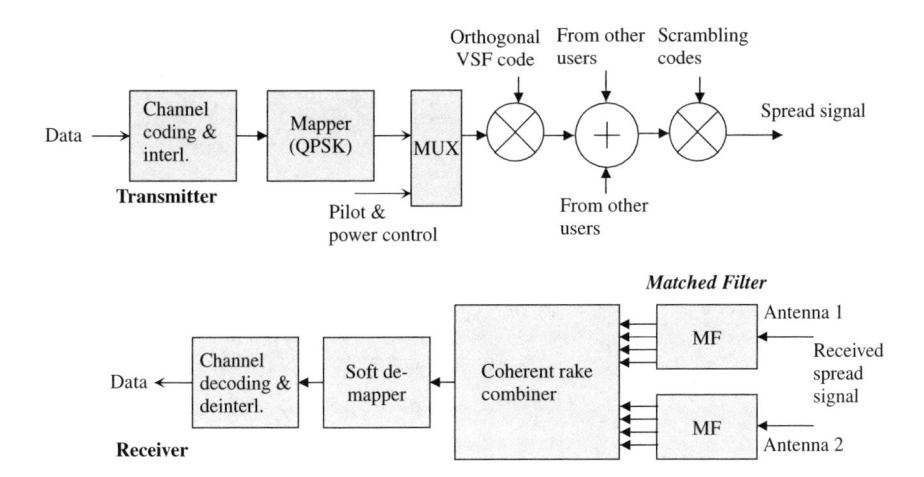

Figure 1-15 Simplified block diagram of a UMTS base station transceiver

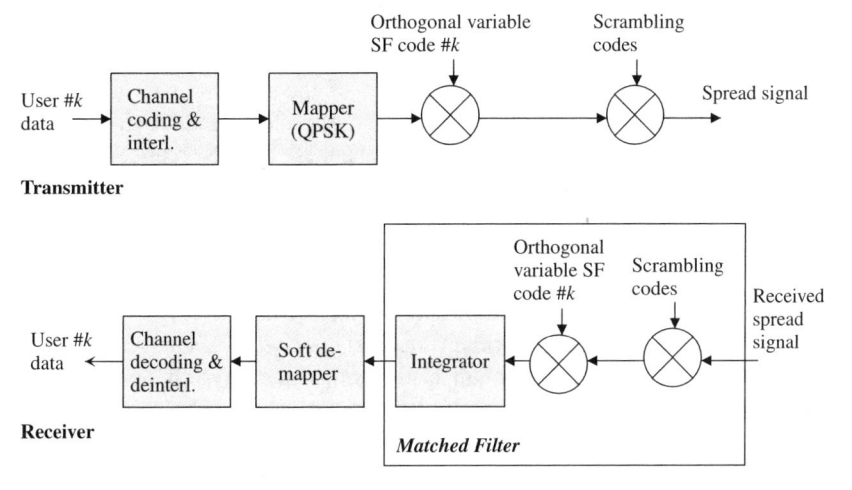

Figure 1-16 Simplified block diagram of a UMTS terminal station transceiver

1.4 Multi-Carrier Spread Spectrum

The success of the spread spectrum techniques for second-generation mobile radio and OFDM for digital broadcasting and wireless LANs motivated many researchers to investigate the combination of both techniques. The combination of DS-CDMA and multi-carrier modulation was proposed in 1993 [6][11][19][21][31][45][50]. Two different realizations of multiple access exploiting multi-carrier spread spectrums are detailed in this section.

1.4.1 Principle of Various Schemes

The first realization is referred to as MC-CDMA, also known as OFDM-CDMA. The second realization is termed as MC-DS-CDMA. In both schemes, the different users

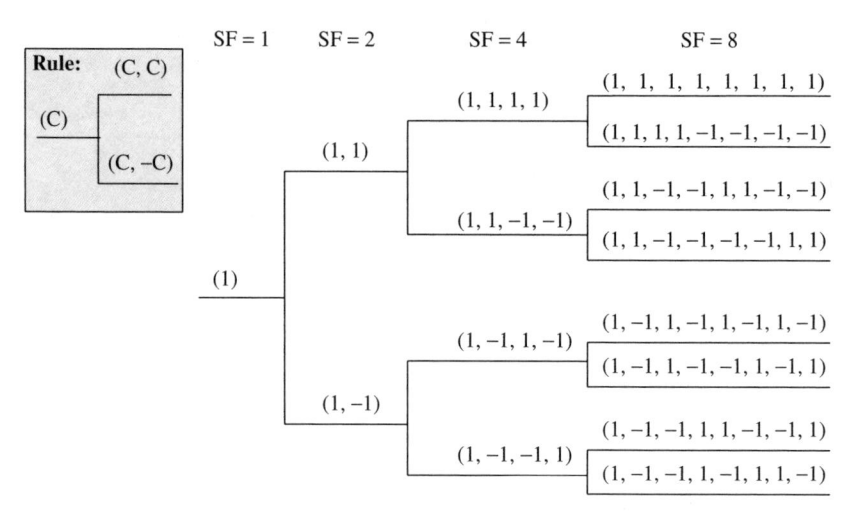

Figure 1-17 Variable length orthogonal spreading code generation

share the same bandwidth at the same time and separate the data by applying different user specific spreading codes, i.e., the separation of the users signals is carried out in the code domain. Moreover, both schemes apply multi-carrier modulation to reduce the symbol rate and, thus, the amount of ISI per sub-channel. This ISI reduction is significant in spread spectrum systems where high chip rates occur.

The difference between MC-CDMA and MC-DS-CMDA is the allocation of the chips to the sub-channels and OFDM symbols. This difference is illustrated in Figures 1-18 and 1-19. The principle of MC-CDMA is to map the chips of a spread data symbol in frequency direction over several parallel sub-channels while MC-DS-CDMA maps the chips of a spread data symbol in the time direction over several multi-carrier symbols.

MC-CDMA transmits a data symbol of a user simultaneously on several narrowband sub-channels. These sub-channels are multiplied by the chips of the user-specific spreading code, as illustrated in Figure 1-18. Multi-carrier modulation is realized by using the low-complex OFDM operation. Since the fading on the narrowband sub-channels can

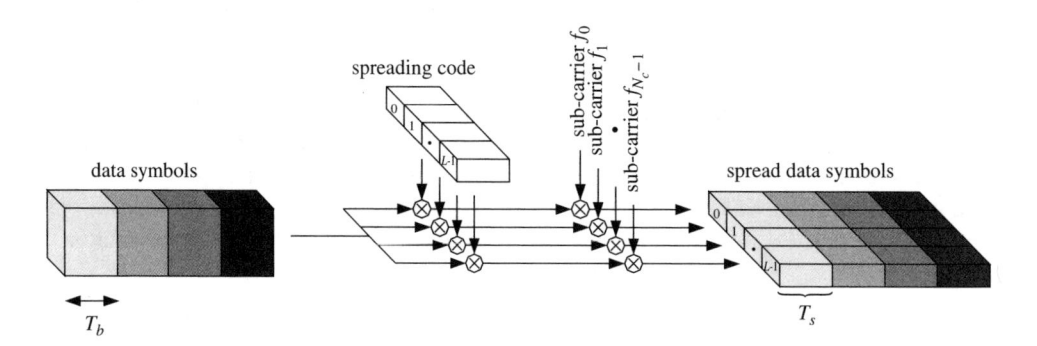

Figure 1-18 MC-CDMA signal generation for one user

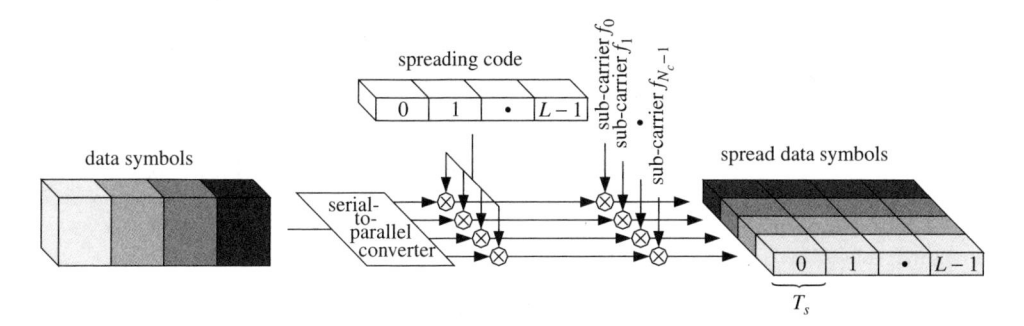

Figure 1-19 MC-DS-CDMA signal generation for one user

be considered flat, simple equalization with one complex-valued multiplication per sub-channel can be realized. MC-CDMA offers a flexible system design, since the spreading code length does not have to be chosen equal to the number of sub-carriers, allowing adjustable receiver complexities. This flexibility is described in detail in Chapter 2.

MC-DS-CDMA serial-to-parallel converts the high-rate data symbols into parallel low-rate sub-streams before spreading the data symbols on each sub-channel with a user-specific spreading code in time direction, which corresponds to direct sequence spreading on each sub-channel. The same spreading codes can be applied on the different sub-channels. The principle of MC-DS-CDMA is illustrated in Figure 1-19.

MC-DS-CDMA systems have been proposed with different multi-carrier modulation schemes, also without OFDM, such that within the description of MC-DS-CDMA the general term multi-carrier symbol instead of OFDM symbol is used. The MC-DS-CDMA schemes can be subdivided in schemes with broadband sub-channels and schemes with narrowband sub-channels. Systems with broadband sub-channels typically apply only few numbers of sub-channels, where each sub-channel can be considered as a classical DS-CDMA system with reduced data rate and ISI, depending on the number of parallel DS-CDMA systems. MC-DS-CDMA systems with narrowband sub-channels typically use high numbers of sub-carriers and can be efficiently realized by using the OFDM operation. Since each sub-channel is narrowband and spreading is performed in time direction, these schemes can only achieve a time diversity gain if no additional measures such as coding or interleaving are applied.

Both multi-carrier spread spectrum concepts are described in detail in Chapter 2.

1.4.2 Advantages and Drawbacks

In Table 1-7, the main advantages and drawbacks of MC-CDMA and MC-DS-CDMA are summarized.

A first conclusion from this table can be derived:

— The high spectral efficiency and the low receiver complexity of MC-CDMA makes it a good candidate for the downlink of a cellular system.
— The low PAPR property of MC-DS-CDMA makes it more appropriate for the uplink of a multiuser system.

Table 1-7 Advantages and drawbacks of MC-CDMA and MC-DS-CDMA

MC-CDMA		MC-DS-CDMA	
Advantages	Disadvantages	Advantages	Disadvantages
– Simple implementation with HT and FFT – Low complex receivers – High spectral efficiency – High frequency diversity gain due to spreading in frequency direction	– High PAPR especially in the uplink – Synchronous transmission	– Low PAPR in the uplink – High time diversity gain due to spreading in time direction	– ISI and/or ICI can occur, resulting in more complex receivers – Less spectral efficient if other multi-carrier modulation schemes than OFDM are used

1.4.3 Examples of Future Application Areas

Multi-carrier spread spectrum concepts have been developed for a wide variety of applications.

Cellular mobile radio: Due to the high spectral efficiency of MC-CDMA, it is a promising candidate for the high rate downlink with peak data rates in the order of 100 Mbit/s for the fourth generation of mobile radio systems [2]. In the uplink, where data rates in the order of several 20 Mbit/s are considered, MC-DS-CDMA seems to be a promising candidate since it has a lower PAPR compared to MC-CDMA, thus increasing the power efficiency of the mobile terminal. In [20] a further concept of MC-CDMA system for mobile cellular system has been proposed.

DVB-T return link: The DVB-T interactive point to multi-point (PMP) network is intended to offer a variety of services requiring different data rates [15]. Therefore, the multiple access scheme needs to be flexible in terms of data rate assignment to each subscriber. As in the downlink terrestrial channel, its return channels suffer especially from high multipath propagation delays. A derivative of MC-CDMA, namely OFDMA, is already adopted in the standard. Several orthogonal sub-carriers are assigned to each terminal station. However, the assignment of these sub-carriers during the time is hopped following a given spreading code.

MMDS/LMDS (FWA): The aim of microwave/local multi-point distribution systems (MMDS/LMDS) or fixed broadband wireless access (FWA) systems is to provide wireless high speed services with, e.g., IP/ATM to fixed positioned terminal stations with a coverage area from 2 km up to 20 km. In order to maintain reasonably low RF costs and good penetration of the radio signals for residential applications, the FWA systems typically use below 10 GHz carrier frequencies, e.g., the MMDS band (2.5–2.7 GHz) or around 5 GHz. As in the DVB-T return channel, OFDMA with frequency hopping for FWA below 10 GHz is proposed [17][27]. However, for microwave frequencies above 10 GHz, e.g., LMDS, the main channel impairment will be the high amount of CCI due to the dense frequency reuse in a cellular environment. In [32] a system architecture based on MC-CDMA for FWA/LMDS applications is proposed. The suggested system provides a high capacity, is quite robust against multipath effects, and can offer service coverage not only to subscribers with LOS but also to subscribers who do not have LOS.

Aeronautical communications: An increase in air traffic will lead to bottlenecks in air traffic handling en route and on ground. Airports have been identified as one of the most capacity-restricted factors in the future if no counter-measures are taken. New digital standards should replace current analog air traffic control systems. Different concepts for future air traffic control based on multi-carrier spread spectrum have been proposed [23][24].

More potential application fields for multi-carrier spread spectrum are in wireless indoor communications [50] and broadband underwater acoustic communications [35].

1.5 References

[1] Adachi F., Sawahashi M. and Suda H., "Wideband CDMA for next generation mobile communications systems," *IEEE Communications Magazine*, vol. 26, pp. 56–69, June 1988.
[2] Atarashi H., Maeda N., Abeta S. and Sawahashi M., "Broadband packet wireless access based on VSF-OFCDM and MC/DS-CDMA," in *Proc. IEEE International Symposium on Personal, Indoor and Mobile Radio Communications (PIMRC 2002)*, Lisbon, Portugal, pp. 992–997, Sept. 2002.
[3] Baier A., Fiebig U.-C., Granzow W., Koch W., Teder P. and Thielecke J., "Design study for a CDMA-based third-generation mobile radio system, "*IEEE Journal on Selected Areas in Communications*, vol. 12, pp. 733–734, May 1994.
[4] Berruto E., Gudmundson M., Menolascino R., Mohr W. and Pizarroso M., "Research activities on UMTS radio interface, network architectures, and planning," *IEEE Communications Magazine*, vol. 36, pp. 82–95, Feb. 1998.
[5] Bingham J.A.C., "Multicarrier modulation for data transmission: An idea whose time has come," *IEEE Communications Magazine*, vol. 28, pp. 5–14, May 1990.
[6] Chouly A., Brajal A. and Jourdan S., "Orthogonal multicarrier techniques applied to direct sequence spread spectrum CDMA systems," in *Proc. IEEE Global Telecommunications Conference (GLOBECOM'93)*, Houston, USA, pp. 1723–1728, Nov./Dec. 1993.
[7] CODIT, "Final propagation model," *Report R2020/TDE/PS/DS/P/040/b1*, 1994.
[8] COST 207, "Digital land mobile radio communications," *Final Report*, 1989.
[9] COST 231, "Digital mobile radio towards future generation systems," *Final Report*, 1996.
[10] COST 259, "Wireless flexible personalized communications," *Final Report*, L.M. Correira (ed.), John Wiley & Sons, 2001.
[11] DaSilva V. and Sousa E.S., "Performance of orthogonal CDMA codes for quasi-synchronous communication systems," in *Proc. IEEE International Conference on Universal Personal Communications (ICUPC'93)*, Ottawa, Canada, pp. 995–999, Oct. 1993.
[12] Dinan E.H. and Jabbari B. "Spreading codes for direct sequence CDMA and wideband CDMA cellular networks," *IEEE Communications Magazine*, vol. 26, pp. 48–54, June 1988.
[13] Dixon R.C., *Spread Spectrum Systems*. New York: John Wiley & Sons, 1976.
[14] Engels M. (ed.), *Wireless OFDM Systems: How to Make Them Work*. Boston: Kluwer Academic Publishers, 2002.
[15] ETSI DVB-RCT (EN 301 958), "Interaction channel for digital terrestrial television (RCT) incorporating multiple access OFDM," Sophia Antipolis, France, March 2001.
[16] ETSI DVB-T (EN 300 744), "Digital video broadcasting (DVB); framing structure, channel coding and modulation for digital terrestrial television," Sophia Antipolis, France, July 1999.
[17] ETSI HIPERMAN (Draft TS 102 177), "High performance metropolitan local area networks, Part 1: Physical layer," Sophia Antipolis, France, Feb. 2003.
[18] ETSI UMTS (TR 101 112), "Universal mobile telecommunications system (UMTS)," Sophia Antipolis, France, 1998.

[19] Fazel K., "Performance of CDMA/OFDM for mobile communication system," in *Proc. IEEE International Conference on Universal Personal Communications (ICUPC'93)*, Ottawa, Canada, pp. 975–979, Oct. 1993.
[20] Fazel K., Kaiser S. and Schnell M., "A flexible and high performance cellular mobile communications system based on multi-carrier SSMA," *Wireless Personal Communications*, vol. 2, nos. 1 & 2, pp. 121–144, 1995.
[21] Fazel K. and Papke L., "On the performance of convolutionally-coded CDMA/OFDM for mobile communication system," in *Proc. IEEE International Symposium on Personal, Indoor and Mobile Radio Communications (PIMRC'93)*, Yokohama, Japan, pp. 468–472, Sept. 1993.
[22] Fettweis G., Bahai A.S. and Anvari K., "On multi-carrier code division multiple access (MC-CDMA) modem design," in *Proc. IEEE Vehicular Technology Conference (VTC'94)*, Stockholm, Sweden, pp. 1670–1674, June 1994.
[23] Haas E., Lang H. and Schnell M., "Development and implementation of an advanced airport data link based on multi-carrier communications," *European Transactions on Telecommunications (ETT)*, vol. 13, no. 5, pp. 447–454, Sept./Oct. 2002.
[24] Haindl B., "Multi-carrier CDMA for air traffic control air/ground communication," in *Proc. International Workshop on Multi-Carrier Spread-Spectrum & Related Topics (MC-SS 2001)*, Oberpfaffenhofen, Germany, pp. 77–84, Sept. 2001.
[25] Hara H. and Prasad R., "Overview of multicarrier CDMA," *IEEE Communications Magazine*, vol. 35, pp. 126–133, Dec. 1997.
[26] Heiskala J. and Terry J., *OFDM Wireless LANs: A Theoretical and Practical Guide*. Indianapolis: SAMS, 2002.
[27] IEEE 802.16ab-01/01, "Air interface for fixed broadband wireless access systems – Part A: Systems between 2 and 11 GHz," IEEE 802.16, June 2000.
[28] Joint Technical Committee (JTC) on Wireless Access, *Final Report on RF Channel Characterization*, JTC(AIR)/93.09.23-238R2, Sep. 1993.
[29] Kaiser S., *Multi-Carrier CDMA Mobile Radio Systems – Analysis and Optimization of Detection, Decoding, and Channel Estimation*. Düsseldorf: VDI-Verlag, Fortschritt-Berichte VDI, series 10, no. 531, 1998, PhD thesis.
[30] Ketchum J.W. and Proakis J.G., "Adaptive algorithms for estimating and suppressing narrow band interference in PN spread spectrum systems," *IEEE, Transactions on Communications*, vol. 30, pp. 913–924, May 1982.
[31] Kondo S. and Milstein L.B., "On the use of multicarrier direct sequence spread spectrum systems," in *Proc. IEEE Military Communications Conference (MILCOM'93)*, Boston, USA, pp. 52–56, Oct. 1993.
[32] Li J. and Kaverhard M., "Multicarrier orthogonal-CDMA for fixed wireless access applications," *International Journal of Wireless Information Network*, vol. 8, no. 4, pp. 189–201, Oct. 2001.
[33] Medbo J. and Schramm P., "Channel models for HIPERLAN/2 in different indoor scenarios," *Technical Report ETSI EP BRAN*, 3ERI085B, March 1998.
[34] Milstein L.B., "Interference rejection techniques in spread spectrum communications," *Proceedings of the IEEE*, vol. 76, pp. 657–671, June 1988.
[35] Ormondroyd R.F., Lam W.K. and Davies J., "A multi-carrier spread spectrum approach to broadband underwater acoustic communications," in *Proc. International Workshop on Multi-Carrier Spread Spectrum & Related Topics (MC-SS'99)*, Oberpfaffenhofen, Germany, pp. 63–70, Sept. 1999.
[36] Parsons D., *The Mobile Radio Propagation Channel*. New York: John Wiley & Sons, 1992.
[37] Petroff A. and Withington P., "Time modulated ultra-wideband (TM-UWB) overview," in *Proc. Wireless Symposium 2000*, San Jose, USA, Feb. 2000.
[38] Pickholtz R.L., Milstein L.B. and Schilling D.L., "Spread spectrum for mobile communications", *IEEE Transactions on Vehicular Technology*, vol. 40, no. 2, pp. 313–322, May 1991.
[39] Pickholtz R.L., Schilling D.L. and Milstein L.B., "Theory of spread spectrum communications – a tutorial," *IEEE Transactions on Communications*, vol. 30, pp. 855–884, May 1982.
[40] Proakis J.G., *Digital Communications*. New York: McGraw-Hill, 1995.
[41] Sarwate D.V. and Pursley M.B., "Crosscorrelation properties of pseudo-random and related sequences," *Proceedings of the IEEE*, vol. 88, pp. 593–619, May 1998.
[42] TIA/EIA/IS-95, "Mobile station-base station compatibility standard for dual mode wideband spread spectrum cellular system," July 1993.

References

[43] Turin G.L., "Introduction to spread spectrum anti-multipath techniques and their application to urban digital radio," *Proceedings of the IEEE*, vol. 68, pp. 328–353, March 1980.

[44] UTRA, *Submission of Proposed Radio Transmission Technologies*, SMG2, 1998.

[45] Vandendorpe L., "Multitone direct sequence CDMA system in an indoor wireless environment," in *Proc. IEEE First Symposium of Communications and Vehicular Technology*, Delft, The Netherlands, pp. 4.1.1–4.1.8, Oct. 1993.

[46] van Nee R. and Prasad R., *OFDM for Wireless Multimedia Communications*. Boston: Artech House Publishers, 2000.

[47] Viterbi A.J., "Spread spectrum communications – myths and realities," *IEEE Communications Magazine*, pp. 11–18, May 1979.

[48] Viterbi A.J., *CDMA: Principles of Spread Spectrum Communication*. Reading: Addison-Wesley, 1995.

[49] Weinstein S.B. and Ebert P.M., "Data transmission by frequency-division multiplexing using the discrete Fourier transform," *IEEE Transactions on Communication Technology*, vol. 19, pp. 628–634, Oct. 1971.

[50] Yee N., Linnartz J.P., and Fettweis G., "Multi-carrier CDMA in indoor wireless radio networks," in *Proc. IEEE International Symposium on Personal, Indoor and Mobile Radio Communications (PIMRC'93)*, Yokohama, Japan, pp. 109–113, Sept. 1993.

2

MC-CDMA and MC-DS-CDMA

In this chapter, the different concepts of the combination of multi-carrier transmission with spread spectrum, namely MC-CDMA and MC-DS-CDMA are analyzed. Several single-user and multiuser detection strategies and their performance in terms of BER and spectral efficiency in a mobile communications system are examined.

2.1 MC-CDMA

2.1.1 Signal Structure

The basic MC-CDMA signal is generated by a serial concatenation of classical DS-CDMA and OFDM. Each chip of the direct sequence spread data symbol is mapped onto a different sub-carrier. Thus, with MC-CDMA the chips of a spread data symbol are transmitted in parallel on different sub-carriers, in contrast to a serial transmission with DS-CDMA. The number of simultaneously active users[1] in an MC-CDMA mobile radio system is K.

Figure 2-1 shows multi-carrier spectrum spreading of one complex-valued data symbol $d^{(k)}$ assigned to user k. The rate of the serial data symbols is $1/T_d$. For brevity, but without loss of generality, the MC-CDMA signal generation is described for a single data symbol per user as far as possible, such that the data symbol index can be omitted. In the transmitter, the complex-valued data symbol $d^{(k)}$ is multiplied with the user specific spreading code

$$\mathbf{c}^{(k)} = (c_0^{(k)}, c_1^{(k)}, \ldots, c_{L-1}^{(k)})^T \qquad (2.1)$$

of length $L = P_G$. The chip rate of the serial spreading code $\mathbf{c}^{(k)}$ before serial-to-parallel conversion is

$$\frac{1}{T_c} = \frac{L}{T_d} \qquad (2.2)$$

---

[1] Values and functions related to user k are marked by the index $^{(k)}$, where k may take on the values $0, \ldots, K-1$.

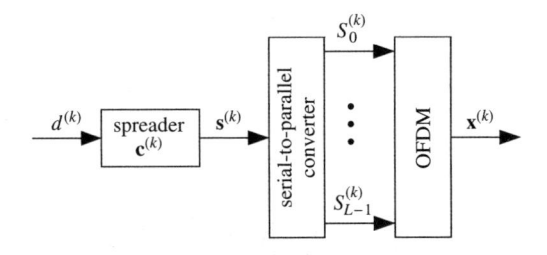

Figure 2-1 Multi-carrier spread spectrum signal generation

and it is L times higher than the data symbol rate $1/T_d$. The complex-valued sequence obtained after spreading is given in vector notations by

$$\mathbf{s}^{(k)} = d^{(k)}\mathbf{c}^{(k)} = (S_0^{(k)}, S_1^{(k)}, \ldots, S_{L-1}^{(k)})^T. \quad (2.3)$$

A multi-carrier spread spectrum signal is obtained after modulating the components $S_l^{(k)}, l = 0, \ldots, L-1$, in parallel onto L sub-carriers. With multi-carrier spread spectrum, each data symbol is spread over L sub-carriers. In cases where the number of sub-carriers N_c of one OFDM symbol is equal to the spreading code length L, the OFDM symbol duration with multi-carrier spread spectrum including a guard interval results in

$$T_s' = T_g + LT_c. \quad (2.4)$$

In this case one data symbol per user is transmitted in one OFDM symbol.

2.1.2 Downlink Signal

In the synchronous downlink, it is computationally efficient to add the spread signals of the K users before the OFDM operation as depicted in Figure 2-2. The superposition of the K sequences $\mathbf{s}^{(k)}$ results in the sequence

$$\mathbf{s} = \sum_{k=0}^{K-1} \mathbf{s}^{(k)} = (S_0, S_1, \ldots, S_{L-1})^T. \quad (2.5)$$

An equivalent representation for $\mathbf{s}$ in the downlink is

$$\mathbf{s} = \mathbf{Cd}, \quad (2.6)$$

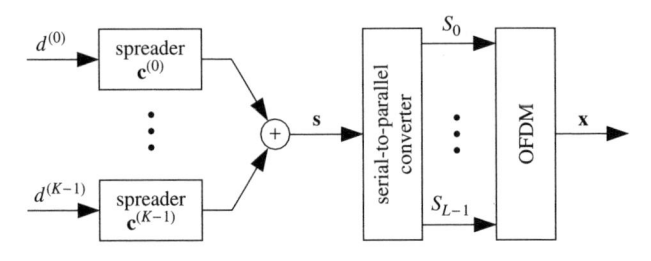

Figure 2-2 MC-CDMA downlink transmitter

where
$$\mathbf{d} = (d^{(0)}, d^{(1)}, \ldots, d^{(K-1)})^T \quad (2.7)$$

is the vector with the transmitted data symbols of the K active users and $\mathbf{C}$ is the spreading matrix given by
$$\mathbf{C} = (\mathbf{c}^{(0)}, \mathbf{c}^{(1)}, \ldots, \mathbf{c}^{(K-1)}). \quad (2.8)$$

The MC-CDMA downlink signal is obtained after processing the sequence $\mathbf{s}$ in the OFDM block according to (1.26). By assuming that the guard time is long enough to absorb all echoes, the received vector of the transmitted sequence $\mathbf{s}$ after inverse OFDM and frequency deinterleaving is given by
$$\mathbf{r} = \mathbf{H}\mathbf{s} + \mathbf{n} = (R_0, R_1, \ldots, R_{L-1})^T, \quad (2.9)$$

where $\mathbf{H}$ is the $L \times L$ channel matrix and $\mathbf{n}$ is the noise vector of length L. The vector $\mathbf{r}$ is fed to the data detector in order to get a hard or soft estimate of the transmitted data. For the description of the multiuser detection techniques, an equivalent notation for the received vector $\mathbf{r}$ is introduced,
$$\mathbf{r} = \mathbf{A}\mathbf{d} + \mathbf{n} = (R_0, R_1, \ldots, R_{L-1})^T. \quad (2.10)$$

The system matrix $\mathbf{A}$ for the downlink is defined as
$$\mathbf{A} = \mathbf{HC}. \quad (2.11)$$

2.1.3 Uplink Signal

In the uplink, the MC-CDMA signal is obtained directly after processing the sequence $\mathbf{s}^{(k)}$ of user k in the OFDM block according to (1.26). After inverse OFDM and frequency deinterleaving, the received vector of the transmitted sequences $\mathbf{s}^{(k)}$ is given by
$$\mathbf{r} = \sum_{k=0}^{K-1} \mathbf{H}^{(k)} \mathbf{s}^{(k)} + \mathbf{n} = (R_0, R_1, \ldots, R_{L-1})^T, \quad (2.12)$$

where $\mathbf{H}^{(k)}$ contains the coefficients of the sub-channels assigned to user k. The uplink is assumed to be synchronous in order to achieve the high spectral efficiency of OFDM. The vector $\mathbf{r}$ is fed to the data detector in order to get a hard or soft estimate of the transmitted data. The system matrix
$$\mathbf{A} = (\mathbf{a}^{(0)}, \mathbf{a}^{(1)}, \ldots, \mathbf{a}^{(K-1)}) \quad (2.13)$$

comprises K user-specific vectors
$$\mathbf{a}^{(k)} = \mathbf{H}^{(k)} \mathbf{c}^{(k)} = (H_{0,0}^{(k)} c_0^{(k)}, H_{1,1}^{(k)} c_1^{(k)}, \ldots, H_{L-1,L-1}^{(k)} c_{L-1}^{(k)})^T. \quad (2.14)$$

2.1.4 Spreading Techniques

The spreading techniques in MC-CDMA schemes differ in the selection of the spreading code and the type of spreading. As well as a variety of spreading codes, different strategies exist to map the spreading codes in time and frequency direction with MC-CDMA. Finally, the constellation points of the transmitted signal can be improved by modifying the phase of the symbols to be distinguished by the spreading codes.

2.1.4.1 Spreading Codes

Various spreading codes exist which can be distinguished with respect to orthogonality, correlation properties, implementation complexity and peak-to-average power ratio (PAPR). The selection of the spreading code depends on the scenario. In the synchronous downlink, orthogonal spreading codes are of advantage, since they reduce the multiple access interference compared to non-orthogonal sequences. However, in the uplink, the orthogonality between the spreading codes gets lost due to different distortions of the individual codes. Thus, simple PN sequences can be chosen for spreading in the uplink. If the transmission is asynchronous, Gold codes have good cross-correlation properties. In cases where pre-equalization is applied in the uplink, orthogonality can be achieved at the receiver antenna, such that in the uplink orthogonal spreading codes can also be of advantage.

Moreover, the selection of the spreading code has influence on the PAPR of the transmitted signal (see Chapter 4). Especially in the uplink, the PAPR can be reduced by selecting, e.g., Golay or Zadoff–Chu codes [8][35][36][39][52]. Spreading codes applicable in MC-CDMA systems are summarized in the following.

Walsh-Hadamard codes: Orthogonal Walsh–Hadamard codes are simple to generate recursively by using the following Hadamard matrix generation,

$$\mathbf{C}_L = \begin{bmatrix} \mathbf{C}_{L/2} & \mathbf{C}_{L/2} \\ \mathbf{C}_{L/2} & -\mathbf{C}_{L/2} \end{bmatrix}, \quad \forall L = 2^m, \quad m \geq 1, \quad \mathbf{C}_1 = 1. \tag{2.15}$$

The maximum number of available orthogonal spreading codes is L which determines the maximum number of active users K.

The Hadamard matrix generation described in (2.15) can also be used to perform an L-ary Walsh–Hadamard modulation which in combination with PN spreading can be applied in the uplink of an MC-CDMA systems [11][12].

Fourier codes: The columns of an FFT matrix can also be considered as spreading codes, which are orthogonal to each other. The chips are defined as

$$c_l^{(k)} = e^{-j2\pi l k/L}. \tag{2.16}$$

Thus, if Fourier spreading is applied in MC-CDMA systems, the FFT for spreading and the IFFT for the OFDM operation cancels out if the FFT and IFFT are the same size, i.e., the spreading is performed over all sub-carriers [7]. Thus, the resulting scheme is a single-carrier system with cyclic extension and frequency domain equalizer. This scheme has a dynamic range of single-carrier systems. The computational efficient implementation of the more general case where the FFT spreading is performed over groups of sub-carriers which are interleaved equidistantly is described in [8]. A comparison of the amplitude distributions between Hadamard codes and Fourier codes shows that Fourier codes result in an equal or lower peak-to-average power ratio [9].

Pseudo noise (PN) spreading codes: The property of a PN sequence is that the sequence appears to be noise-like if the construction is not known at the receiver. They are typically generated by using shift registers. Often used PN sequences are maximum-length shift register sequences, known as m-sequences. A sequence has a length of

$$n = 2^m - 1 \tag{2.17}$$

bits and is generated by a shift register of length m with linear feedback [40]. The sequence has a period length of n and each period contains 2^{m-1} ones and $2^{m-1} - 1$ zeros, i.e., it is a balanced sequence.

Gold codes: PN sequences with better cross-correlation properties than m-sequences are the so-called Gold sequences [40]. A set of n Gold sequences is derived from a preferred pair of m-sequences of length $L = 2^n - 1$ by taking the modulo-2 sum of the first preferred m-sequence with the n cyclically shifted versions of the second preferred m-sequence. By including the two preferred m-sequences, a family of $n + 2$ Gold codes is obtained. Gold codes have a three-valued cross correlation function with values $\{-1, -t(m), t(m) - 2\}$ where

$$t(m) = \begin{cases} 2^{(m+1)/2} + 1 & \text{for } m \text{ odd} \\ 2^{(m+2)/2} + 1 & \text{for } m \text{ even} \end{cases} \tag{2.18}$$

Golay codes: Orthogonal Golay complementary codes can recursively be obtained by

$$\mathbf{C}_L = \begin{bmatrix} \mathbf{C}_{L/2} & \overline{\mathbf{C}}_{L/2} \\ \mathbf{C}_{L/2} & -\overline{\mathbf{C}}_{L/2} \end{bmatrix}, \quad \forall L = 2^m, \quad m \geq 1, \quad \mathbf{C}_1 = 1, \tag{2.19}$$

where the complementary matrix $\overline{\mathbf{C}}_L$ is defined by reverting the original matrix $\mathbf{C}_L$. If

$$\mathbf{C}_L = \begin{bmatrix} \mathbf{A}_L & \mathbf{B}_L \end{bmatrix}, \tag{2.20}$$

and $\mathbf{A}_L$ and $\mathbf{B}_L$ are $L \times L/2$ matrices, then

$$\overline{\mathbf{C}}_L = [\mathbf{A}_L \quad -\mathbf{B}_L]. \tag{2.21}$$

Zadoff–Chu codes: The Zadoff–Chu codes have optimum correlation properties and are a special case of generalized chirp-like sequences. They are defined as

$$c_l^{(k)} = \begin{cases} e^{j2\pi k(ql+l^2/2)/L} & \text{for } L \text{ even} \\ e^{j2\pi k(ql+l(l+1)/2)/L} & \text{for } L \text{ odd} \end{cases}, \tag{2.22}$$

where q is any integer, and k is an integer, prime with L. If L is a prime number, a set of Zadoff–Chu codes is composed of $L - 1$ sequences. Zadoff–Chu codes have an optimum periodic autocorrelation function and a low constant magnitude periodic cross-correlation function.

Low-rate convolutional codes: Low-rate convolutional codes can be applied in CDMA systems as spreading codes with inherent coding gain [50]. These codes have been applied as alternative to the use of a spreading code followed by a convolutional code. In MC-CDMA systems, low-rate convolutional codes can achieve good performance results for moderate numbers of users in the uplink [30][32][46]. The application of low-rate convolutional codes is limited to very moderate numbers of users since, especially in the downlink, signals are not orthogonal between the users, resulting in possibly severe multiple access interference. Therefore, they cannot reach the high spectral efficiency of MC-CDMA systems with separate coding and spreading.

2.1.4.2 Peak-to-Average Power Ratio (PAPR)

The variation of the envelope of a multi-carrier signal can be defined by the peak-to-average power ratio (PAPR) which is given by

$$PAPR = \frac{\max |x_v|^2}{\frac{1}{N_c} \sum_{v=0}^{N_c-1} |x_v|^2}. \quad (2.23)$$

The values x_v, $v = 0, \ldots, N_c - 1$, are the time samples of an OFDM symbol. An additional measure to determine the envelope variation is the crest factor (CF) which is

$$CF = \sqrt{PAPR}. \quad (2.24)$$

By appropriately selecting the spreading code, it is possible to reduce the PAPR of the multi-carrier signal [4][36][39]. This PAPR reduction can be of advantage in the uplink where low power consumption is required in the terminal station.

Uplink PAPR
The uplink signal assigned to user k results in

$$x_v = x_v^{(k)}. \quad (2.25)$$

The PAPR for different spreading codes can be upper-bounded for the uplink by [35]

$$PAPR \leq \frac{2 \max \left\{ \left| \sum_{l=0}^{L-1} c_l^{(k)} e^{j2\pi l t / T_s} \right|^2 \right\}}{L}, \quad (2.26)$$

assuming that $N_c = L$. Table 2-1 summarizes the PAPR bounds for MC-CDMA uplink signals with different spreading codes.

The PAPR bound for Golay codes and Zadoff–Chu codes is independent of the spreading code length. When N_c is a multiple of L, the PAPR of the Walsh-Hadamard code is upper-bounded by $2N_c$.

Downlink PAPR
The time samples of a downlink multi-carrier symbol assuming synchronous transmission are given as

$$x_v = \sum_{k=0}^{K-1} x_v^{(k)}. \quad (2.27)$$

Table 2-1 PAPR bounds of MC-CDMA uplink signals; $N_c = L$

Spreading code	PAPR
Walsh–Hadamard	$\leqslant 2L$
Golay	$\leqslant 4$
Zadoff–Chu	2
Gold	$\leqslant 2 \left(t(m) - 1 - \dfrac{t(m)+2}{L} \right)$

The PAPR of an MC-CDMA downlink signal with K users and $N_c = L$ can be upper-bounded by [35]

$$PAPR \leqslant \frac{2 \max \left\{ \sum_{k=0}^{K-1} \left| \sum_{l=0}^{L-1} c_l^{(k)} e^{j2\pi l t/T_s} \right|^2 \right\}}{L}. \qquad (2.28)$$

2.1.4.3 One- and Two-Dimensional Spreading

Spreading in MC-CDMA systems can be carried out in frequency direction, time direction or two-dimensional in time and frequency direction. An MC-CDMA system with spreading only in the time direction is equal to an MC-DS-CDMA system. Spreading in two dimensions exploits time and frequency diversity and is an alternative to the conventional approach with spreading in frequency or time direction only. A two-dimensional spreading code is a spreading code of length L where the chips are distributed in the time and frequency direction. Two-dimensional spreading can be performed by a two-dimensional spreading code or by two cascaded one-dimensional spreading codes. An efficient realization of two-dimensional spreading is to use a one-dimensional spreading code followed by a two-dimensional interleaver as illustrated in Figure 2-3 [23]. With two cascaded one-dimensional spreading codes, spreading is first carried out in one dimension with the first spreading code of length L_1. In the next step, the data-modulated chips of the first spreading code are again spread with the second spreading code in the second dimension. The length of the second spreading code is L_2. The total spreading length with two cascaded one-dimensional spreading codes results in

$$L = L_1 L_2. \qquad (2.29)$$

If the two cascaded one-dimensional spreading codes are Walsh–Hadamard codes, the resulting two-dimensional code is again a Walsh–Hadamard code with total length L. For large L, two-dimensional spreading can outperform one-dimensional in an uncoded MC-CDMA system [13][42].

Two-dimensional spreading for maximum diversity gain is efficiently realized by using a sufficiently long spreading code with $L \geqslant D_O$, where D_O is the maximum achievable two-dimensional diversity (see Section 1.1.7). The spread sequence of length L has to be appropriately interleaved in time and frequency, such that all chips of this sequence are faded independently as far as possible.

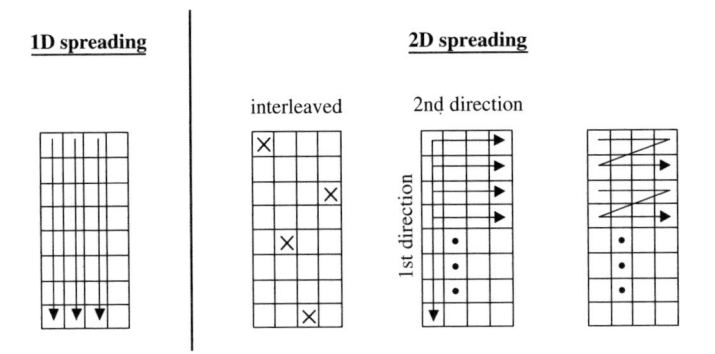

Figure 2-3 1D and 2D spreading schemes

Another approach with two-dimensional spreading is to locate the chips of the two-dimensional spreading code as close together as possible in order to get all chips similarly faded and, thus, preserve orthogonality of the spreading codes at the receiver as far as possible [3][38]. Due to reduced multiple access interference, low complex receivers can be applied. However, the diversity gain due to spreading is reduced such that powerful channel coding is required. If the fading over all chips of a spreading code is flat, the performance of conventional OFDM without spreading is the lower bound for this spreading approach; i.e., the BER performance of an MC-CDMA system with two-dimensional spreading and Rayleigh fading which is flat over the whole spreading sequence results in the performance of OFDM with $L = 1$ shown in Figure 1-3. One- or two-dimensional spreading concepts with interleaving of the chips in time and/or frequency are lower-bounded by the diversity performance curves in Figure 1-3 which are assigned to the chosen spreading code length L.

2.1.4.4 Rotated Constellations

With spreading codes like Walsh–Hadamard codes, the achievable diversity gain degrades, if the signal constellation points of the resulting spread sequence **s** in the downlink concentrate their energy in less than L sub-channels, which in the worst case is only in one sub-channel while the signal on all other sub-channels is zero. Here we consider a full loaded scenario with $K = L$. The idea of rotated constellations [8] is to guarantee the existence of M^L distinct points at each sub-carrier for a transmitted alphabet size of M and a spreading code length of L and that all points are nonzero. Thus, if all except one sub-channel are faded out, detection of all data symbols is still possible.

With rotated constellations, the L data symbols are rotated before spreading such that the data symbol constellations are different for each of the L data symbols of the transmit symbol vector **s**. This can be achieved by rotating the phase of the transmit symbol alphabet of each of the L spread data symbols by a fraction proportional to $1/L$. The rotation factor for user k is

$$r^{(k)} = e^{j2\pi k/(M_{rot}L)}, \qquad (2.30)$$

where M_{rot} is a constant whose choice depends on the symbol alphabet. For example, $M_{rot} = 2$ for BPSK and $M_{rot} = 4$ for QPSK. For M-PSK modulation, the constant

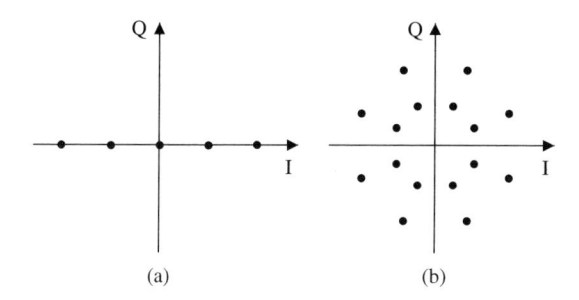

Figure 2-4 Constellation points after Hadamard spreading a) nonrotated, b) rotated, both for BPSK and $L = 4$

$M_{rot} = M$. The constellation points of the Walsh-Hadamard spread sequence **s** with BPSK modulation with and without rotation is illustrated in Figure 2-4 for a spreading code length of $L = 4$.

Spreading with rotated constellations can achieve better performance than the use of nonrotated spreading sequences. The performance improvements strongly depend on the chosen symbol mapping scheme. Large symbol alphabets reduce the degree of freedom for placing the points in a rotated signal constellation and decrease the gains. Moreover, the performance improvements with rotated constellations strongly depend on the chosen detection techniques. For higher-order symbol mapping schemes, relevant performance improvements require the application of powerful multiuser detection techniques. The achievable performance improvements in SNR with rotated constellations can be in the order of several dB at a BER of 10^{-3} for an uncoded MC-CDMA system with QPSK in fading channels.

2.1.5 Detection Techniques

Data detection techniques can be classified as either single-user detection (SD) or multiuser detection (MD). The approach using SD detects the user signal of interest by not taking into account any information about multiple access interference. In MC-CDMA mobile radio systems, SD is realized by one tap equalization to compensate for the distortion due to flat fading on each sub-channel, followed by user-specific despreading. As in OFDM, the one tap equalizer is simply one complex-valued multiplication per sub-carrier. If the spreading code structure of the interfering signals is known, the multiple access interference could not be considered in advance as noise-like, yielding SD to be suboptimal. The suboptimality of SD can be overcome with MD where the *a priori* knowledge about the spreading codes of the interfering users is exploited in the detection process.

The performance improvements with MD compared to SD are achieved at the expense of higher receiver complexity. The methods of MD can be divided into interference cancellation (IC) and joint detection. The principle of IC is to detect the information of the interfering users with SD and to reconstruct the interfering contribution in the received signal before subtracting the interfering contribution from the received signal and detecting the information of the desired user. The optimal detector applies joint detection with maximum likelihood detection. Since the complexity of maximum likelihood detection grows exponentially with the number of users, its use is limited in practice to applications

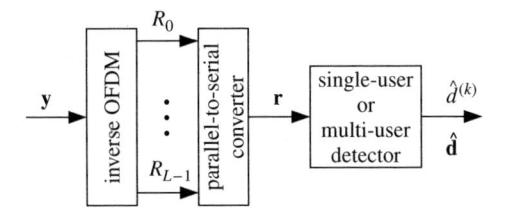

Figure 2-5 MC-CDMA receiver in the terminal station

with a small number of users. Simpler joint detection techniques can be realized by using block linear equalizers.

An MC-CDMA receiver in the terminal station of user k is depicted in Figure 2-5.

2.1.5.1 Single-User Detection

The principle of single-user detection is to detect the user signal of interest by not taking into account any information about the multiple access interference. A receiver with single-user detection of the data symbols of user k is shown in Figure 2-6.

After inverse OFDM the received sequence $\mathbf{r}$ is equalized by employing a bank of adaptive one-tap equalizers to combat the phase and amplitude distortions caused by the mobile radio channel on the sub-channels. The one tap equalizer is simply realized by one complex-valued multiplication per sub-carrier. The received sequence at the output of the equalizer has the form

$$\mathbf{u} = \mathbf{G}\,\mathbf{r} = (U_0, U_1, \ldots, U_{L-1})^T. \tag{2.31}$$

The diagonal equalizer matrix

$$\mathbf{G} = \begin{pmatrix} G_{0,0} & 0 & \cdots & 0 \\ 0 & G_{1,1} & & 0 \\ \vdots & & \ddots & \vdots \\ 0 & 0 & \cdots & G_{L-1,L-1} \end{pmatrix} \tag{2.32}$$

of dimension $L \times L$ represents the L complex-valued equalizer coefficients of the sub-carriers assigned to $\mathbf{s}$. The complex-valued output $\mathbf{u}$ of the equalizer is despread by correlating it with the conjugate complex user-specific spreading code $\mathbf{c}^{(k)*}$. The complex-valued soft decided value at the output of the despreader is

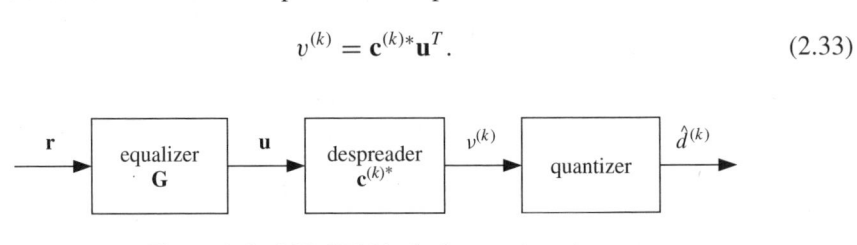

$$v^{(k)} = \mathbf{c}^{(k)*}\mathbf{u}^T. \tag{2.33}$$

Figure 2-6 MC-CDMA single-user detection

The hard decided value of a detected data symbol is given by

$$\hat{d}^{(k)} = Q\{v^{(k)}\}, \qquad (2.34)$$

where $Q\{\cdot\}$ is the quantization operation according to the chosen data symbol alphabet.

The term equalizer is generalized in the following, since the processing of the received vector **r** according to typical diversity combining techniques is also investigated using the SD scheme shown in Figure 2-6.

Maximum Ratio Combining (MRC): MRC weights each sub-channel with its respective conjugate complex channel coefficient, leading to

$$G_{l,l} = H_{l,l}^*, \qquad (2.35)$$

where $H_{l,l}, l = 0, \ldots, L-1$, are the diagonal components of **H**. The drawback of MRC in MC-CDMA systems in the downlink is that it destroys the orthogonality between the spreading codes and, thus, additionally enhances the multiple access interference. In the uplink, MRC is the most promising single-user detection technique since the spreading codes do not superpose in an orthogonal fashion at the receiver and maximization of the signal-to-interference ratio is optimized.

Equal Gain Combining (EGC): EGC compensates only for the phase rotation caused by the channel by choosing the equalization coefficients as

$$G_{l,l} = \frac{H_{l,l}^*}{|H_{l,l}|}. \qquad (2.36)$$

EGC is the simplest single-user detection technique, since it only needs information about the phase of the channel.

Zero Forcing (ZF): ZF applies channel inversion and can eliminate multiple access interference by restoring the orthogonality between the spread data in the downlink with an equalization coefficient chosen as

$$G_{l,l} = \frac{H_{l,l}^*}{|H_{l,l}|^2}. \qquad (2.37)$$

The drawback of ZF is that for small amplitudes of $H_{l,l}$ the equalizer enhances noise.

Minimum Mean Square Error (MMSE) Equalization: Equalization according to the MMSE criterion minimizes the mean square value of the error

$$\varepsilon_l = S_l - G_{l,l} R_l \qquad (2.38)$$

between the transmitted signal and the output of the equalizer. The mean square error

$$J_l = E\{|\varepsilon_l|^2\} \qquad (2.39)$$

can be minimized by applying the orthogonality principle, stating that the mean square error J_l is minimum if the equalizer coefficient $G_{l,l}$ is chosen such that the error ε_l is orthogonal to the received signal R_l^*, i.e.,

$$E\{\varepsilon_l R_l^*\} = 0. \qquad (2.40)$$

The equalization coefficient based on the MMSE criterion for MC-CDMA systems results in

$$G_{l,l} = \frac{H_{l,l}^*}{|H_{l,l}|^2 + \sigma^2}. \tag{2.41}$$

The computation of the MMSE equalization coefficients requires knowledge about the actual variance of the noise σ^2. For very high SNRs, the MMSE equalizer becomes identical to the ZF equalizer. To overcome the additional complexity for the estimation of σ^2, a low-complex suboptimum MMSE equalization can be realized [21].

With suboptimum MMSE equalization, the equalization coefficients are designed such that they perform optimally only in the most critical cases for which successful transmission should be guaranteed. The variance σ^2 is set equal to a threshold λ at which the optimal MMSE equalization guarantees the maximum acceptable BER. The equalization coefficient with suboptimal MMSE equalization results in

$$G_{l,l} = \frac{H_{l,l}^*}{|H_{l,l}|^2 + \lambda} \tag{2.42}$$

and requires only information about $H_{l,l}$. The value λ has to be determined during the system design.

A controlled equalization can be applied in the receiver, which performs slightly worse than suboptimum MMSE equalization [23]. Controlled equalization applies zero forcing on sub-carriers where the amplitude of the channel coefficients exceeds a predefined threshold a_{th}. All other sub-carriers apply equal gain combining in order to avoid noise amplification.

In the uplink **G** and **H** are user-specific.

2.1.5.2 Multiuser Detection

Maximum Likelihood Detection

The optimum multiuser detection technique exploits the maximum *a posteriori* (MAP) criterion or the maximum likelihood criterion, respectively. In this section, two optimum maximum likelihood detection algorithms are shown, namely the maximum likelihood sequence estimation (MLSE), which optimally estimates the transmitted data sequence $\mathbf{d} = (d^{(0)}, d^{(1)}, \ldots, d^{(K-1)})^T$ and the maximum likelihood symbol-by-symbol estimation (MLSSE), which optimally estimates the transmitted data symbol $d^{(k)}$. It is straightforward that both algorithms can be extended to a MAP sequence estimator and to a MAP symbol-by-symbol estimator by taking into account the *a priori* probability of the transmitted sequence and symbol, respectively. When all possible transmitted sequences and symbols, respectively, are equally probable *a priori*, the estimator based on the MAP criterion and the one based on the maximum likelihood criterion are identical. The possible transmitted data symbol vectors are $\mathbf{d}_\mu$, $\mu = 0, \ldots, M^K - 1$, where M^K is the number of possible transmitted data symbol vectors and M is the number of possible realizations of $d^{(k)}$.

Maximum Likelihood Sequence Estimation (MLSE): MLSE minimizes the sequence error probability, i.e., the data symbol vector error probability, which is equivalent to maximizing the conditional probability $P\{\mathbf{d}_\mu|\mathbf{r}\}$ that $\mathbf{d}_\mu$ was transmitted given the received vector $\mathbf{r}$. The estimate of $\mathbf{d}$ obtained with MLSE is

$$\hat{\mathbf{d}} = \arg\max_{\mathbf{d}_\mu} P\{\mathbf{d}_\mu|\mathbf{r}\}, \tag{2.43}$$

with arg denoting the argument of the function. If the noise N_l is additive white Gaussian, (2.43) is equivalent to finding the data symbol vector $\mathbf{d}_\mu$ that minimizes the squared Euclidean distance

$$\Delta^2(\mathbf{d}_\mu, \mathbf{r}) = \|\mathbf{r} - \mathbf{A}\,\mathbf{d}_\mu\|^2 \tag{2.44}$$

between the received and all possible transmitted sequences. The most likely transmitted data vector is

$$\hat{\mathbf{d}} = \arg\min_{\mathbf{d}_\mu} \Delta^2(\mathbf{d}_\mu, \mathbf{r}). \tag{2.45}$$

MLSE requires the evaluation of M^K squared Euclidean distances for the estimation of the data symbol vector $\hat{\mathbf{d}}$.

Maximum Likelihood Symbol-by-Symbol Estimation (MLSSE): MLSSE minimizes the symbol error probability, which is equivalent to maximizing the conditional probability $P\{d_\mu^{(k)}|\mathbf{r}\}$ that $d_\mu^{(k)}$ was transmitted given the received sequence $\mathbf{r}$. The estimate of $d^{(k)}$ obtained by MLSSE is

$$\hat{d}^{(k)} = \arg\max_{d_\mu^{(k)}} P\{d_\mu^{(k)}|\mathbf{r}\}. \tag{2.46}$$

If the noise N_l is additive white Gaussian the most likely transmitted data symbol is

$$\hat{d}^{(k)} = \arg\max_{d_\mu^{(k)}} \sum_{\forall \mathbf{d}_\mu \text{ with same realization of } d_\mu^{(k)}} \exp\left(-\frac{1}{\sigma^2}\Delta^2(\mathbf{d}_\mu, \mathbf{r})\right). \tag{2.47}$$

The increased complexity with MLSSE compared to MLSE can be observed in the comparison of (2.47) with (2.45). An advantage of MLSSE compared to MLSE is that MLSSE inherently generates reliability information for detected data symbols which can be exploited in a subsequent soft decision channel decoder.

Block Linear Equalizer
The block linear equalizer is a suboptimum, low-complex multiuser detector which requires knowledge about the system matrix $\mathbf{A}$ in the receiver. Two criteria can be applied to use this knowledge in the receiver for data detection.

Zero Forcing Block Linear Equalizer: Joint detection applying a zero forcing block linear equalizer delivers at the output of the detector the soft decided data vector

$$\mathbf{v} = (\mathbf{A}^H\mathbf{A})^{-1}\mathbf{A}^H\mathbf{r} = (v^{(0)}, v^{(1)}, \ldots, v^{(K-1)})^T, \tag{2.48}$$

where $(\cdot)^H$ is the Hermitian transposition.

MMSE Block Linear Equalizer: An MMSE block linear equalizer delivers at the output of the detector the soft decided data vector

$$\mathbf{v} = (\mathbf{A}^H\mathbf{A} + \sigma^2\mathbf{I})^{-1}\mathbf{A}^H\mathbf{r} = (v^{(0)}, v^{(1)}, \ldots, v^{(K-1)})^T. \tag{2.49}$$

Hybrid combinations of block linear equalizers and interference cancellation schemes (see the next section) are possible, resulting in block linear equalizers with decision feedback.

Interference Cancellation

The principle of interference cancellation is to detect and subtract interfering signals from the received signal before detection of the wanted signal. It can be applied to reduce intra-cell and inter-cell interference. Most detection schemes focus on intra-cell interference, which will be further discussed in this section. Interference cancellation schemes can use signals for reconstruction of the interference either obtained at the detector output (see Figure 2-7), or at the decoder output (see Figure 2-8).

Both schemes can be applied in several iterations. Values and functions related to the iteration j are marked by an index $^{[j]}$, where j may take on the values $j = 1, \ldots, J_{it}$, and J_{it} is the total number of iterations. The initial detection stage is indicated by the index $^{[0]}$. Since the interference is detected more reliably at the output of the channel decoder than at the output of the detector, the scheme with channel decoding included in the iterative process outperforms the other scheme. Interference cancellation distinguishes between parallel and successive cancellation techniques. Combinations of parallel and successive interference cancellation are also possible.

Parallel Interference Cancellation (PIC): The principle of PIC is to detect and subtract all interfering signals in parallel before detection of the wanted signal. PIC is suitable for

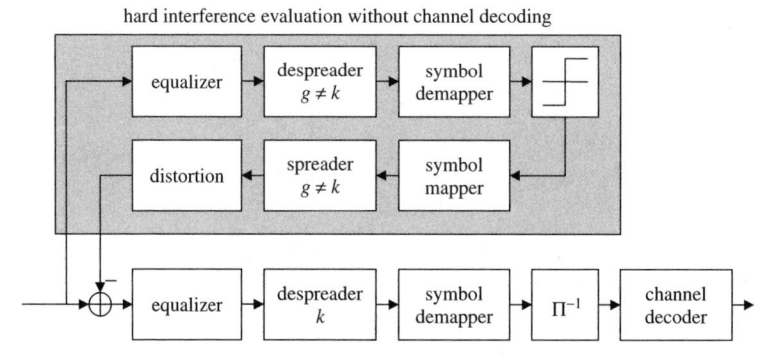

Figure 2-7 Hard interference cancellation scheme

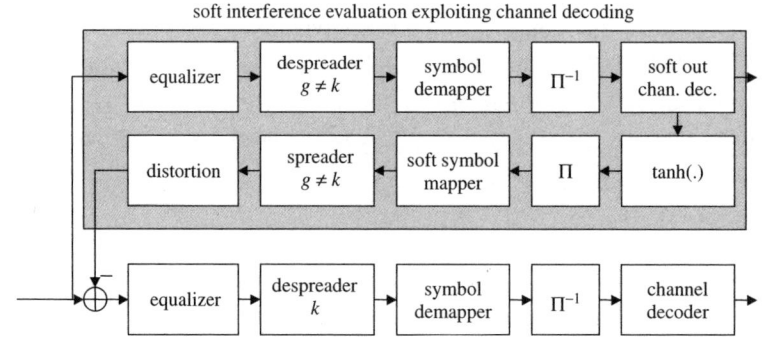

Figure 2-8 Soft interference cancellation scheme

systems where the interfering signals have similar power. In the initial detection stage, the data symbols of all K active users are detected in parallel by single-user detection. That is,

$$\hat{d}^{(k)[0]} = Q\{\mathbf{c}^{(k)*}\mathbf{G}^{(k)[0]}\mathbf{r}^T\}, \quad k = 0, \ldots, K-1, \tag{2.50}$$

where $\mathbf{G}^{(k)[0]}$ denotes the equalization coefficients assigned to the initial stage. The following detection stages work iteratively by using the decisions of the previous stage to reconstruct the interfering contribution in the received signal. The obtained interference is subtracted, i.e., cancelled from the received signal, and the data detection is performed again with reduced multiple access interference. Thus, the second and further detection stages apply

$$\hat{d}^{(k)[j]} = Q\left\{\mathbf{c}^{(k)*}\mathbf{G}^{(k)[j]}\left(\mathbf{r} - \sum_{\substack{g=0 \\ g \neq k}}^{K-1} \mathbf{H}^{(g)} d^{(g)[j-1]} \mathbf{c}^{(g)}\right)^T\right\}, \quad j = 1, \ldots, J_{it}. \tag{2.51}$$

where, except for the final stage, the detection has to be applied for all K users.

PIC can be applied with different detection strategies in the iterations. Starting with EGC in each iteration [15] various combinations have been proposed [6][22][27]. Very promising results are obtained with MMSE equalization adapted in the first iteration to the actual system load and in all further iterations to MMSE equalization adapted to the single-user case [21]. The application of MRC seems theoretically to be of advantage for the second and further detection stages, since MRC is the optimum detection technique in the multiple access interference free case, i.e., in the single-user case. However, if one or more decision errors are made, MRC has a poor performance [22].

Successive Interference Cancellation (SIC): SIC detects and subtracts the interfering signals in the order of the interfering signal power. First, the strongest interferer is cancelled, before the second strongest interferer is detected and subtracted, i.e.,

$$\hat{d}^{(k)[j]} = Q\{\mathbf{c}^{(k)*}\mathbf{G}^{(k)[j]}(\mathbf{r} - \mathbf{H}^{(g)}(d^{(g)[j-1]}\mathbf{c}^{(g)}))^T\}, \tag{2.52}$$

where g is the strongest interferer in the iteration j, $j = 1, \ldots, J_{it}$. This procedure is continued until a predefined stop criteria. SIC is suitable for systems with large power variations between the interferers [6].

Soft-Interference Cancellation: Interference cancellation can use reliability information about the detected interference in the iterative process. These schemes can be without [37] and with [18][25] channel decoding in the iterative process, and are termed soft interference cancellation. If reliability information about the detected interference is taken into account in the cancellation scheme, the performance of the iterative scheme can be improved since error propagation can be reduced compared to schemes with hard decided feedback. The block diagram of an MC-CDMA receiver with soft interference cancellation is illustrated in Figure 2-8. The data of the desired user k are detected by applying interference cancellation with reliability information. Before detection of user k's data in the lowest path of Figure 2-8 with an appropriate single-user detection technique, the contributions of the $K-1$ interfering users g, $g = 0, \ldots, K-1$, and $g \neq k$ is detected with single-user detection and subtracted from the received signal. The principle of parallel or successive interference cancellation or combinations of both can be applied within a soft interference cancellation scheme.

In the following, we focus on the contribution of the interfering user g with $g \neq k$. The soft decided values $\mathbf{w}^{(g)[j]}$ are obtained after single-user detection, symbol demapping, and deinterleaving. The corresponding log-likelihood ratios (LLRs) for channel decoding are given by the vector $\mathbf{l}^{(g)[j]}$. LLRs are the optimum soft decided values which can be exploited in a Viterbi decoder (see Section 2.1.7). From the subsequent soft-in/soft-out channel decoder, besides the output of the decoded source bits, reliability information in the form of LLRs of the coded bits can be obtained. These LLRs are given by the vector

$$\mathbf{l}_{out}^{(g)[j]} = (\Gamma_{0,out}^{(g)[j]}, \Gamma_{1,out}^{(g)[j]}, \ldots, \Gamma_{L_b-1,out}^{(g)[j]})^T. \tag{2.53}$$

In contrast to the LLRs of the coded bits at the input of the soft-in/soft-out channel decoder, the LLRs of the coded bits at the output of the soft-in/soft-out channel decoder

$$\Gamma_{\kappa,out}^{(g)[j]} = \ln\left(\frac{P\{b_\kappa^{(g)} = +1 | \mathbf{w}^{(g)[j]}\}}{P\{b_\kappa^{(g)} = -1 | \mathbf{w}^{(g)[j]}\}}\right), \quad \kappa = 0, \ldots, L_b-1, \tag{2.54}$$

are the estimates of all the other soft decided values in the sequence $\mathbf{w}^{(g)[j]}$ about this coded bit, and not only of one received soft decided value $w_\kappa^{(g)[j]}$. For brevity, the index κ is omitted since the focus is on the LLR of one coded bit in the sequel. To avoid error propagation, the average value of coded bit $b^{(g)}$ is used, which is the so-called soft bit $w_{out}^{(g)[j]}$ [18]. The soft bit is defined as

$$w_{out}^{(g)[j]} = E\{b^{(g)} | \mathbf{w}^{(g)[j]}\}$$
$$= (+1)P\{b^{(g)} = +1 | \mathbf{w}^{(g)[j]}\} + (-1)P\{b^{(g)} = -1 | \mathbf{w}^{(g)[j]}\}. \tag{2.55}$$

With (2.54), the soft bit results in

$$w_{out}^{(g)[j]} = \tanh\left(\frac{\Gamma_{out}^{(g)[j]}}{2}\right). \tag{2.56}$$

The soft bit $w_{out}^{(g)[j]}$ can take on values in the interval $[-1, +1]$. After interleaving, the soft bits are soft symbol mapped such that the reliability information included in the soft bits is not lost. The obtained complex-valued data symbols are spread with the user-specific spreading code and each chip is predistorted with the channel coefficient assigned to the sub-carrier that the chip has been transmitted on. The total reconstructed multiple access interference is subtracted from the received signal $\mathbf{r}$. After canceling the interference, the data of the desired user k are detected using single-user detection. However, in contrast to the initial detection stage, in further stages, the equalizer coefficients given by the matrix $\mathbf{G}^{(k)[j]}$ and the LLRs given by the vector $\mathbf{l}^{(k)[j]}$ after soft interference cancellation are adapted to the quasi multiple access interference-free case.

2.1.6 Pre-Equalization

If information about the actual channel is *a priori* known at the transmitter, pre-equalization can be applied at the transmitter such that the signal at the receiver appears non-distorted and an estimation of the channel at the receiver is not necessary. Information about the channel state can, for example, be made available in TDD schemes if the TDD slots are short enough such that the channel of an up- and a subsequent downlink slots can be considered as constant and the transceiver can use the channel state information obtained from previously received data.

An application scenario of pre-equalization in a TDD mobile radio system would be that the terminal station sends pilot symbols in the uplink which are used in the base station for channel estimation and detection of the uplink data symbols. The estimated channel state is used for pre-equalization of the downlink data to be transmitted to the terminal station. Thus, no channel estimation is necessary in the terminal station which reduces its complexity. Only the base station has to estimate the channel, i.e., the complexity can be shifted to the base station.

A further application scenario of pre-equalization in a TDD mobile radio system would be that the base station sends pilot symbols in the downlink to the terminal station, which performs channel estimation. In the uplink, the terminal station applies pre-equalization with the intention to get quasi-orthogonal user signals at the base station receiver antenna. This results in a high spectral efficiency in the uplink, since MAI can be avoided. Moreover, a complex uplink channel estimation is not necessary.

The accuracy of pre-equalization can be increased by using prediction of the channel state in the transmitter where channel state information from the past is filtered.

Pre-equalization is performed by multiplying the symbols on each sub-channel with an assigned pre-equalization coefficient before transmission [20][33][41][43]. The selection criteria for the equalization coefficients is to compensate the channel fading as far as possible, such that the signal at the receiver antenna seems to be only affected by AWGN. In Figure 2-9, an OFDM transmitter with pre-equalization is illustrated which results with a spreading operation in an MC-SS transmitter.

2.1.6.1 Downlink

In a multi-carrier system in the downlink (e.g., SS-MC-MA) the pre-equalization operation is given by

$$\bar{s} = \overline{G} s, \qquad (2.57)$$

where the source symbols S_l before pre-equalization are represented by the vector s and $\overline{G}$ is the diagonal $L \times L$ pre-equalization matrix with elements $\overline{G}_{l,l}$. In the case of spreading L corresponds to the spreading code length and in the case of OFDM (OFDMA, MC-TDMA), L is equal to the number of sub-carriers N_c. The pre-equalized sequence $\bar{s}$ is fed to the OFDM operation and transmitted.

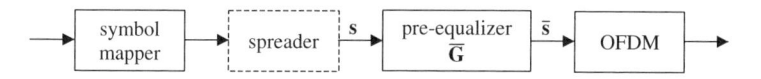

Figure 2-9 OFDM or MC-SS transmitter with pre-equalization

In the receiver, the signal after inverse OFDM operation results in

$$\mathbf{r} = \mathbf{H}\bar{\mathbf{s}} + \mathbf{n}$$
$$= \mathbf{H}\overline{\mathbf{G}}\mathbf{s} + \mathbf{n} \tag{2.58}$$

where $\mathbf{H}$ represents the channel matrix with the diagonal components $H_{l,l}$ and $\mathbf{n}$ represents the noise vector. It can be observed from (2.58) that by choosing

$$\overline{G}_{l,l} = \frac{1}{H_{l,l}} \tag{2.59}$$

the influence of the fading channel can be compensated and the signal is only disturbed by AWGN. In practice, this optimum technique cannot be realized since this would require transmission with very high power on strongly faded sub-channels. Thus, in the following section we focus on pre-equalization with power constraint where the total transmission power with pre-equalization is equal to the transmission power without pre-equalization [33].

The condition for pre-equalization with power constraint is

$$\sum_{l=0}^{L-1} |\overline{G}_{l,l} S_l|^2 = \sum_{l=0}^{L-1} |S_l|^2. \tag{2.60}$$

When assuming that all symbols S_l are transmitted with same power, the condition for pre-equalization with power constraint becomes

$$\sum_{l=0}^{L-1} |\overline{G}_{l,l}|^2 = \sum_{l=0}^{L-1} |G_{l,l} C|^2 = L, \tag{2.61}$$

where $G_{l,l}$ is the pre-equalization coefficient without power constraint and C is a normalizing factor which keeps the transmit power constant. The factor C results in

$$C = \sqrt{\frac{L}{\sum_{l=0}^{L-1} |G_{l,l}|^2}}. \tag{2.62}$$

By applying the equalization criteria introduced in Section 2.1.5.1, the following pre-equalization coefficients are obtained.

Maximum Ratio Combining (MRC)

$$\overline{G}_{l,l} = H_{l,l}^* \sqrt{\frac{L}{\sum_{l=0}^{L-1} |H_{l,l}|^2}}. \tag{2.63}$$

Equal Gain Combining (EGC)

$$\overline{G}_{l,l} = \frac{H_{l,l}^*}{|H_{l,l}|}. \tag{2.64}$$

Zero Forcing (ZF)

$$\overline{G}_{l,l} = \frac{H_{l,l}^*}{|H_{l,l}|^2} \sqrt{\frac{L}{\sum_{l=0}^{L-1} \frac{1}{|H_{l,l}|^2}}}. \qquad (2.65)$$

Quasi Minimum Mean Square Error (MMSE) Pre-Equalization

$$\overline{G}_{l,l} = \frac{H_{l,l}^*}{|H_{l,l}|^2 + \sigma^2} \sqrt{\frac{L}{\sum_{l=0}^{L-1} \left|\frac{H_{l,l}^*}{|H_{l,l}|^2 + \sigma^2}\right|^2}}. \qquad (2.66)$$

We call this technique quasi MMSE pre-equalization, since this is an approximation. The optimum technique requires a very high computational complexity, due to the power constraint condition.

As with the single-user detection techniques presented in Section 2.1.5.1, controlled pre-equalization can be applied. Controlled pre-equalization applies zero forcing pre-equalization on sub-carriers where the amplitude of the channel coefficients exceeds a predefined threshold a_{th}. All other sub-carriers apply equal gain combining for pre-equalization.

2.1.6.2 Uplink

In an MC-CDMA uplink scenario, pre-equalization is performed in the terminal station of user k according to

$$\overline{\mathbf{s}}^{(k)} = \overline{\mathbf{G}}^{(k)} \mathbf{s}^{(k)}. \qquad (2.67)$$

The received signal at the base station after inverse OFDM operation results in

$$\mathbf{r} = \sum_{k=0}^{K-1} \mathbf{H}^{(k)} \overline{\mathbf{s}}^{(k)} + \mathbf{n}$$

$$= \sum_{k=0}^{K-1} \mathbf{H}^{(k)} \overline{\mathbf{G}}^{(k)} \mathbf{s}^{(k)} + \mathbf{n} \qquad (2.68)$$

The pre-equalization techniques presented in (2.63) to (2.66) are applied in the uplink individually for each terminal station, i.e., $\overline{G}_{l,l}^{(k)}$ and $H_{l,l}^{(k)}$ have to be applied instead of $\overline{G}_{l,l}$ and $H_{l,l}$, respectively.

Finally, knowledge about the channel in the transmitter can be exploited, not only to perform pre-equalization, but also to apply adaptive modulation per sub-carrier in order to increase the capacity of the system (see Chapter 4).

2.1.7 Soft Channel Decoding

Channel coding with bit interleaving is an efficient technique to combat degradation due to fading, noise, interference, and other channel impairments. The basic idea of channel coding is to introduce controlled redundancy into the transmitted data that is exploited at the receiver to correct channel-induced errors by means of forward error correction (FEC). Binary convolutional codes are chosen as channel codes in current mobile radio, digital broadcasting, WLAN and WLL systems, since there exist very simple decoding algorithms based on the Viterbi algorithm that can achieve a soft decision decoding gain. Moreover, convolutional codes are used as component codes for Turbo codes, which have become part of 3G mobile radio standards. A detailed channel coding description is given in Chapter 4.

Many of the convolutional codes that have been developed for increasing the reliability in the transmission of information are effective when errors caused by the channel are statistically independent. Signal fading due to time-variant multipath propagation often causes the signal to fall below the noise level, thus resulting in a large number of errors called burst errors. An efficient method for dealing with burst error channels is to interleave the coded bits in such a way that the bursty channel is transformed into a channel with independent errors. Thus, a code designed for independent errors or short bursts can be used. Code bit interleaving has become an extremely useful technique in 2G and 3G digital cellular systems, and can for example be realized as a block, diagonal, or random interleaver.

A block diagram of channel encoding and user-specific spreading in an MC-CDMA transmitter assigned to user k is shown in Figure 2-10. The block diagrams are the same for up- and downlinks. The input sequence of the convolutional encoder is represented by the source bit vector

$$\mathbf{a}^{(k)} = (a_0^{(k)}, a_1^{(k)}, \ldots, a_{L_a-1}^{(k)})^T \qquad (2.69)$$

of length L_a. The code word is the discrete time convolution of $\mathbf{a}^{(k)}$ with the impulse response of the convolutional encoder. The memory M_c of the code determines the complexity of the convolutional decoder, given by 2^{M_c} different memory realizations, also called states, for binary convolutional codes. The output of the channel encoder is a coded bit sequence of length L_b which is represented by the coded bit vector

$$\mathbf{b}^{(k)} = (b_0^{(k)}, b_1^{(k)}, \ldots, b_{L_b-1}^{(k)})^T. \qquad (2.70)$$

(a) channel encoding and user-specific spreading

(b) single-user detection and soft decision channel decoding

(c) multiuser detection and soft decision channel decoding

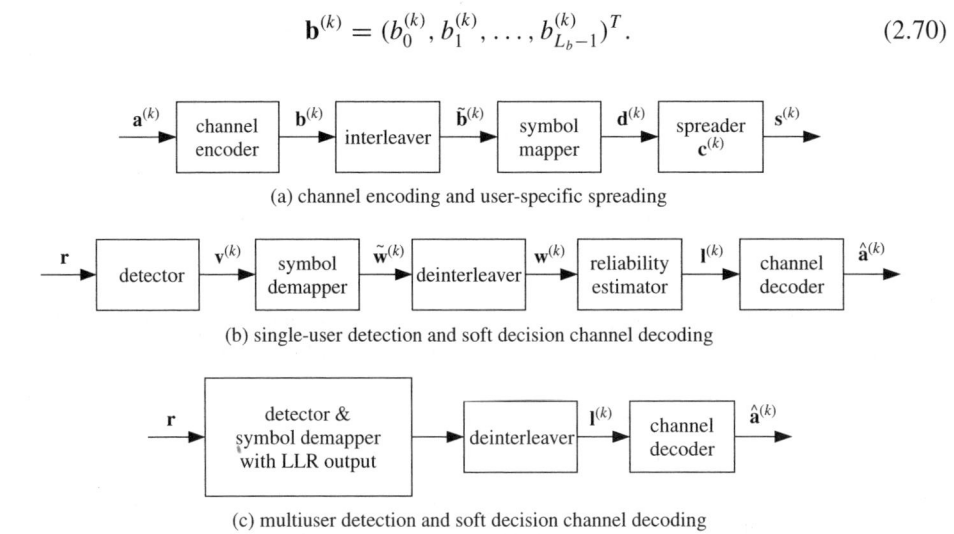

Figure 2-10 Channel encoding and decoding in MC-CDMA systems

The channel code rate is defined as the ratio

$$R = \frac{L_a}{L_b}. \tag{2.71}$$

The interleaved coded bit vector $\tilde{\mathbf{b}}^{(k)}$ is passed to a symbol mapper, where $\tilde{\mathbf{b}}^{(k)}$ is mapped into a sequence of L_d complex-valued data symbols, i.e.,

$$\mathbf{d}^{(k)} = (d_0^{(k)}, d_1^{(k)}, \ldots, b_{L_d-1}^{(k)})^T. \tag{2.72}$$

A data symbol index κ, $\kappa = 0, \ldots, L_d - 1$, is introduced to distinguish the different data symbols $d_\kappa^{(k)}$ assigned to $\mathbf{d}^{(k)}$. Each data symbol is multiplied with the spreading code $\mathbf{c}^{(k)}$ according to (2.3) and processed as described in Section 2.1.

With single-user detection, the L_d soft decided values at the output of the detector are given by the vector

$$\mathbf{v}^{(k)} = (v_0^{(k)}, v_1^{(k)}, \ldots, v_{L_d-1}^{(k)})^T. \tag{2.73}$$

The L_d complex-valued, soft decided values of $\mathbf{v}^{(k)}$ assigned to the data symbols of $\mathbf{d}^{(k)}$ are mapped on to L_b real-valued, soft decided values represented by $\tilde{\mathbf{w}}^{(k)}$ assigned to the coded bits of $\tilde{\mathbf{b}}^{(k)}$. The output of the symbol demapper after deinterleaving is written as the vector

$$\mathbf{w}^{(k)} = (w_0^{(k)}, w_1^{(k)}, \ldots, w_{L_b-1}^{(k)})^T. \tag{2.74}$$

Based on the vector $\mathbf{w}^{(k)}$, LLRs of the detected coded bits are calculated. The vector

$$\mathbf{l}^{(k)} = (\Gamma_0^{(k)}, \Gamma_1^{(k)}, \ldots, \Gamma_{L_b-1}^{(k)})^T \tag{2.75}$$

of length L_b represents the LLRs assigned to the transmitted coded bit vector $\mathbf{b}^{(k)}$. Finally, the sequence $\mathbf{l}^{(k)}$ is soft decision-decoded by applying the Viterbi algorithm. At the output of the channel decoder, the detected source bit vector

$$\hat{\mathbf{a}}^{(k)} = (\hat{a}_0^{(k)}, \hat{a}_1^{(k)}, \ldots, \hat{a}_{L_a-1}^{(k)})^T \tag{2.76}$$

is obtained.

Before presenting the coding gains of different channel coding schemes applied in MC-CDMA systems, the calculation of LLRs in fading channels is given generally for MC modulated transmission systems. Based on this introduction, the LLRs for MC-CDMA systems are derived. The LLRs for MC-CDMA systems with single-user detection and with joint detection are in general applicable for the up- and downlink. In the uplink, only the user index $^{(k)}$ has to be assigned to the individual channel fading coefficients of the corresponding users.

2.1.7.1 Log-Likelihood Ratio for OFDM Systems

The LLR is defined as

$$\Gamma = \ln\left(\frac{p(w|b = +1)}{p(w|b = -1)}\right), \tag{2.77}$$

which is the logarithm of the ratio between the likelihood function $p(w|b = +1)$ and $p(w|b = -1)$. The LLR can take on values in the interval $[-\infty, +\infty]$. With flat fading on the sub-carriers and in the presence of AWGN, the log-likelihood ratio for OFDM systems results in

$$\Gamma = \frac{4|H_{l,l}|}{\sigma^2} w. \tag{2.78}$$

2.1.7.2 Log-Likelihood Ratio for MC-CDMA Systems

Since in MC-CDMA systems a coded bit $b^{(k)}$ is transmitted in parallel on L sub-carriers, where each sub-carrier may be affected by both independent fading and multiple access interference, the LLR for OFDM systems is not applicable for MC-CDMA systems. The LLR for MC-CDMA systems is presented in the next section.

Single-User Detection
A received MC-CDMA data symbol after single-user detection results in the soft decided value

$$v^{(k)} = \sum_{l=0}^{L-1} C_l^{(k)*} G_{l,l} \left(H_{l,l} \sum_{g=0}^{K-1} d^{(g)} C_l^{(g)} + N_l \right) = \underbrace{d^{(k)} \sum_{l=0}^{L-1} \left|C_l^{(k)}\right|^2 G_{l,l} H_{l,l}}_{\text{desired symbol}}$$

$$+ \underbrace{\sum_{\substack{g=0 \\ g \neq k}}^{K-1} d^{(g)} \sum_{l=0}^{L-1} G_{l,l} H_{l,l} C_l^{(g)} C_l^{(k)*}}_{\text{MAI}} + \underbrace{\sum_{l=0}^{L-1} N_l G_{l,l} C_l^{(k)*}}_{\text{noise}} \tag{2.79}$$

Since a frequency interleaver is applied, the L complex-valued fading factors $H_{l,l}$ affecting $d^{(k)}$ can be assumed to be independent. Thus, for sufficiently long spreading codes, the multiple access interference can be considered to be additive zero-mean Gaussian noise according to the central limit theorem. The noise term can also be considered as additive zero-mean Gaussian noise. The attenuation of the transmitted data symbol $d^{(k)}$ is the magnitude of the sum of the equalized channel coefficients $G_{l,l} H_{l,l}$ of the L sub-carriers used for the transmission of $d^{(k)}$, weighted with $|C_l^{(k)}|^2$. The symbol demapper delivers the real-valued soft decided value $w^{(k)}$. According to (2.78), the LLR for MC-CDMA systems can be calculated as

$$\Gamma^{(k)} = \frac{2 \left| \sum_{l=0}^{L-1} |C_l^{(k)}|^2 G_{l,l} H_{l,l} \right|}{\sigma_{MAI}^2 + \sigma_{noise}^2} w^{(k)}. \tag{2.80}$$

Since the variances are assigned to real-valued noise, $w^{(k)}$ is multiplied by a factor of 2. When applying Walsh–Hadamard codes as spreading codes, the property can be exploited that the product $C_l^{(g)} C_l^{(k)*}$, $l = 0, \ldots, L-1$, in half of the cases equals -1 and in the other half equals $+1$ if $g \neq k$. Furthermore, when assuming that the realizations $b^{(k)} = +1$ and $b^{(k)} = -1$ are equally probable, the LLR for MC-CDMA systems with single-user detection results in [24][26]

$$\Gamma^{(k)} = \frac{2\left|\sum_{l=0}^{L-1} G_{l,l} H_{l,l}\right|}{(K-1)\left(\frac{1}{L}\sum_{l=0}^{L-1}|G_{l,l}H_{l,l}|^2 - \left|\frac{1}{L}\sum_{l=0}^{L-1} G_{l,l}H_{l,l}\right|^2\right) + \frac{\sigma^2}{2}\sum_{l=0}^{L-1}|G_{l,l}|^2} w^{(k)}. \quad (2.81)$$

When MMSE equalization is used in MC-CDMA systems, (2.81) can be approximated by [24]

$$\Gamma^{(k)} = \frac{4}{L\sigma^2}\sum_{l=0}^{L-1}|H_{l,l}|w^{(k)}, \quad (2.82)$$

since the variance of $G_{l,l}H_{l,l}$ reduces such that only the noise remains relevant.

The gain with soft decision decoding compared to hard decision decoding in MC-CDMA systems with single-user detection depends on the spreading code length and is in the order of 4 dB for small L (e.g., $L = 8$) and reduces to 3 dB with increasing L (e.g., $L = 64$). This shows the effect that the spreading averages the influence of the fading on a data symbol. When using LLRs instead of the soft decided information $w^{(k)}$, the performance further improves up to 1 dB [23].

Maximum Likelihood Detection

The LLR for coded MC-CDMA mobile radio systems with joint detection based on MLSSE is given by

$$\Gamma^{(k)} = \ln\left(\frac{p(\mathbf{r}|b^{(k)} = +1)}{p(\mathbf{r}|b^{(k)} = -1)}\right) \quad (2.83)$$

and is inherently delivered in the symbol-by-symbol estimation process presented in Section 2.1.5.2. The set of all possible transmitted data vectors $\mathbf{d}_\mu$ where the considered coded bit $b^{(k)}$ of user k is equal to $+1$ is denoted by $D_+^{(k)}$. The set of all possible data symbols where $b^{(k)}$ is equal to -1 is denoted by $D_-^{(k)}$. The LLR for MC-CDMA systems with MLSSE results in [24][26]

$$\Gamma^{(k)} = \ln\left(\frac{\sum_{\forall \mathbf{d}_\mu \in D_+^{(k)}} \exp\left(-\frac{1}{\sigma^2}\Delta^2(\mathbf{d}_\mu, \mathbf{r})\right)}{\sum_{\forall \mathbf{d}_\mu \in D_-^{(k)}} \exp\left(-\frac{1}{\sigma^2}\Delta^2(\mathbf{d}_\mu, \mathbf{r})\right)}\right), \quad (2.84)$$

where $\Delta^2(\mathbf{d}_\mu, \mathbf{r})$ is the squared Euclidean distance according to (2.44).

For coded MC-CDMA systems with joint detection based on MLSE, the sequence estimation process cannot provide reliability information on the detected, coded bits. However, an appropriate approximation for the LLR with MLSE is given by [16]

$$\Gamma^{(k)} \approx \frac{1}{\sigma^2}(\Delta^2(\mathbf{d}_{\mu-}, \mathbf{r}) - \Delta^2(\mathbf{d}_{\mu+}, \mathbf{r})). \quad (2.85)$$

The indices μ_- and μ_+ mark the smallest squared Euclidean distances $\Delta^2(\mathbf{d}_{\mu-}, \mathbf{r})$ and $\Delta^2(\mathbf{d}_{\mu+}, \mathbf{r})$ where $b^{(k)}$ is equal to -1 and $b^{(k)}$ is equal to $+1$, respectively.

Interference Cancellation

MC-CDMA receivers using interference cancellation exploit the LLRs derived for single-user detection in each detection stage, where in the second and further stages the term representing the multiple access interference in the LLRs can approximately be set to zero.

2.1.8 Flexibility in System Design

The MC-CDMA signal structure introduced in Section 2.1.1 enables the realization of powerful receivers with low complexity due to the avoidance of ISI and ICI in the detection process. Moreover, the spreading code length L has not necessarily to be equal to the number of sub-carriers N_c in an MC-CDMA system, which enables a flexible system design and can further reduce the complexity of the receiver. The three MC-CDMA system modifications presented in the following are referred to as M-Modification, Q-Modification, and $M\&Q$-Modification [15][16][23]. These modifications can be applied in the up- and in the downlink of a mobile radio system.

2.1.8.1 Parallel Data Symbols (M-Modification)

As depicted in Figure 2-11, the M-Modification increases the number of sub-carriers N_c while maintaining constant the overall bandwidth B, the spreading code length L and the maximum number of active users K. The OFDM symbol duration increases and the loss in spectral efficiency due to the guard interval decreases. Moreover, the tighter sub-carrier spacing enables one to guarantee flat fading per sub-channel in propagation scenarios with small coherence bandwidth. With the M-Modification, each user transmits simultaneously $M > 1$ data symbols per OFDM symbol.

The total number of sub-carriers of the modified MC-CDMA system is

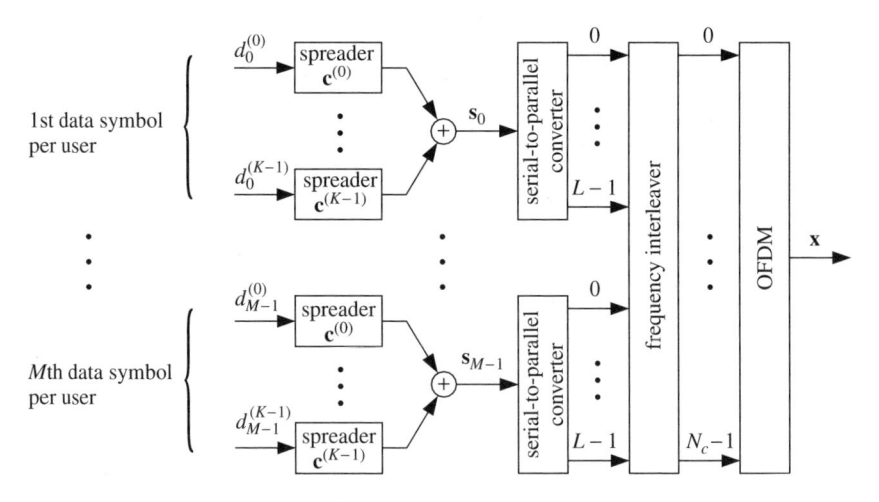

$$N_c = ML. \tag{2.86}$$

Figure 2-11 M-Modification

Each user exploits the total of N_c sub-carriers for data transmission. The OFDM symbol duration (including the guard interval) increases to

$$T'_s = T_g + MLT_c, \qquad (2.87)$$

where it can be observed that the loss in spectral efficiency due to the guard interval decreases with increasing M. The maximum number of active users is still $K = L$.

The data symbol index m, $m = 0, \ldots, M - 1$, is introduced in order to distinguish the M simultaneously transmitted data symbols $d_m^{(k)}$ of user k. The number M is upper-limited by the coherence time $(\Delta t)_c$ of the channel. To optimally exploit frequency diversity, the components of the sequences $\mathbf{s}_m$, $m = 0, \ldots, M - 1$, transmitted in the same OFDM symbol, are interleaved over the frequency. The interleaving is carried out prior to OFDM.

2.1.8.2 Parallel User Groups (Q-Modification)

With an increasing number of active users K the number of required spreading codes and, thus, the spreading code length L, increases. Since L and K determine the complexity of the receiver, both values have to be kept as small as possible. The Q-Modification introduces an OFDMA component (see Chapter 3) on sub-carrier level and with that reduces the receiver complexity by reducing the spreading code length per user, while maintaining constant the maximum number of active users K and the number of sub-carriers N_c. The MC-CDMA transmitter with Q-Modification is shown in Figure 2-12 where Q different user groups transmit simultaneously in one OFDM symbol. Each user group has a specific set of sub-carriers for transmission which avoids interference between different user groups. Assuming that each user group applies spreading codes of length L, the total number of sub-carriers is

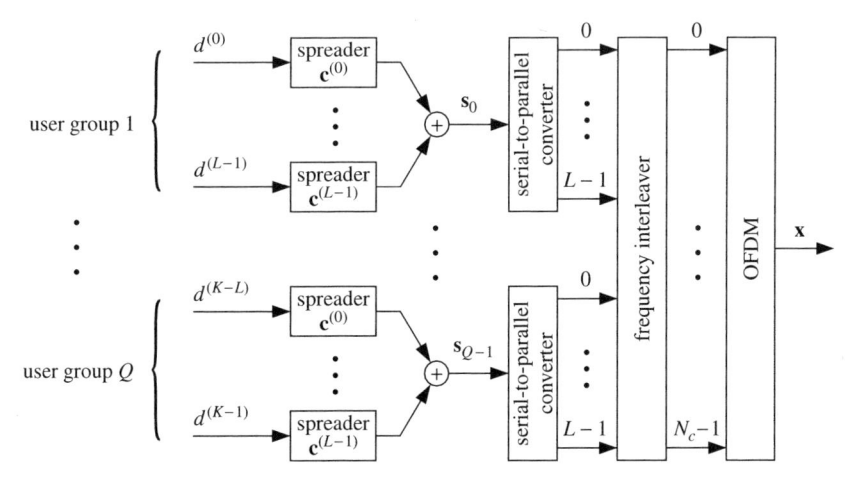

$$N_c = QL, \qquad (2.88)$$

Figure 2-12 Q-Modification

where each user exploits a subset of L sub-carriers for data transmission. Depending on the coherence bandwidth $(\Delta f)_c$ of the channel, it can be sufficient to apply spreading codes with $L \ll N_c$ to obtain the full diversity gain [17][23].

To optimally exploit the frequency diversity of the channel, the components of the spread sequences $\mathbf{s}_q$, $q = 0, \ldots, Q - 1$, transmitted in the same OFDM symbol are interleaved over the frequency. The interleaving is carried out prior to OFDM. The OFDM symbol duration (including the guard interval) is

$$T'_s = T_g + QLT_c. \tag{2.89}$$

Only one set of L spreading codes of length L is required within the whole MC-CDMA system. This set of spreading codes can be used in each subsystem. An adaptive sub-carrier allocation can also increase the capacity of the system [2][10].

2.1.8.3 $M \& Q$-Modification

$M \& Q$-Modification combines the flexibility of M- and Q-Modification. The transmission of M data symbols per user and, additionally, the splitting of the users in Q independent user groups according to $M \& Q$-Modification is illustrated in Figure 2-13.

The total number of sub-carriers used is

$$N_c = MQL, \tag{2.90}$$

where each user only exploits a subset of ML sub-carriers for data transmission due to the OFDMA component introduced by Q-Modification. The total OFDM symbol duration (including the guard interval) results in

$$T'_s = T_g + MQLT_c. \tag{2.91}$$

A frequency interleaver scrambles the information of all subsystems prior to OFDM to guarantee an optimum exploitation of the frequency diversity offered by the mobile radio channel.

M-, Q-, and $M \& Q$-Modification are also suitable for the uplink of an MC-CDMA mobile radio system. For Q- and $M \& Q$-Modification in the uplink only the inputs of the frequency interleaver of the user group of interest are connected in the transmitter; all other inputs are set to zero.

Finally, it should be noted that an MC-CDMA system with its basic implementation or with any of the three modifications presented in this section could support an additional TDMA component in the up- and downlink, since the transmission is synchronized on OFDM symbols.

2.1.9 Performance Analysis
2.1.9.1 System Parameters

The parameters of the MC-CDMA system analyzed in this section are summarized in Table 2-2. Orthogonal Walsh–Hadamard codes are used for spreading. The spreading code length in a subsystem is $L = 8$. Unless otherwise stated, cases with fully loaded systems are considered. QPSK, 8-PSK and 16-QAM with Gray encoding are applied for

MC-CDMA

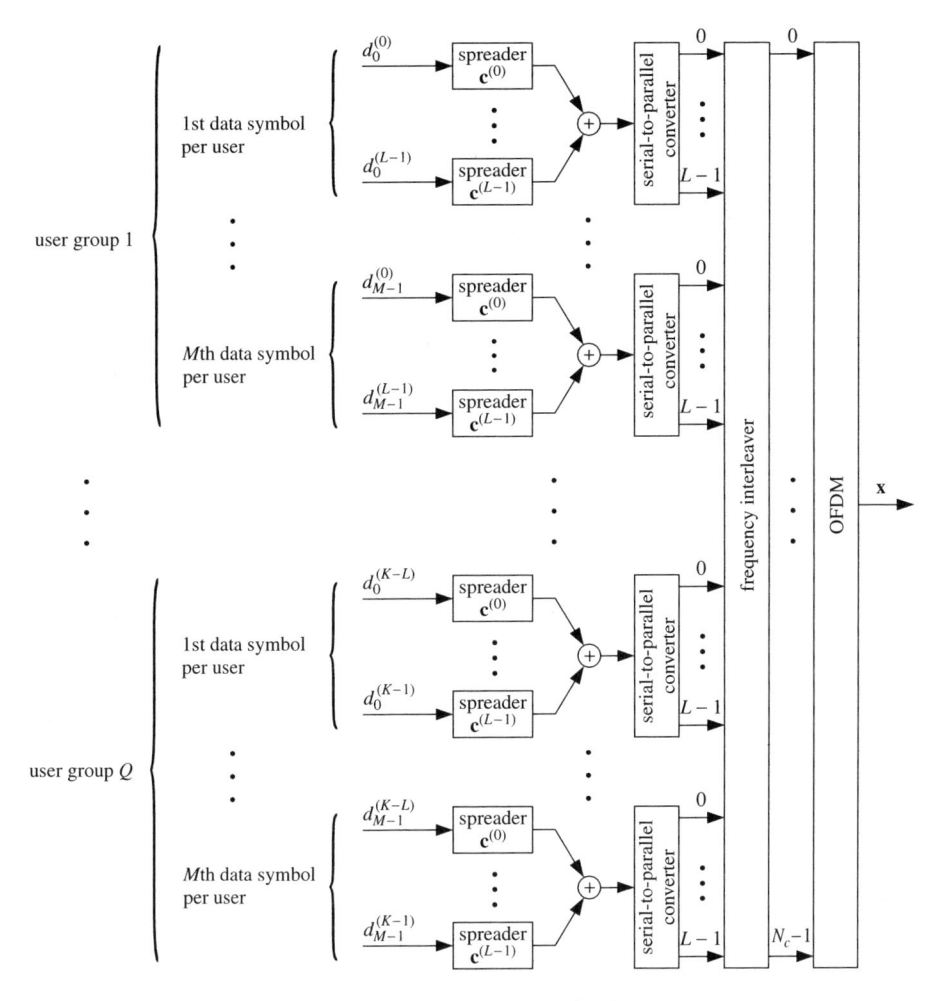

Figure 2-13 $M \& Q$-Modification

data symbol mapping. Moreover, the guard interval of the reference system is chosen such that ISI and ICI are eliminated. The mobile radio channel is implemented as an uncorrelated Rayleigh fading channel, described in detail in Section 1.1.6.

The performance of the MC-CDMA reference system presented in this section is applicable to any MC-CDMA system with an arbitrary transmission bandwidth B, an arbitrary number of subsystems Q, and an arbitrary number of data symbols M transmitted per user in an OFDM symbol, resulting in an arbitrary number of sub-carriers. The number of sub-carriers within a subsystem has to be 8, the amplitudes of the channel fading have to be Rayleigh-distributed and have to be uncorrelated on the sub-carriers of a subsystem due to appropriate frequency interleaving. The loss in SNR due to the guard interval is not taken into account in the results. The intention is that the loss in SNR due to the guard interval can be calculated individually for each specified guard interval. So, the results presented can be adapted to any guard interval.

Table 2-2 MC-CDMA system parameters

Parameter	Value/Characteristics
Spreading codes	Walsh–Hadamard codes
Spreading code length L	8
System load	Fully loaded
Symbol mapping	QPSK, 8-PSK, 16-QAM
FEC codes	Convolutional codes with memory 6
FEC code rate R and FEC-Decoder	4/5, 2/3, 1/2, 1/3 with Viterbi decoder
Channel estimation & synchronization	Perfect
Mobile radio channel	Uncorrelated Rayleigh fading channel

2.1.9.2 Synchronous Downlink

The BER versus the SNR per bit for single-user detection techniques MRC, EGC, ZF and MMSE equalization in an MC-CDMA system without FEC coding is depicted in Figure 2-14. The results show that with a fully loaded system the MMSE equalization outperforms the other single-user detection techniques. ZF equalization restores the orthogonality between the user signals and avoids MAI. However, it introduces noise amplification. EGC avoids noise amplification but does not counteract the MAI caused by the loss of the orthogonality between the user signals, resulting in a high error floor. The worst performance is obtained with MRC which additionally enhances the MAI. As reference, the matched filter bound (lower bound) for the MC-CDMA system is given. Analytical approaches to evaluate the performance of MC-CDMA systems with MRC and EGC are given in [51], with ZF equalization in [47] and with MMSE equalization in [22].

Figure 2-15 shows the BER versus the SNR per bit for the multiuser detection techniques parallel IC, MLSE, and MLSSE applied in an MC-CDMA system without FEC coding. The performance of parallel IC with adapted MMSE equalization is presented for two detection stages. The significant performance improvements with parallel IC are obtained after the first iteration. The optimum joint detection techniques MLSE and MLSSE perform almost identically and outperform the other detection techniques. The SNR degradation with the optimum detection techniques compared to the matched filter bound (lower bound) is caused by the superposition of orthogonal Walsh–Hadamard codes, resulting in sequences of length L which can contain up to $L-1$ zeros. Sequences with many zeros perform worse in the fading channel due to the reduced diversity gain. These diversity losses can be reduced by applying rotated constellations, as described in Section 2.1.4.4. An upper bound of the BER for MC-CDMA systems applying joint detection with MLSE and MLSSE for the uncorrelated Rayleigh channel is derived in [16] and for the uncorrelated Rice fading channel in [22]. Analytical approaches to determine the performance of MC-CDMA systems with interference cancellation are shown in [22][27].

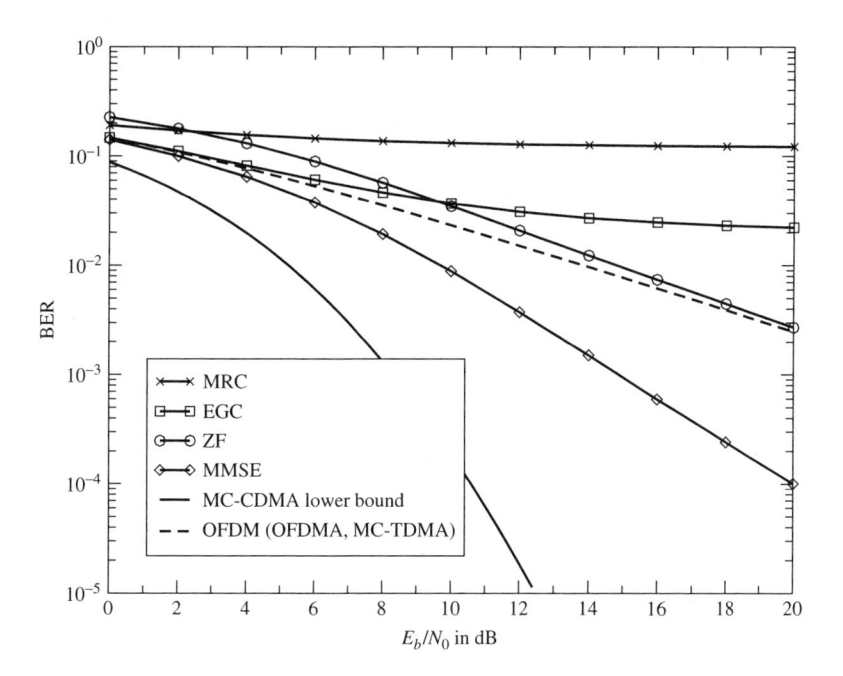

Figure 2-14 BER versus SNR for MC-CDMA with different single-user detection techniques; fully loaded system; no FEC coding; QPSK; Rayleigh fading

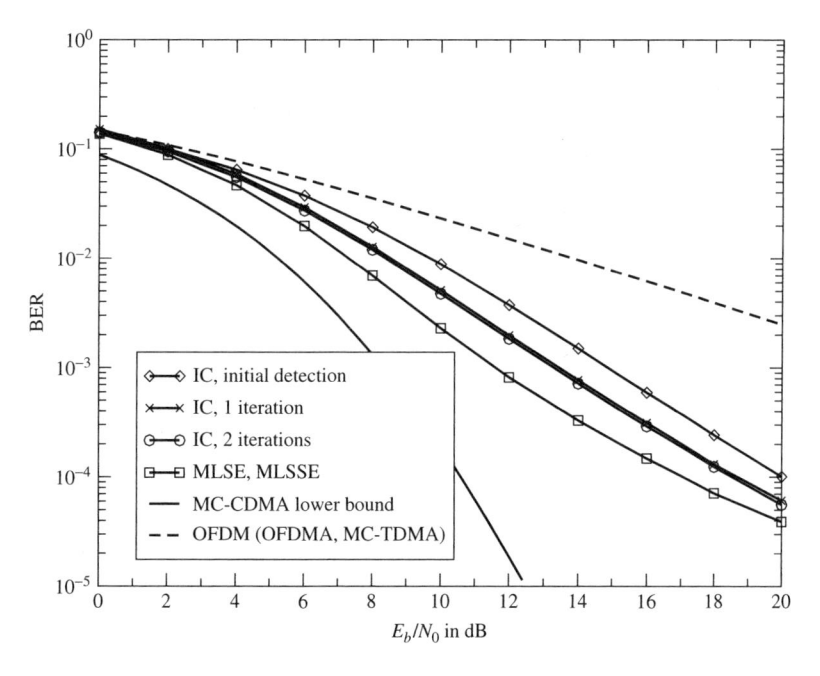

Figure 2-15 BER versus SNR for MC-CDMA with different multiuser detection techniques; fully loaded system; no FEC coding; QPSK; Rayleigh fading

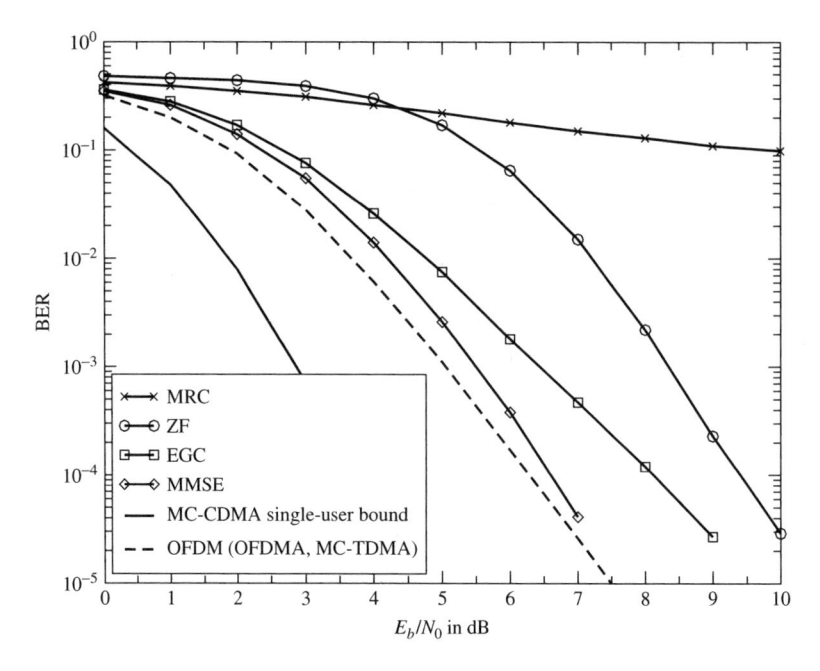

Figure 2-16 FEC coded BER versus SNR for MC-CDMA with different single-user detection techniques; fully loaded system; channel code rate $R = 1/2$; QPSK; Rayleigh fading

The FEC coded BER versus the SNR per bit for single-user detection with MRC, EGC, ZF and MMSE equalization in MC-CDMA systems is presented in Figure 2-16. It can be observed that rate 1/2 coded OFDM (OFDMA, MC-TDMA) systems slightly outperform rate 1/2 coded MC-CDMA systems with MMSE equalization when considering cases with full system load in a single cell. Furthermore, the performance of coded MC-CDMA systems with simple EGC requires only about a 1 dB higher SNR to reach the BER of 10^{-3} compared to more complex MC-CDMA systems with MMSE equalization. With a fully loaded system, the single-user detection technique MRC is not of interest in practice.

The FEC coded BER versus the SNR per bit for multiuser detection with soft IC, MLSE, MLSSE, and single-user detection with MMSE equalization is shown in Figure 2-17 for code rate 1/2. Coded MC-CDMA systems with the soft IC detection technique outperform coded OFDM (OFDMA, MC-TDMA) systems and MC-CDMA systems with MLSE/MLSSE. The performance of the initial stage with soft IC is equal to the performance with MMSE equalization. Promising results are obtained with soft IC already after the first iteration.

The FEC coded BER versus the SNR per bit for different symbol mapping schemes in MC-CDMA systems with soft IC and in OFDM (OFDMA, MC-TDMA) systems is shown in Figure 2-18 for code rate 2/3. Coded MC-CDMA systems with the soft IC detection technique outperform coded OFDM (OFDMA, MC-TDMA) systems for all symbol mapping schemes at lower BERs due to the steeper slope obtained with MC-CDMA.

Finally, the spectral efficiency of MC-CDMA with soft IC and of OFDM (OFDMA, MC-TDMA) versus the SNR is shown in Figure 2-19. The results are given for the code

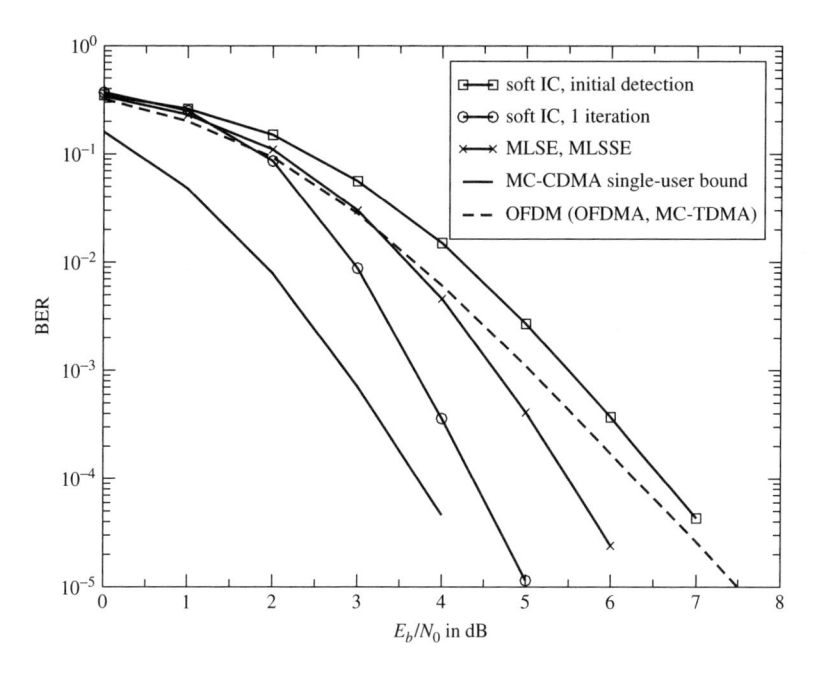

Figure 2-17 FEC coded BER versus SNR for MC-CDMA with different multiuser detection techniques; fully loaded system; channel code rate $R = 1/2$; QPSK; Rayleigh fading

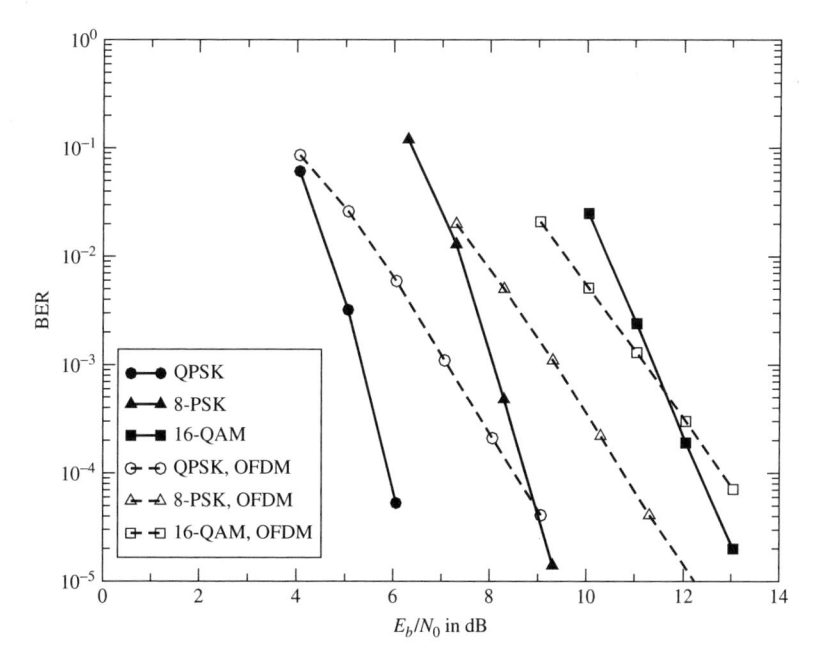

Figure 2-18 FEC coded BER versus SNR for MC-CDMA with different symbol mapping schemes; fully loaded system; channel code rate $R = 2/3$; Rayleigh fading

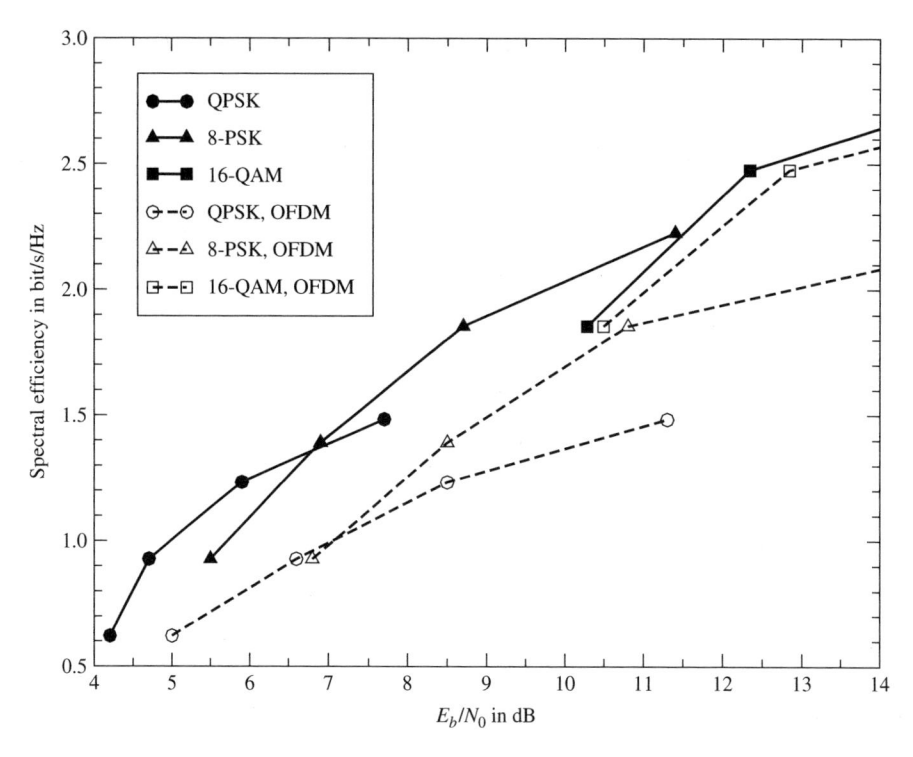

Figure 2-19 Spectral efficiency of MC-CDMA and OFDM (OFDMA, MC-TDMA); fully loaded system; Rayleigh fading; BER = 10^{-4}

rates 1/3, 1/2, 2/3, and 4/5 and are shown for a BER of 10^{-4}. The curves in Figure 2-19 show that MC-CDMA with soft IC can outperform OFDM (OFDMA, MC-TDMA).

Figure 2-19 represents the most important results regarding spectral/power efficiency in a cellular system in favor of MC-CDMA schemes. These curves lead to the following conclusions:

— for a given coverage, the transmitted data rate can be augmented by at least 40% compared to MC-TDMA or OFDMA, or
— for a given data rate, about 2.5 dB can be gained in SNR. The 2.5 dB extension in power will give a higher coverage for an MC-CDMA system.

2.1.9.3 Synchronous Uplink

The parameters used for the synchronous uplink are the same as for the downlink presented in the previous section. Orthogonal spreading codes outperform other codes such as Gold codes in the synchronous MC-CDMA uplink scenario which motivates the choice of Walsh–Hadamard codes also in the uplink. Each user has an uncorrelated Rayleigh fading channel. Due to the loss of orthogonality of the spreading codes at the receiver antenna, MRC is the optimum single-user detection technique in the uplink (see Section 2.1.5.1).

The performance of an MC-CDMA system with different loads and MRC in the synchronous uplink is shown in Figure 2-20. It can be observed that due to the loss of orthogonality between the user signals in the uplink only moderate numbers of active users can be handled with single-user detection.

The performance of MC-CDMA in the synchronous uplink can be significantly improved by applying multiuser detection techniques. Various concepts have been investigated in the literature. In the uplink, the performance of MLSE and MLSSE closely approximates the single user bound (1 user curve in Figure 2-20) since here the Walsh–Hadamard codes do not superpose orthogonally and the maximum diversity can be exploited [43]. The performance degradation of a fully loaded MC-CDMA system with MLSE/MLSSE compared to the single-user bound is about 1 dB in SNR.

Moreover, suboptimum multiuser detection techniques have also been investigated for MC-CDMA in the uplink, which benefit from reduced complexity in the receiver. Interference cancellation schemes are analyzed in [1] and [29] and joint detection schemes in [5] and [45].

To take advantage of MC-CDMA with nearly orthogonal user separation at the receiver antenna, the pre-equalization techniques presented in Section 2.1.6 can be applied. The parameters of the TDD MC-CDMA system under investigation are presented in Table 2-3.

In Figure 2-21, the BER versus the SNR for an MC-CDMA system with different pre-equalization techniques in the uplink is shown. The system is fully loaded. It can be

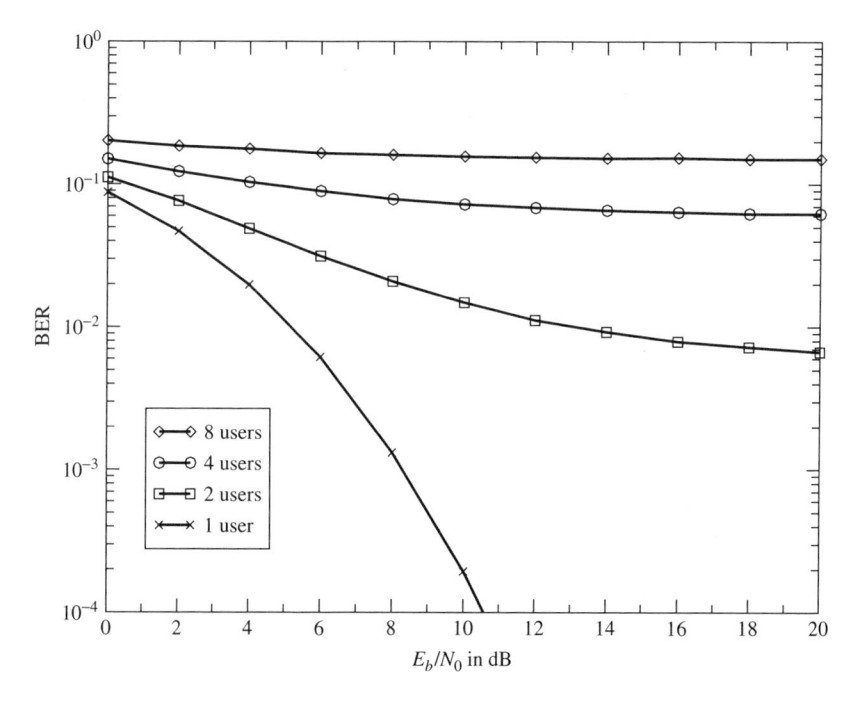

Figure 2-20 BER versus SNR for MC-CDMA in the synchronous uplink; MRC; no FEC coding; QPSK; $L = 8$; Rayleigh fading

Table 2-3 Parameters of the TDD MC-CDMA uplink system with pre-equalization

Parameter	Value/Characteristics
Bandwidth	20 MHz
Carrier frequency	5.2 GHz
Number of sub-carriers	256
Spreading codes	Walsh-Hadamard codes
Spreading code length L	16
Symbol mapping	QPSK
FEC coding	Non
Mobile radio channel	Indoor fading channel with $T_g > \tau_{max}$
Max. Doppler frequency	26 Hz

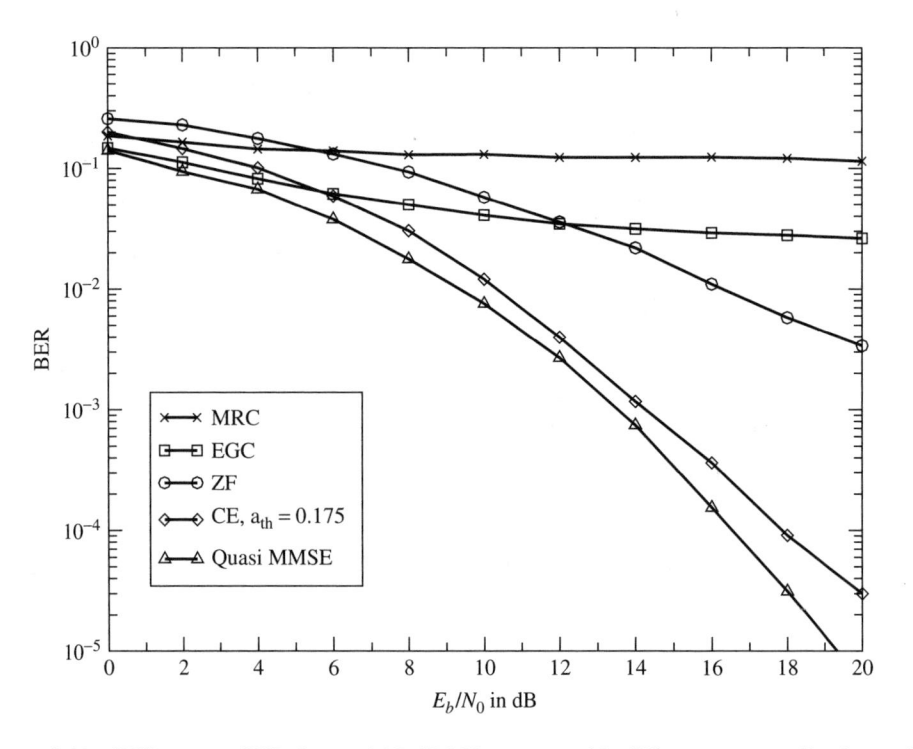

Figure 2-21 BER versus SNR for an MC-CDMA system with different pre-equalization techniques in the uplink; fully loaded system; no FEC coding

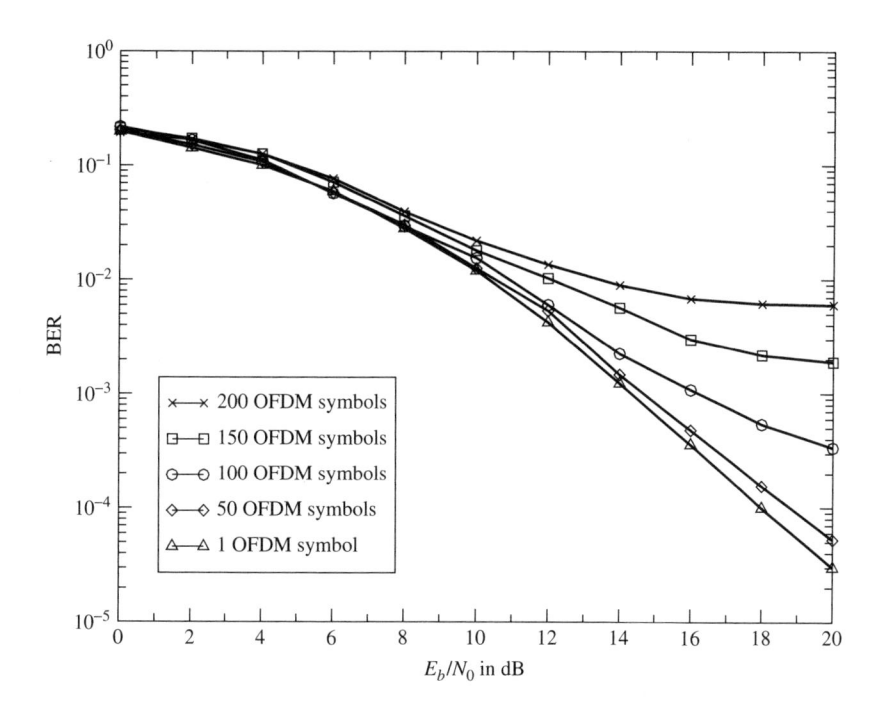

Figure 2-22 BER versus SNR for an MC-CDMA system with controlled pre-equalization and different frame lengths in the uplink; fully loaded system; no FEC coding

observed that with pre-equalization a fully loaded MC-CDMA system can achieve promising results in the uplink. Quasi MMSE pre-equalization and controlled pre-equalization with a threshold of $a_{th} = 0.175$ outperform other pre-equalization techniques.

When assuming that the information about the uplink channel for pre-equalization is only available at the beginning of each transmission frame, the performance of the system degrades with increasing frame duration due to the time variation of the channel. A typical scenario would be that at the beginning of each frame a feedback channel provides the transmitter with the required channel state information. Of importance for the selection of a proper frame duration is that it is smaller than the coherence time of the channel. The influence of the frame length for an MC-CDMA system with controlled pre-equalization is shown in Figure 2-22. The Doppler frequency is 26 Hz and the OFDM symbol duration is 13.6 μs.

In Figure 2-23, the performance of an MC-CDMA system with controlled pre-equalization and an update of the channel coefficients at the beginning of each OFDM frame is shown for different system loads. An OFDM frame consists of 200 OFDM symbols.

2.2 MC-DS-CDMA

2.2.1 Signal Structure

The MC-DS-CDMA signal is generated by serial-to-parallel converting data symbols into N_c sub-streams and applying DS-CDMA on each individual sub-stream. Thus, with

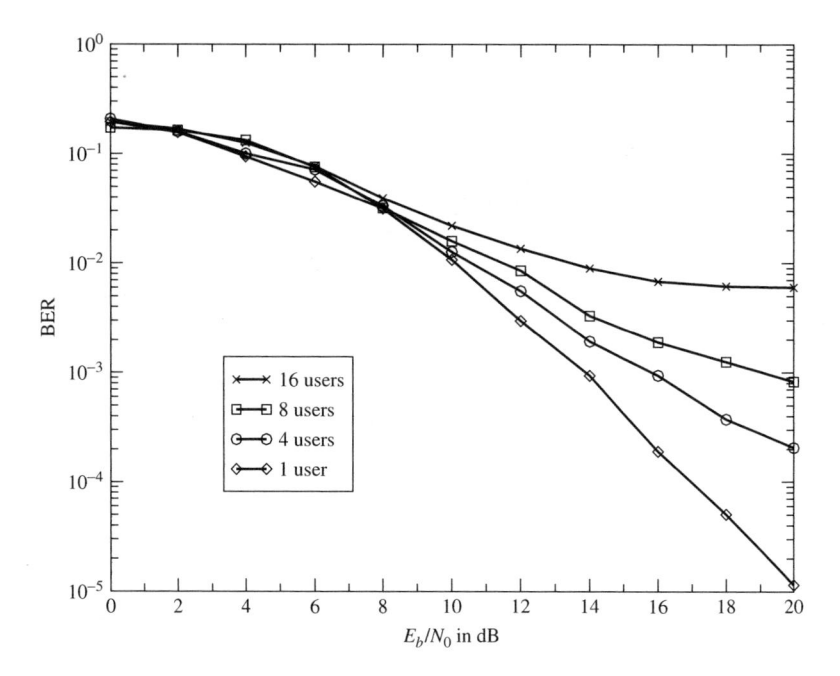

Figure 2-23 BER versus SNR in an MC-CDMA system with controlled pre-equalization and different system loads in the uplink; frame length is equal to 200 OFDM symbols; no FEC coding

MC-DS-CDMA, each data symbol is spread in bandwidth within its sub-channel, but in contrast to MC-CDMA or DS-CDMA not over the whole transmission bandwidth for $N_c > 1$. An MC-DS-CDMA system with one sub-carrier is identical to a single-carrier DS-CDMA system. MC-DS-CDMA systems can be distinguished in systems where the sub-channels are narrowband and the fading per sub-channel appears flat and in systems with broadband sub-channels where the fading is frequency-selective per sub-channel. The fading over the whole transmission bandwidth can be frequency-selective in both cases. The complexity of the receiver with flat fading per sub-channel is comparable to that of an MC-CDMA receiver, when OFDM is assumed for multi-carrier modulation. As soon as the fading per sub-channel is frequency-selective and ISI occurs, more complex detectors have to be applied. MC-DS-CDMA is of special interest for the asynchronous uplink of mobile radio systems, due to its close relation to asynchronous single-carrier DS-CDMA systems. On one hand, a synchronization of users can be avoided, however, on the other hand, the spectral efficiency of the system decreases due to asynchronism.

Figure 2-24 shows the generation of a multi-carrier direct sequence spread spectrum signal. The data symbol rate is $1/T_d$. A sequence of N_c complex-valued data symbols $d_n^{(k)}$, $n = 0, \ldots, N_c - 1$, of user k is serial-to-parallel converted onto N_c sub-streams. The data symbol rate on each sub-stream becomes $1/(N_c T_d)$. Within a single sub-stream, a data symbol is spread with the user-specific spreading code

$$c^{(k)}(t) = \sum_{l=0}^{L-1} c_l^{(k)} p_{T_c}(t - lT_s) \qquad (2.92)$$

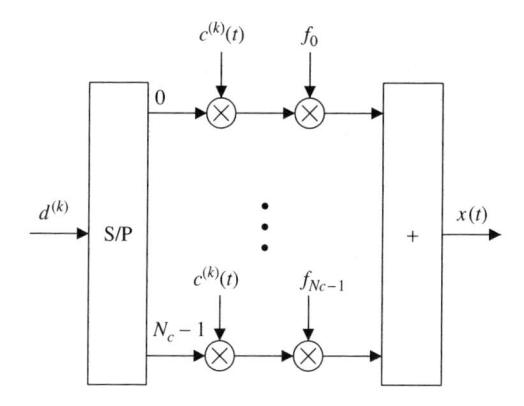

Figure 2-24 MC-DS-CDMA transmitter

of length L. The pulse form of the chips is given by $p_{T_c}(t)$. For the description of the MC-DS-CDMA signal, the continuous time representation is chosen, since MC-DS-CDMA systems are of interest for the asynchronous uplink. Here, OFDM might not necessarily be the best choice of multi-carrier modulation technique. The duration of a chip within a sub-stream is

$$T_c = T_s = \frac{N_c T_d}{L}. \tag{2.93}$$

With multi-carrier direct sequence spread spectrum, each data symbol is spread over L multi-carrier symbols, each of duration T_s. The complex-valued sequence obtained after spreading is given by

$$x^{(k)}(t) = \sum_{n=0}^{N_c-1} d_n^{(k)} c^{(k)}(t) e^{j2\pi f_n t}, \quad 0 \leqslant t < LT_s. \tag{2.94}$$

The nth sub-carrier frequency is

$$f_n = \frac{(1+\alpha)n}{T_s}, \tag{2.95}$$

where $0 \leqslant \alpha \leqslant 1$. The choice of α depends on the chosen chip form $p_{T_c}(t)$ and is typically chosen such that the N_c parallel sub-channels are disjoint. In the case of OFDM, α is equal to 0 and $p_{T_c}(t)$ has a rectangular form.

A special case of MC-DS-CDMA systems is obtained when the sub-carrier spacing is equal to $1/(N_c T_s)$. The tight sub-carrier spacing allows the use of longer spreading codes to better reduce multiple access interference; however, it results in an overlap of the signal spectra of the sub-carriers and introduces ICI. This special case of MC-DS-CDMA is referred to as multitone CDMA (MT-CDMA) [48]. An MT-CDMA signal is generated by first modulating a block of N_c data symbols on N_c sub-carriers applying OFDM before spreading the resulting signal with a code of length $N_c L$, where L is the spreading code length of conventional MC-DS-CDMA. Due to the N_c times increased sub-channel bandwidth with MT-CDMA, each sub-channel is broadband and more complex receivers are required.

2.2.2 Downlink Signal

In the synchronous downlink, the signals of K users are superimposed in the transmitter. The resulting transmitted MC-DS-CDMA signal is

$$x(t) = \sum_{k=0}^{K-1} x^{(k)}(t). \quad (2.96)$$

The signal received at a terminal station is given by

$$y(t) = x(t) \otimes h(t) + n(t). \quad (2.97)$$

Since the downlink can be synchronized, ISI and ICI can be avoided by choosing an appropriate guard interval. Moreover, if narrowband sub-channels are achieved in the system design, a low-complex implementation of an MC-DS-CDMA system exploiting the advantages of OFDM can be realized.

2.2.3 Uplink Signal

In the uplink, the MC-DS-CDMA signal transmitted by user k is $x^{(k)}(t)$. The channel output assigned to user k is given by the convolution of $x^{(k)}(t)$ with the channel impulse response $h^{(k)}(t)$, i.e.,

$$y^{(k)}(t) = x^{(k)}(t) \otimes h^{(k)}(t). \quad (2.98)$$

The received signal of all K users at the base station including the additive noise signal $n(t)$ results in

$$y(t) = \sum_{k=0}^{K-1} y^{(k)}(t - \tau^{(k)}) + n(t). \quad (2.99)$$

The user-specific delay relative to the first arriving signal is given by $\tau^{(k)}$. If all users are synchronized, then $\tau^{(k)} = 0$ for all K users.

As soon as each sub-channel can be considered as narrowband, i.e., the sub-channel bandwidth is smaller than the coherence bandwidth $(\Delta f)_c$, the fading per sub-channel is frequency nonselective and low-complex detection techniques compared to broadband sub-channels can be realized. Narrowband sub-channels are achieved by choosing a sufficiently large number of sub-carriers relative to the bandwidth B. A rough approximation for the minimum number of sub-carriers is given by

$$N_c \geq \tau_{\max} B. \quad (2.100)$$

The overall transmission bandwidth is given by B, and $\tau_{\max}$ is the maximum delay of the mobile radio channel.

2.2.4 Spreading

Since MC-DS-CDMA is of interest for the asynchronous uplink, spreading codes such as PN or Gold codes described in Section 2.1.4.1 are of interest for this scenario. As for asynchronous single-carrier DS-CDMA systems, good auto- and cross-correlation properties are required. In the case of a synchronous downlink, orthogonal codes are preferable.

2.2.5 Detection Techniques

MC-DS-CDMA systems with broadband sub-channels can be split into N_c classical broadband DS-CDMA systems. Thus, single- and multiuser detection techniques known for DS-CDMA can be applied in each data stream, which are in detail described in Section 1.3.1.2 and in [49]. Deep analysis for MC-DS-CDMA systems with broadband sub-channels and pre-rake diversity combining techniques have been carried out in [19] and [34].

MC-DS-CDMA is typically applied for asynchronous uplink scenarios. The assigned single-user detector for MC-DS-CDMA with narrowband sub-channels can be realized by a spreading code correlator on each sub-channel. Figure 2-25 shows the single-user detector with N_c correlators.

Modified MC-DS-CDMA systems transmit the same spread data symbol in parallel on p, $0 < p \leqslant N_c$, sub-carriers in order to achieve an additional diversity gain. However, this reduces the spectral efficiency of the system by the factor of p. In the receiver, the detected data of the p sub-channels are combined with either EGC or MRC [28][44].

In the case of synchronous uplink and downlink transmission with narrowband sub-channels, the same detection techniques described in Section 2.1.5 for MC-CDMA can also be applied for MC-DS-CDMA.

2.2.6 Performance Analysis

The BER performance of an MC-DS-CDMA system in the synchronous and asynchronous uplink is analyzed in this section. The transmission bandwidth is 5 MHz and the carrier frequency is 5 GHz. Compared to the MC-CDMA parameters, longer spreading codes of length $L = 31$ with Gold codes and $L = 32$ with Walsh–Hadamard codes are chosen with MC-DS-CDMA due to the spreading in time direction. The number of sub-carriers is identical to the spreading code length, i.e., $N_c = L$. QPSK symbol mapping is chosen. No FEC coding scheme is applied. The mobile radio channels of the individual users are modeled by the COST 207 rural area (RA) channel model and the Doppler frequency is 1.15 kHz (250 km/h). A high Doppler frequency is of advantage for MC-DS-CDMA, since it offers high time diversity.

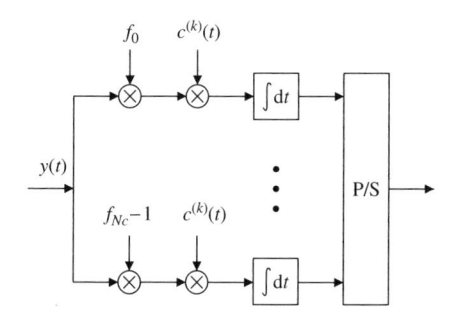

Figure 2-25 MC-DS-CDMA correlation detector

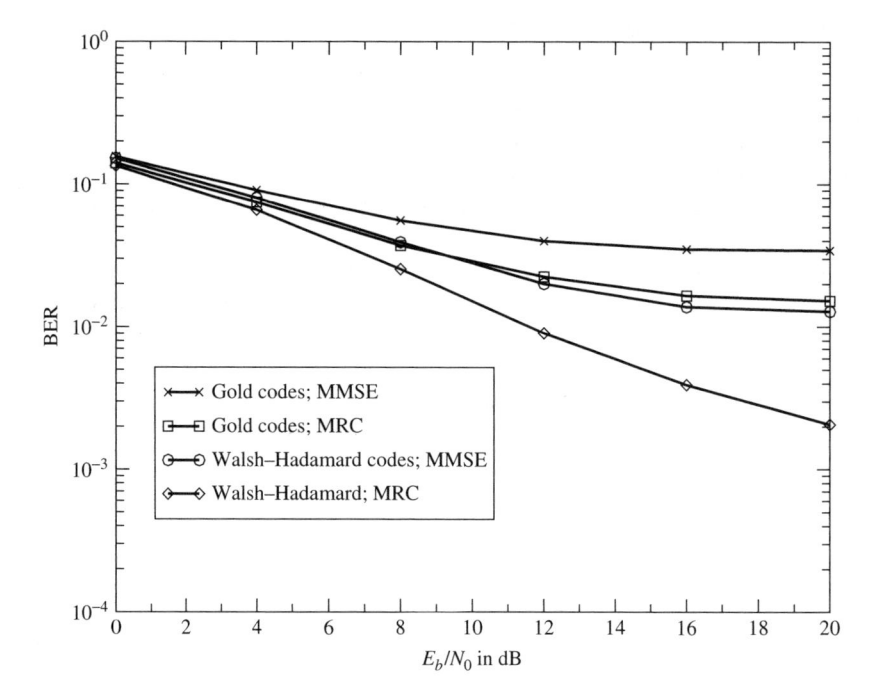

Figure 2-26 BER versus SNR for an MC-DS-CDMA system with different spreading codes and detection techniques; synchronous uplink; $K = 8$ users; QPSK; COST 207 RA fading channel

2.2.6.1 Synchronous Uplink

Figure 2-26 shows the BER versus the SNR per bit for an MC-DS-CDMA system with different spreading codes and detection techniques. The number of active users is 8. It can be observed that Walsh–Hadamard codes outperform Gold codes in the synchronous uplink. Moreover, the single-user detection technique MRC outperforms MMSE equalization. All curves show a quite high error floor due to multiple access interference.

The influence of the system load is presented in Figure 2-27 for a system with Walsh–Hadamard codes and MRC. The MC-DS-CDMA system can only handle moderate numbers of users in the synchronous uplink due to the limitation by the multiple access inference.

To overcome the limitation with single-user detection techniques in the uplink, more complex multiuser detection techniques have to be used which have been analyzed, e.g., in [31] for MC-DS-CDMA in the synchronous uplink.

2.2.6.2 Asynchronous Uplink

The BER performance of an asynchronous MC-DS-CDMA in the uplink is shown in Figure 2-28 for different spreading codes and system loads. MRC has been chosen as detection technique, since it is the optimum single-user detection technique in the uplink (see Section 2.1.5.1). It can be observed that Gold codes outperform Walsh–Hadamard

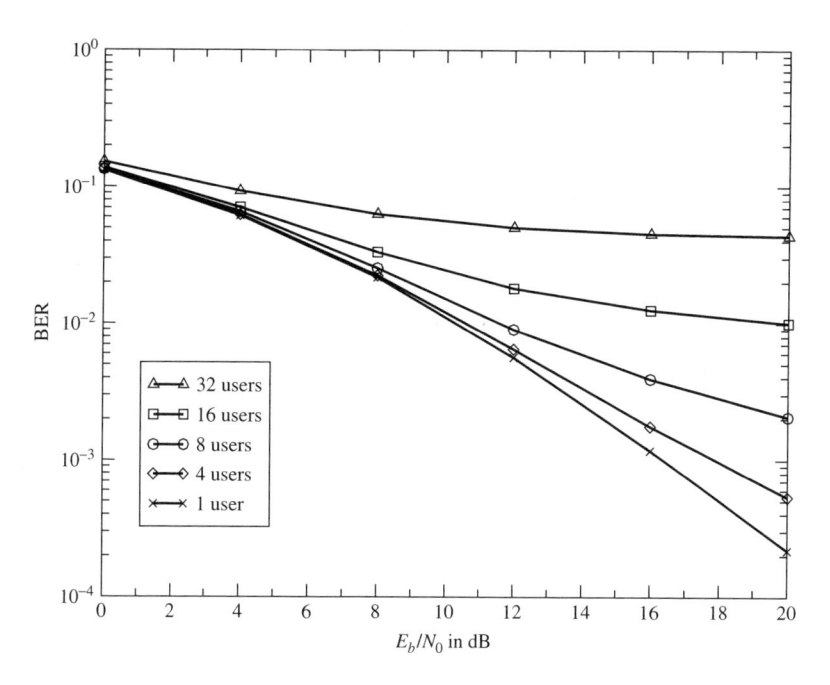

Figure 2-27 BER versus SNR for an MC-DS-CDMA system with different loads; synchronous uplink; Walsh–Hadamard codes; MRC; QPSK; COST 207 RA fading channel

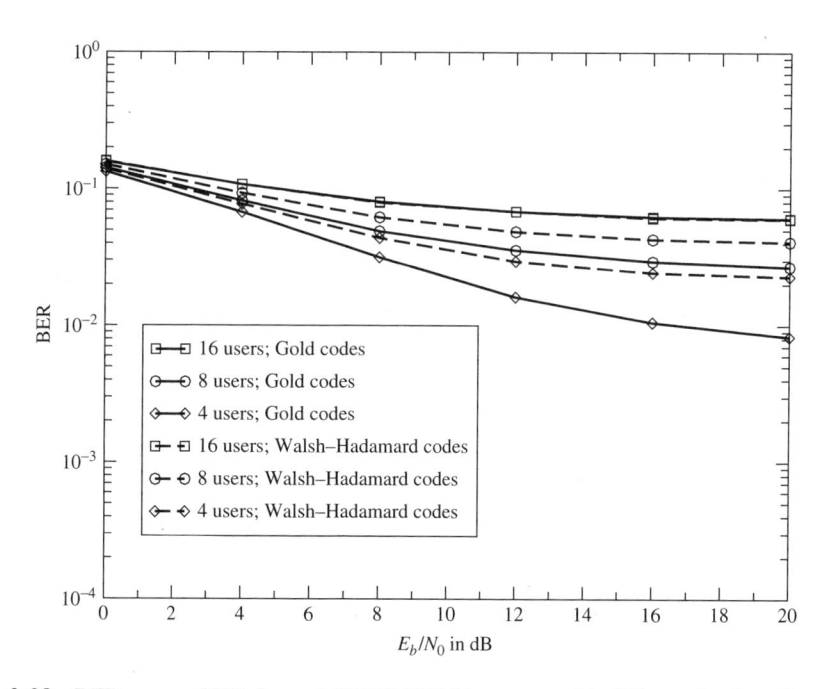

Figure 2-28 BER versus SNR for an MC-DS-CDMA system with different loads and spreading codes in the asynchronous uplink; MRC; QPSK; COST 207 RA fading channel

codes due to better cross-correlation properties, which are of importance in the asynchronous case. The considered MC-DS-CDMA system shows a high error floor due to multiple access interference, which is higher than in the synchronous case.

These results show that MC-DS-CDMA in the asynchronous uplink needs more complex multiuser detectors to handle larger numbers of users. Investigations into interference cancellation have been carried out in [14].

2.3 References

[1] Akther M.S., Asenstorfer J., Alexander P.D. and Reed M.C., "Performance of multi-carrier CDMA with iterative detection," in *Proc. IEEE International Conference on Universal Personal Communications (ICUPC'98)*, Florence, Italy, pp. 131–135, Oct. 1998.

[2] Al-Susa E. and Cruickshank D., "An adaptive orthogonal multicarrier multiuser CDMA technique for a broadband mobile communication system," in *Proc. International Workshop on Multi-Carrier Spread-Spectrum & Related Topics (MC-SS 2001)*, Oberpfaffenhofen, Germany, pp. 45–52, Sept. 2001.

[3] Atarashi H., Maeda N., Abeta S. and Sawahashi M., "Broadband packet wireless access based on VSF-OFCDM and MC/DS-CDMA," in *Proc. IEEE International Symposium on Personal, Indoor and Mobile Radio Communications (PIMRC 2002)*, Lisbon, Portugal, pp. 992–997, Sept. 2002.

[4] Aue V. and Fettweis G., "Multi-carrier spread spectrum modulation with reduced dynamic range," in *Proc. IEEE Vehicular Technology Conference (VTC'96)*, Atlanta, USA, pp. 914–917, May/June 1996.

[5] Bader F., Zazo S. and Borrallo J.M., "Decorrelation MUD for MC-CDMA in an uplink transmission mode," in *Proc. International Workshop on Multi-Carrier Spread-Spectrum & Related Topics (MC-SS 2001)*, Oberpfaffenhofen, Germany, pp. 173–180, Sept. 2001.

[6] Baudais J.-Y. Helard J.-F. and Citerne J., "An improved linear MMSE detection technique for multi-carrier CDMA systems: Comparison and combination with interference cancellation schemes," *European Transactions on Telecommunications (ETT)*, vol. 11, pp. 547–554, Nov./Dec. 2000.

[7] Brüninghaus K. and Rohling H., "On the duality of multi-carrier spread spectrum and single-carrier transmission," in *Proc. International Workshop on Multi-Carrier Spread-Spectrum (MC-SS'97)*, Oberpfaffenhofen, Germany, pp. 187–194, April 1997.

[8] Bury A., *Efficient Multi-Carrier Spread Spectrum Transmission*. Düsseldorf: VDI-Verlag, Fortschritt-Berichte VDI, series 10, no. 685, 2001, PhD thesis.

[9] Bury A. and Lindner J., "Comparison of amplitude distributions for Hadamard spreading and Fourier spreading in multi-carrier code division multiplexing," in *Proc. IEEE Global Telecommunications Conference (GLOBECOM 2000)*, San Francisco, USA, pp. 857–860, Nov./Dec. 2000.

[10] Costa E., Haas H. and Schulz E., "Optimization of capacity assignment in MC-CDMA transmission systems," in *Proc. International Workshop on Multi-Carrier Spread-Spectrum & Related Topics (MC-SS 2001)*, Oberpfaffenhofen, Germany, pp. 217–224, Sept. 2001.

[11] Dekorsy A. and Kammeyer K.-D., "A new OFDM-CDMA uplink concept with M-ary orthogonal modulation," *European Transactions on Telecommunications (ETT)*, vol. 10, pp. 377–389, July/Aug. 1999.

[12] Dekorsy A. and Kammeyer K.-D., "Serial code concatenation with complex valued Walsh-Hadamard codes applied to OFDM-CDMA," in *Proc. International Workshop on Multi-Carrier Spread-Spectrum & Related Topics (MC-SS 2001)*, Oberpfaffenhofen, Germany, pp. 131–138, Sept. 2001.

[13] Egle J., Reinhardt M. and Lindner J., "Equalization and coding for extended MC-CDMA over time and frequency selective channels," in *Proc. International Workshop on Multi-Carrier Spread-Spectrum (MC-SS'97)*, Oberpfaffenhofen, Germany, pp. 127–134, April 1997.

[14] Fang L. and Milstein L.B., "Successive interference cancellation in multicarrier DS/CDMA," *IEEE Transactions on Communications*, vol. 48, pp. 1530–1540, Sept. 2000.

[15] Fazel K., "Performance of CDMA/OFDM for mobile communication system," in *Proc. IEEE International Conference on Universal Personal Communications (ICUPC'93)*, Ottawa, Canada, pp. 975–979, Oct. 1993.

[16] Fazel K. and Papke L., "On the performance of convolutionally-coded CDMA/OFDM for mobile communication system," in *Proc. IEEE International Symposium on Personal, Indoor and Mobile Radio Communications (PIMRC'93)*, Yokohama, Japan, pp. 468–472, Sept. 1993.

[17] Fujii M., Shimizu R., Suzuki S., Itami M. and Itoh K., "A study on downlink capacity of FD-MC/CDMA for channels with frequency selective fading," in *Proc. International Workshop on Multi-Carrier Spread Spectrum & Related Topics (MC-SS 2001)*, Oberpfaffenhofen, Germany, pp. 139–146, Sept. 2001.

[18] Hagenauer J., "Forward error correcting for CDMA systems," in *Proc. IEEE International Symposium on Spread Spectrum Techniques & Applications (ISSSTA'96)*, Mainz, Germany, pp. 566–569, Sept. 1996.

[19] Jarot S.P.W. and Nakagawa M., "Investigation on using channel information of MC-CDMA for pre-rake diversity combining in TDD/CDMA system," in *Proc. International Workshop on Multi-Carrier Spread-Spectrum & Related Topics (MC-SS 2001)*, Oberpfaffenhofen, Germany, pp. 265–272, Sept. 2001.

[20] Jeong D.G. and Kim M.J., "Effects of channel estimation error in MC-CDMA/TDD systems," in *Proc. IEEE Vehicular Technology Conference (VTC 2000-Spring)*, Tokyo, Japan, pp. 1773–1777, May 2000.

[21] Kaiser S., "On the performance of different detection techniques for OFDM-CDMA in fading channels," in *Proc. IEEE Global Telecommunications Conference (GLOBECOM'95)*, Singapore, pp. 2059–2063, Nov. 1995.

[22] Kaiser S., "Analytical performance evaluation of OFDM-CDMA mobile radio systems," in *Proc. European Personal and Mobile Communications Conference (EPMCC'95)*, Bologna, Italy, pp. 215–220, Nov. 1995.

[23] Kaiser S., *Multi-Carrier CDMA Mobile Radio Systems – Analysis and Optimization of Detection, Decoding, and Channel Estimation*. Düsseldorf: VDI-Verlag, Fortschritt-Berichte VDI, series 10, no. 531, 1998, PhD thesis.

[24] Kaiser S., "OFDM code division multiplexing in fading channels," *IEEE Transactions on Communications*, vol. 50, pp. 1266–1273, Aug. 2002.

[25] Kaiser S. and Hagenauer J., "Multi-carrier CDMA with iterative decoding and soft-interference cancellation," in *Proc. IEEE Global Telecommunications Conference (GLOBECOM'97)*, Phoenix, USA, pp. 6–10, Nov. 1997.

[26] Kaiser S. and Papke L., "Optimal detection when combining OFDM-CDMA with convolutional and Turbo channel coding," in *Proc. IEEE International Conference on Communications (ICC'96)*, Dallas, USA, pp. 343–348, June 1996.

[27] Kalofonos D.N. and Proakis J.G., "Performance of the multistage detector for a MC-CDMA system in a Rayleigh fading channel," in *Proc. IEEE Global Telecommunications Conference (GLOBECOM'96)*, London, UK, pp. 1784–1788, Nov. 1997.

[28] Kondo S. and Milstein L.B., "Performance of multi-carrier DS-CDMA systems," *IEEE Transactions on Communications*, vol. 44, pp. 238–246, Feb. 1996.

[29] Kühn V., "Combined MMSE-PIC in coded OFDM-CDMA systems," in *Proc. IEEE Global Telecommunications Conference (GLOBECOM 2001)*, San Antonio, USA, Nov. 2001.

[30] Kühn V., Dekorsy A. and Kammeyer K.-D., "Channel coding aspects in an OFDM-CDMA system," in *Proc. ITG Conference on Source and Channel Coding*, Munich, Germany, pp. 31–36, Jan. 2000.

[31] Liu H. and Yin H., "Receiver design in multi-carrier direct-sequence CDMA communications," *IEEE Transactions on Communications*, vol. 49, pp. 1479–1487, Aug. 2001.

[32] Maxey J.J. and Ormondroyd R.F., "Multi-carrier CDMA using convolutional coding and interference cancellation over fading channels," in *Proc. International Workshop on Multi-Carrier Spread-Spectrum (MC-SS'97)*, Oberpfaffenhofen, Germany, pp. 89–96, April 1997.

[33] Mottier D. and Castelain D., "SINR-based channel pre-compensation for uplink multi-carrier CDMA systems," in *Proc. IEEE International Symposium on Personal, Indoor and Mobile Radio Communications (PIMRC 2002)*, Lisbon, Portugal, Sept. 2002.

[34] Nakagawa M. and Esmailzadeh R., "Time division duplex-CDMA," in *Proc. International Workshop on Multi-Carrier Spread-Spectrum & Related Topics (MC-SS 2001)*, Oberpfaffenhofen, Germany, pp. 13–21, Sept. 2001

[35] Nobilet S., Helard J.-F. and Mottier D., "Spreading sequences for uplink and downlink MC-CDMA systems: PAPR and MAI minimization," *European Transactions on Telecommunications (ETT)*, vol. 13, pp. 465–474, Sept./Oct. 2002.

[36] Ochiai H. and Imai H., "Performance of OFDM-CDMA with simple peak power reduction," *European Transactions on Telecommunications (ETT)*, vol. 10, pp. 391–398, July/Aug. 1999.

[37] Ochiai H. and Imai H., "Performance of downlink MC-CDMA with simple interference cancellation," in *Proc. International Workshop on Multi-Carrier Spread-Spectrum & Related Topics (MC-SS'99)*, Oberpfaffenhofen, Germany, pp. 211–218, Sept. 1999.

[38] Persson A., Ottosson T. and Ström E., "Time-frequency localized CDMA for downlink multi-carrier systems," in *Proc. IEEE International Symposium on Spread Spectrum Techniques & Applications (ISSSTA 2002)*, Prague, Czech Republic, pp. 118–122, Sept. 2002.
[39] Popovic B.M., "Spreading sequences for multicarrier CDMA systems," *IEEE Transactions on Communications*, vol. 47, pp. 918–926, June 1999.
[40] Proakis J., *Digital Communications*. New York: McGraw-Hill, 1995.
[41] Pu Z., You X., Cheng S. and Wang H., "Transmission and reception of TDD multicarrier CDMA signals in mobile communications system," in *Proc. IEEE Vehicular Technology Conference (VTC'99-Spring)*, Houston, USA, pp. 2134–2138, May 1999.
[42] Reinhardt M., Egle J. and Lindner J., "Transformation methods, coding and equalization for time- and frequency-selective channels," *European Transactions on Telecommunications (ETT)*, vol. 11, pp. 555–565, Nov./Dec. 2000.
[43] Schnell M., *Systeminhärente Störungen bei "Spread-Spectrum"-Vielfachzugriffsverfahren für die Mobilfunkübertragung*. Düsseldorf: VDI-Verlag, Fortschritt-Berichte VDI, series 10, no. 505, 1997, PhD thesis.
[44] Sourour E.A. and Nakagawa M., "Performance of orthogonal multi-carrier CDMA in a multipath fading channel," *IEEE Transactions on Communications*, vol. 44, pp. 356–367, Mar. 1996.
[45] Steiner B., "Uplink performance of a multicarrier-CDMA mobile radio system concept," in *Proc. IEEE Vehicular Technology Conference (VTC'97)*, Phoenix, USA, pp. 1902–1906, May 1997.
[46] Stirling-Gallacher R.A. and Povey G.J.R., "Different channel coding strategies for OFDM-CDMA," in *Proc. IEEE Vehicular Technology Conference (VTC'97)*, Phoenix, USA, pp. 845–849, May 1997.
[47] Tomba L. and Krzymien W.A., "Downlink detection schemes for MC-CDMA systems in indoor environments," *IEICE Transactions on Communications*, vol. E79-B, pp. 1351–1360, Sept. 1996.
[48] Vandendorpe L., "Multitone spread spectrum multiple access communications system in a multipath Rician fading channel," *IEEE Transactions on Vehicular Technology*, vol. 44, pp. 327–337, May 1995.
[49] Verdu S., *Multiuser Detection*. Cambridge: Cambridge University Press, 1998.
[50] Viterbi A.J., "Very low rate convolutional codes for maximum theoretical performance of spread-spectrum multiple-access channels," *IEEE Journal on Selected Areas in Communications*, vol. 8, pp. 641–649, May 1990.
[51] Yee N., Linnartz J.-P. and Fettweis G., "Multi-carrier CDMA in indoor wireless radio networks," in *Proc. IEEE International Symposium on Personal, Indoor and Mobile Radio Communications (PIMRC'93)*, Yokohama, Japan, pp. 109–113, Sept. 1993.
[52] Yip K.-W. and Ng T.-S., "Tight error bounds for asynchronous multi-carrier CDMA and their applications," *IEEE Communications Letters*, vol. 2, pp. 295–297, Nov. 1998.

3

Hybrid Multiple Access Schemes

3.1 Introduction

The simultaneous transmission of multiple data streams over the same medium can be achieved with different multiplexing schemes. Most communications systems, such as GSM, DECT, IEEE 802.11a, and HIPERLAN/2, use multiplexing based on either time division, frequency division or a combination of both. Space division multiplexing is applied to further increase the user capacity of the system. The simplest scheme of space division multiplexing is antenna sectorization at the base station where often antennas with 120°/90° beams are used. Recently, multiplexing schemes using code division have gained significant interest and have become part of wireless standards such as UMTS, IS-95, and WLAN. Moreover, code division multiplexing is also a promising candidate for the fourth generation of mobile radio systems.

Time Division Multiplexing: The separation of different data streams with time division multiplexing is carried out by assigning each stream exclusively a certain period of time, i.e., time slot, for transmission. After each time slot, the next data stream transmits in the following time slot. The number of slots assigned to each user can be supervised by the medium access controller (MAC). A MAC frame determines a group of time slots in which all data streams transmit once. The duration of the different time slots can vary according to the requirements of the different data streams. If the different data steams belong to different users, the access scheme is called time division multiple access (TDMA).

Time division multiplexing can be used with both time division duplex (TDD) and frequency division duplex (FDD). However, it is often used in communication systems with TDD duplex transmission, where up- and downlink are separated by the assignment of different time slots. It is adopted in several wireless LAN and WLL systems including IEEE 802.11a and HIPERLAN/2 as well as IEEE 802.16a and HIPERMAN.

Frequency Division Multiplexing: With frequency division multiplexing, the different data streams are separated by exclusively assigning each stream a certain fraction of the frequency band for transmission. In contrast to time division multiplexing, each stream can continuously transmit within its sub-band. The efficiency of frequency division multiplexing strongly depends on the minimum separation of the sub-bands to avoid adjacent channel interference. OFDM is an efficient frequency division multiplexing schemes, which offers minimum spacing of the sub-bands without interference from adjacent channels in the synchronous case.

In multiple access schemes, where different data streams belong to different users, the frequency division multiplexing scheme is known as frequency division multiple access (FDMA).

Frequency division multiplexing is often used in communication systems with FDD, where up- and downlink are separated by the assignment of different frequency bands for each link. They are, for example, used in the mobile radio systems GSM, IS-95, and UMTS FDD Mode.

Code Division Multiplexing: Multiplexing of different data streams can be carried out by multiplying the data symbols of a data stream with a spreading code exclusively assigned to this data stream before superposition with the spread data symbols of the other data streams. All data streams use the same bandwidth at the same time in code division multiplexing. Depending on the application, the spreading codes should as far as possible be orthogonal to each other in order to reduce interference between different data streams.

Multiple access schemes where the user data are separated by code division multiplexing are referred to as code division multiple access (CDMA).

Space Division Multiplexing: The spatial dimension can also be used for the multiplexing of different data streams by transmitting the data streams over different, non-overlapping transmission channels. Space division multiplexing can be achieved using beam-forming or sectorization.

The use of space division multiplexing for multiple access is termed space division multiple access (SDMA).

Hybrid Multiplexing Schemes: The above multiplexing schemes are often combined to hybrid schemes in communication systems like GSM where TDMA and FDMA are applied, or UMTS, where CDMA, TDMA and FDMA are used. These hybrid combinations additionally increase the user capacity and flexibility of the system. For example, the combination of MC-CDMA with DS-CDMA or TDMA offers the possibility to overload an otherwise limited MC-CDMA scheme. The idea is to load the orthogonal MC-CDMA scheme up to its limits and in case of additional users, other non-orthogonal multiple access schemes are superimposed. For small numbers of overload and using efficient interference cancellation schemes nearly all additional multiple access interference caused by the system overlay can be canceled [33].

In this chapter, different hybrid multiple access concepts will be presented and compared to each other.

3.2 Multi-Carrier FDMA

The concept of the combination of spread spectrum and frequency hopping with multi-carrier transmission opened the door for alternative hybrid multiple access solutions such as: *OFDMA* [28], *OFDMA with CDM*, called *SS-MC-MA* [18], and *Interleaved FDMA* [35]. All these schemes are discussed in the following.

3.2.1 Orthogonal Frequency Division Multiple Access (OFDMA)

3.2.1.1 Basic Principle

Orthogonal frequency division multiple access (OFDMA) consists of assigning one or several sub-carrier frequencies to each user (terminal station) with the constraint that the sub-carrier spacing is equal to the OFDM frequency spacing $1/T_s$ (see [28][30][31][32]).

To describe the basic principle of OFDMA we will make the following assumptions:

— one sub-carrier is assigned per user (the generalization for several sub-carriers per user is straightforward) and
— the only source of disturbance is AWGN.

The signal of user k, $k = 0, 1, \ldots, K-1$, where $K = N_c$, has the form

$$s^{(k)}(t) = \text{Re}\{d^{(k)}(t) e^{j2\pi f_k t} e^{j2\pi f_c t}\}, \qquad (3.1)$$

with

$$f_k = \frac{k}{T_s} \qquad (3.2)$$

and f_c representing the carrier frequency. Furthermore, we assume that the frequency f_k is permanently assigned to user k, although in practice frequency assignment could be made upon request. Therefore, an OFDMA system with, e.g., $N_c = 1024$ sub-carriers and adaptive sub-carrier allocation is able to handle thousands of users.

In the following, we consider a permanent channel assignment scheme in which the number of sub-carriers is equal to the number of users. Under this assumption the modulator of the terminal station of user k has the form of an unfiltered modulator with rectangular pulse (e.g., unfiltered QPSK) with carrier $f_k + f_c$. The transmitted data symbols are given by

$$d^{(k)}(t) = \sum_{i=-\infty}^{+\infty} d_i^{(k)} \text{rect}(t - iT_s), \qquad (3.3)$$

where $d_i^{(k)}$ designates the data symbol transmitted by user k during the ith symbol period and $\text{rect}(t)$ is a rectangular pulse shape which spans the time interval $[0, T_s]$.

The received signal before down-conversion of all K users at the base station in the presence of only noise (in the absence of multipath) can be written as

$$q(t) = \sum_{k=0}^{K-1} s^{(k)}(t) + n(t), \qquad (3.4)$$

where $n(t)$ is an additive noise term. After demodulation at the base station using a local oscillator with carrier frequency f_c, we obtain

$$r(t) = \sum_{k=0}^{K-1} r^{(k)}(t) + w(t), \qquad (3.5)$$

where $r^{(k)}(t)$ is the complex envelope of the kth user signal and $w(t)$ is the baseband equivalent noise. This expression can also be written as

$$r(t) = \sum_{i=-\infty}^{\infty} \sum_{k=0}^{K-1} d_i^{(k)}(t) e^{j2\pi f_k t} + w(t), \quad (3.6)$$

where we explicitly find in this expression the information part $d_i^{(k)}(t)$.

The demodulated signal is sampled at a sampling rate of N_c/T_s and a block of N_c regularly spaced signal samples is generated per symbol period T_s. Over the ith symbol period, we generate an N_c-point sequence

$$r_{n,i} = \sum_{k=0}^{K-1} d_i^{(k)} e^{j2\pi kn/N_c} + w_{n,i}, \quad n = 0, \ldots, N_c - 1. \quad (3.7)$$

It is simple to verify that except for a scaling factor $1/N_c$, the above expression is a noisy version of the IDFT of the sequence $d_i^{(k)}$, $k = 0, \ldots, K - 1$. This indicates that the data symbols can be recovered using an N_c-point DFT after sampling. In other words, the receiver at the base station is an OFDM receiver.

As illustrated in Figure 3-1, in the simplest OFDMA scheme (one sub-carrier per user) each user signal is a single-carrier signal. At the base station (of, e.g., fixed wireless access or interactive DVB-T) the received signal, being the sum of K users' signals, acts as an OFDM signal due to its multi-point to point nature. Unlike conventional FDMA which requires K demodulators to handle K simultaneous users, OFDMA requires only a single demodulator, followed by an N_c-point DFT.

Hence, the basic components of an OFDMA transmitter at the terminal station are FEC channel coding, mapping, sub-carrier assignment, and single carrier modulator (or multi-carrier modulator in the case that several sub-carriers are assigned per user).

Since OFDMA is preferably used for the uplink in a multiuser environment, low-order modulation such as QPSK with Gray mapping is preferred. However, basically high-order modulation (e.g., 16- or 64-QAM) can also be employed.

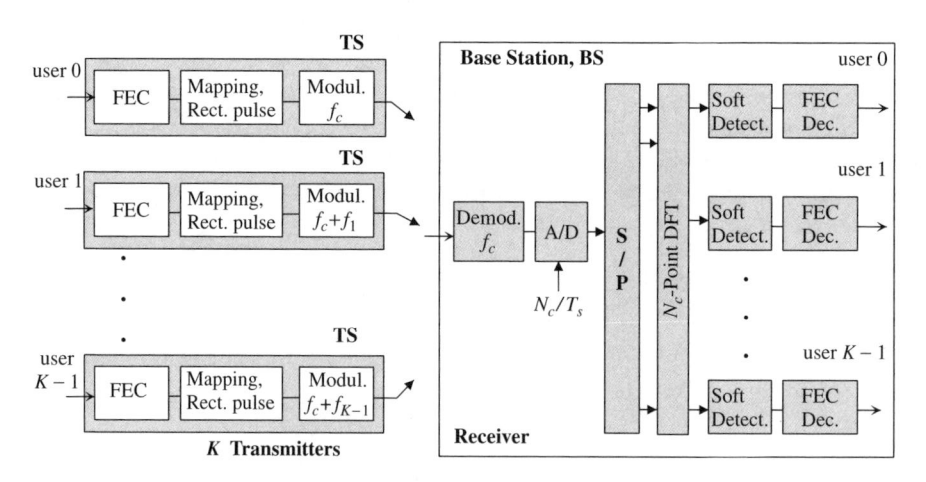

Figure 3-1 Basic principle of OFDMA

The sub-carrier assignment can be fixed or dynamic. In practice, in order to increase the system robustness (to exploit frequency diversity) a dynamic assignment of sub-carriers (i.e., frequency hopping) for each user is preferable. This approach is similar to M- or Q-Modification in MC-CDMA (see Section 2.1.8). For pulse shaping, rectangular shaping is usually used which results for K users in an OFDM-type signal at the receiver side.

In summary, where only one sub-carrier is assigned to a user, the modulator for the user could be a single-carrier modulator. If several carriers are used for a given terminal station, the modulator will be a multi-carrier (OFDM) modulator.

A very accurate clock and carrier synchronization is essential for an OFDMA system, to ensure orthogonality between the K modulated signals originating from different terminal stations. This can be achieved, for instance, by transmitting synchronization signals from the base stations to all terminal stations. Therefore, each terminal station modulator derives its carrier frequency and symbol timing from these common downlink signals.

At the base station the main components of the receiver are the demodulator (including synchronization functions), FFT and channel decoder (with soft decisions). Since in the case of a synchronous system the clock and carrier frequencies are readily available at the base station (see Section 3.2.1.2), very simple carrier and clock recovery circuits are sufficient in the demodulator to extract this information from the received signal [30]. This fact can greatly simplify the OFDM demodulator.

3.2.1.2 Synchronization Sensitivity

As mentioned before, OFDMA requires an accurate carrier spacing between different users and precise symbol clock frequency. Hence, in a synchronous system, the OFDMA transmitter is synchronized (clock and frequency) to the base station downlink signal, received by all terminal stations [3][5][11].

In order to avoid time drift, the symbol clock of the terminal station is locked to the downlink reference clock and on some extra time synchronization messages (e.g. time stamps) transmitted periodically from the base station to all terminal stations. The reference clock in the base station requires a quite high accuracy [3]. Furthermore, the terminal station can synchronize the transmit sub-carriers in phase and frequency to the received downlink channel.

Since the clock and carrier frequencies are readily available at the reception side in the base station, no complex carrier and clock recovery circuits are necessary in the demodulator to extract this information from the received signal [30]. This simplifies the OFDMA demodulator. Although the carrier frequency is locally available, there are phase differences between different user signals and local references. These phase errors can be compensated, for instance, by a phase equalizer which takes the form of a complex multiplier bank with one multiplier per sub-carrier. This phase equalization is not necessary if the transmitted data is differentially encoded and demodulated.

Regarding the sensitivity to the oscillator's phase noise, the OFDMA technique will have the same sensitivity as an OFDM system. Therefore, low noise oscillators are needed, particularly if the number of sub-carriers is high or high-order modulation is used.

If each terminal station is fixed positioned (e.g., return channel of DVB-T), the ranging procedure (i.e., measuring the delay and power of individual signals) and adjusting the phase and the transmit power of the transmitters can be done at the installation and later on periodically in order to cope with drifts which may be due to weather or aging variations and other factors. The ranging information can be transmitted periodically from the base station to all terminal stations within a given frame format [3][5][11].

Phase alignment of different users through ranging cannot be perfect. Residual misalignment can be compensated for using a larger guard interval (cyclic extension).

3.2.1.3 Pulse Shaping

In the basic version of OFDMA, one sub-carrier is assigned to each user. The spectrum of each user is quite narrow, which makes OFDMA more sensitive to narrowband interference. In this section, another variant is described which may lead to increased robustness against narrowband interference.

With rectangular pulse shaping, OFDMA has a $\text{sinc}^2(f)$ shaped spectrum with overlapping sub-channels (see Figure 3-2a). The consequence of this is that a narrowband interferer will affect not only one sub-carrier but several sub-carriers [31]. The robustness of OFDMA to band-limited interference can be increased if the bandwidth of individual sub-channels is strictly limited so that either adjacent sub-channels do not overlap, or each sub-channel spectrum only overlaps with two adjacent sub-channels. The non-overlapping concept is illustrated in Figure 3-2b. As long as the bandwidth of one sub-channel is smaller than $1/T_s$, the narrowband interferer will only affect one sub-channel. As shown in Figure 3-2b, the orthogonality between sub-channels is guaranteed, since there is no overlapping between the spectra of adjacent sub-channels. Here a Nyquist pulse shaping is needed for ISI-free transmission on each sub-carrier, comparable to a conventional single-carrier transmission scheme. This requires oversampling of the received signal and DFT operations at a higher rate than N_c/T_s. In other words, the increased robustness to narrowband interference is achieved at the expense of increased complexity.

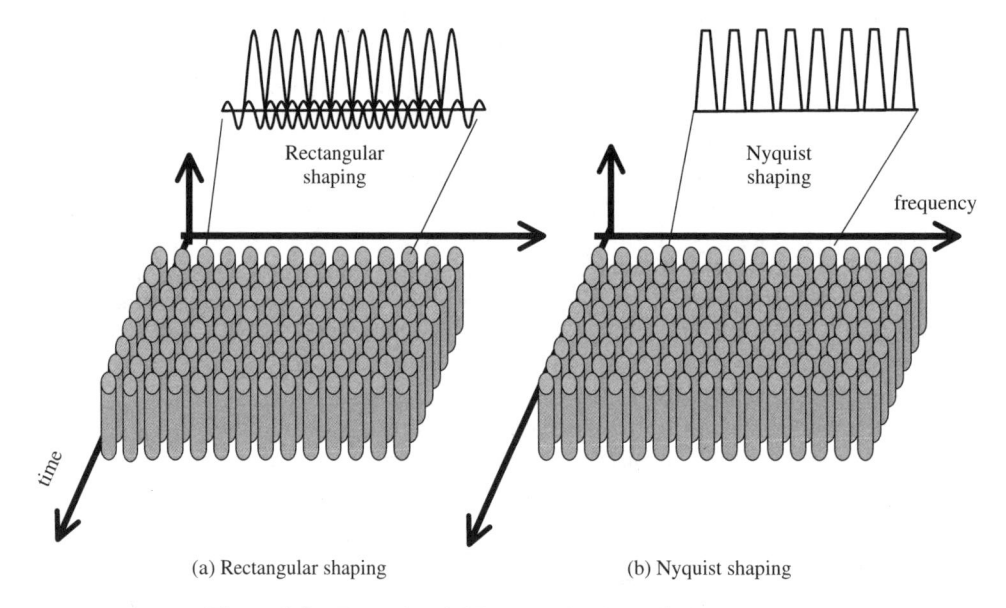

(a) Rectangular shaping

(b) Nyquist shaping

Figure 3-2 Example of OFDMA with band-limited spectra

The Nyquist shaping function $g(t)$ can be implemented with a time-limited square root raised cosine pulse with a roll-off factor α,

$$g(t) = \begin{cases} \dfrac{\sin\left[\dfrac{\pi t}{T'_s}(1-\alpha)\right] + \dfrac{4\alpha t}{T'_s}\cos\left[\dfrac{\pi t}{T'_s}(1+\alpha)\right]}{\dfrac{\pi t}{T'_s}\left[1-\left(\dfrac{4\alpha t}{T'_s}\right)^2\right]} & \text{for } t \in \{-4T'_s, 4T'_s\} \\ 0 & \text{otherwise} \end{cases} \quad (3.8)$$

The relationship between T'_s, T_s and α (roll-off factor) provides the property of the individual separated spectra, where $T'_s = (1+\alpha)T_s$.

3.2.1.4 Frequency Hopping OFDMA

The application of frequency hopping (FH) in an OFDMA system is straightforward. Rather than assigning a fixed particular frequency to a given user, the base station assigns a hopping pattern [2][11][28][36]. In the following it is assumed that N_c sub-carriers are available and that the frequency hopping sequence is periodic and uniformly distributed over the signal bandwidth.

Suppose that the frequency sequence $(f_0, f_7, f_{14}, \ldots)$ is assigned to the first user, the sequence $(f_1, f_8, f_{15}, \ldots)$ to the second user and so on. The frequency assignment to N_c users can be written as

$$f(n, k) = f_{k+(7n \bmod N_c)}, \quad k = 0, \ldots, N_c - 1, \quad (3.9)$$

where $f(n, k)$ designates the sub-carrier frequency assigned to user k at symbol time n.

OFDMA with frequency hopping has a close relationship with MC-CDMA. We know that MC-CDMA is based on spreading the signal bandwidth using direct sequence spreading with processing gain P_G. In OFDMA, frequency assignments can be specified with a code according to a frequency hopping (FH) pattern, where the number of hops can be slow. Both schemes employ OFDM for chip transmission.

3.2.1.5 General OFDMA Transceiver

A general conceptual block diagram of an OFDMA transceiver for the uplink of a multiuser cellular system is illustrated in Figure 3-3. The terminal station is synchronized to the base station. The transmitter of the terminal station extracts from the demodulated downlink received data MAC messages on information about sub-carrier allocation, frequency hopping pattern, power control messages and timing, and further clock and frequency synchronization information. Synchronization of the terminal station is achieved by using the MAC control messages to perform time synchronization and using frequency information issued from the terminal station downlink demodulator (the recovered base station system clock). The MAC control messages are processed by the MAC management block to instruct the terminal station modulator on the transmission resources assigned to

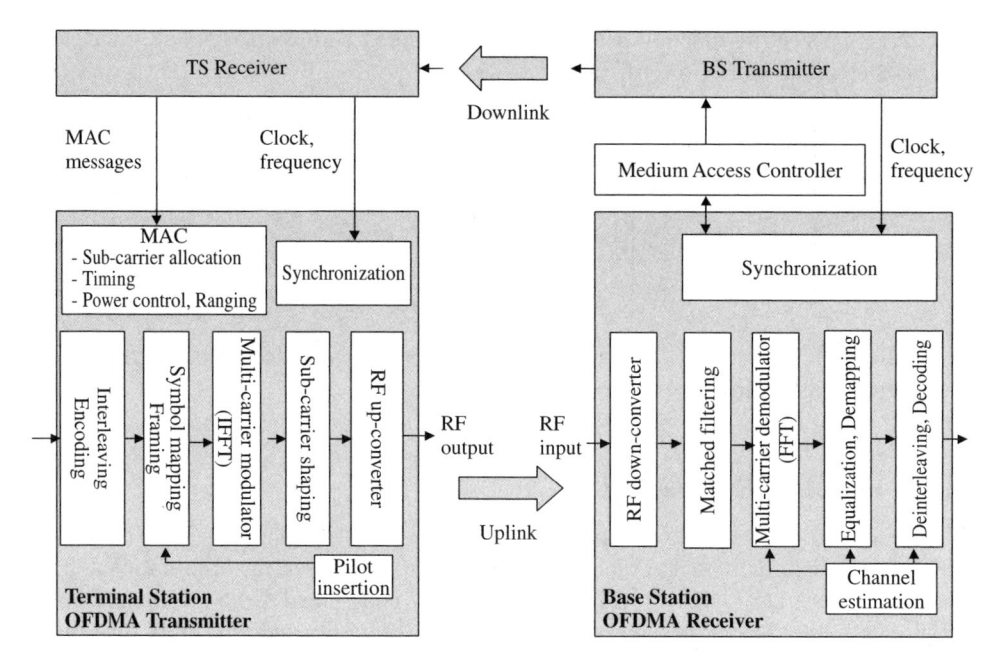

Figure 3-3 General OFDMA conceptual transceiver

it and to tune the access performed to the radio frequency channel. The pilot symbols are inserted to ease the channel estimation task at the base station.

At the base station, the received signals issued by all terminal stations are demodulated by the use of an FFT as an OFDM receiver, assisted by the MAC layer management block.

It should be emphasized that the transmitter and the receiver structure of an OFDMA system is quite similar to an OFDM system. Same components like FFT, channel estimation, equalization and soft channel decoding can be used for both cases.

In order to offer a variety of multimedia services requiring different data rates, the OFDMA scheme needs to be flexible in terms of data rate assignment. This can be achieved by assigning the required number of sub-carriers according to the request of a given user. This method of assignment is part of a MAC protocol at the base station.

Note that if the number of assigned sub-carriers is an integer power of two, the inverse FFT can be used at the terminal station transmitter, which will be equivalent to a conventional OFDM transmitter.

3.2.2 *OFDMA with Code Division Multiplexing: SS-MC-MA*

The extension of OFDMA by code division multiplexing (CDM) results in a multiple access scheme referred to as spread spectrum multi-carrier multiple access (SS-MC-MA) [18][19]. It applies OFDMA for user separation and additionally uses CDM on data symbols belonging to the same user. The CDM component is introduced in order to achieve additional diversity gains. Like MC-CDMA, SS-MC-MA exploits the advantages given by the combination of the spread spectrum technique and multi-carrier modulation. The SS-MC-MA scheme is similar to the MC-CDMA transmitter with M-Modification. Both transmitters are identical except for the mapping of the user data to the subsystems. In SS-MC-MA systems, one user maps L data symbols to one subsystem which this user exclusively uses for transmission. Different users use different subsystems in SS-MC-MA systems. In MC-CDMA systems, M data symbols per user are mapped to M different subsystems where each subsystem is shared by different users. The principle of SS-MC-MA is illustrated for a downlink transmitter in Figure 3-4.

The SS-MC-MA and MC-CDMA systems have the following similarities:

— SS-MC-MA and MC-CDMA systems exploit frequency diversity by spreading each data symbol over L sub-carriers.
— Per subsystem, the same data detection techniques can be applied with both SS-MC-MA and MC-CDMA systems.
— ISI and ICI can be avoided in SS-MC-MA and MC-CDMA systems, resulting in simple data detection techniques.

However, their main differences are:

— In SS-MC-MA systems, CDM is used for the simultaneous transmission of the data of one user on the same sub-carriers, whereas in MC-CDMA systems, CDM is used for the transmission of the data of different users on the same sub-carriers. Therefore, SS-MC-MA is an OFDMA scheme on a sub-carrier level, whereas MC-CDMA is a CDMA scheme.
— MC-CDMA systems have to cope with multiple access interference, which is not present in SS-MC-MA systems. Instead of multiple access interference, SS-MC-MA systems have to cope with self-interference caused by the superposition of signals from the same user.

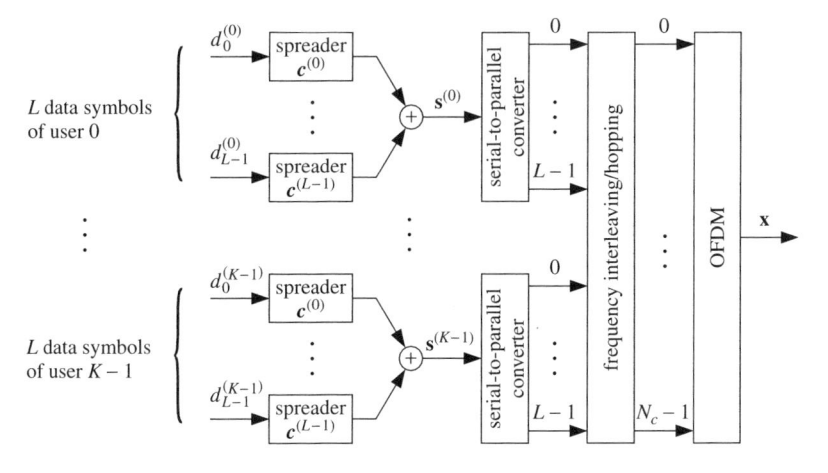

Figure 3-4 SS-MC-MA downlink transmitter

— In SS-MC-MA systems, each sub-carrier is exclusively used by one user, enabling low complex channel estimation especially for the uplink. In MC-CDMA systems, the channel estimation in the uplink has to cope with the superposition of signals from different users which are faded independently on the same sub-carriers, increasing the complexity of the uplink channel estimation.

After this comparative introduction of SS-MC-MA, the uplink transmitter and the assigned receiver are described in detail in this section.

Figure 3-5 shows an SS-MC-MA uplink transmitter with channel coding for the data of user k. The vector

$$\mathbf{d}^{(k)} = (d_0^{(k)}, d_1^{(k)}, \ldots, d_{L-1}^{(k)})^T \qquad (3.10)$$

represents one block of L parallel converted data symbols of user k. Each data symbol is multiplied with another orthogonal spreading code of length L. The $L \times L$ matrix

$$\mathbf{C} = (\mathbf{c}_0, \mathbf{c}_1, \ldots, \mathbf{c}_{L-1}) \qquad (3.11)$$

represents the L different spreading codes $c_l, l = 0, \ldots, L-1$, used by user k. The spreading matrix $\mathbf{C}$ can be the same for all users. The modulated spreading codes are synchronously added, resulting in the transmission vector

$$\mathbf{s}^{(k)} = \mathbf{C}\,\mathbf{d}^{(k)} = (S_0^{(k)}, S_1^{(k)}, \ldots, S_{L-1}^{(k)})^T. \qquad (3.12)$$

To increase the robustness of SS-MC-MA systems, less than L data modulated spreading codes can be added in one transmission vector $\mathbf{s}^{(k)}$.

Comparable to frequency interleaving in MC-CDMA systems, the SS-MC-MA transmitter performs a user-specific frequency mapping such that subsequent chips of $\mathbf{s}^{(k)}$ are interleaved over the whole transmission bandwidth. The user-specific frequency mapping assigns each user exclusively its L sub-carriers, avoiding multiple access interference. The Q-Modification introduced in Section 2.1.8.2 for MC-CDMA systems is inherent in SS-MC-MA systems. M-Modification can, as in MC-CDMA systems, be applied to SS-MC-MA systems by assigning a user more than one subsystem.

OFDM with guard interval is applied in SS-MC-MA systems in the same way as in MC-CDMA systems. In order to perform coherent data detection at the receiver and to

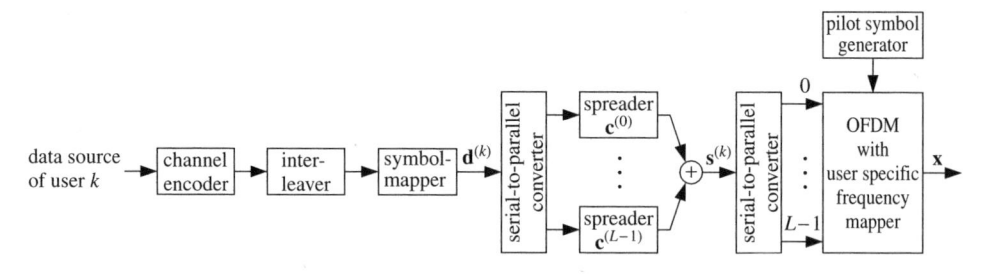

Figure 3-5 SS-MC-MA transmitter of user k

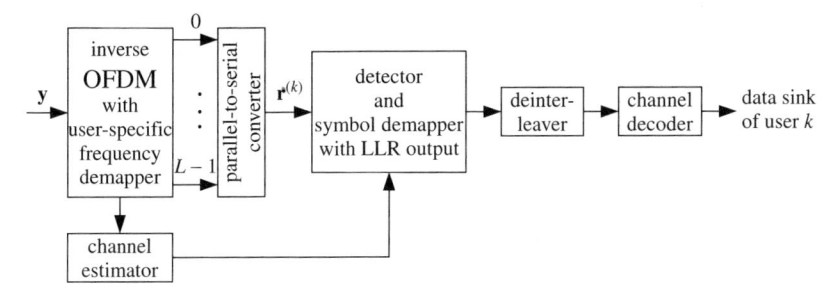

Figure 3-6 SS-MC-MA receiver of user k

guarantee robust time and frequency synchronization, pilot symbols are multiplexed in the transmitted data.

An SS-MC-MA receiver with coherent detection of the data of user k is shown in Figure 3-6. After inverse OFDM with user-specific frequency demapping and extraction of the pilot symbols from the symbols with user data, the received vector

$$\mathbf{r}^{(k)} = \mathbf{H}^{(k)}\mathbf{s}^{(k)} + \mathbf{n}^{(k)} = (R_0^{(k)}, R_1^{(k)}, \ldots, R_{L-1}^{(k)})^T \qquad (3.13)$$

with the data of user k is obtained. The $L \times L$ diagonal matrix $\mathbf{H}^{(k)}$ and the vector $\mathbf{n}^{(k)}$ of length L describe the channel fading and noise, respectively, on the sub-carriers exclusively used by user k.

Any of the single-user or multiuser detection techniques presented for MC-CDMA systems in Section 2.1.5 can be applied for the detection of the data of a single user per subsystem in SS-MC-MA systems. However, SS-MC-MA systems offer (especially in the downlink) the advantage that with multi-symbol detection (equivalent to multiuser detection in MC-CDMA systems) in one estimation step simultaneously L data symbols of a single user are estimated. Compared to MC-CDMA systems, the complexity per data symbol of multi-symbol detection in SS-MC-MA systems reduces by a factor of L in the downlink. With multi-symbol detection, LLRs can inherently be obtained from the detection algorithm which may also include the symbol demapping. After deinterleaving and decoding of the LLRs, the detected source bits of user k are obtained.

A promising future mobile radio system may use MC-CDMA in the downlink and SS-MC-MA in the uplink. This combination achieves for both links a high spectral efficiency and flexibility. Furthermore, in both links the same hardware can be used, only the user data have to be mapped differently [16]. Alternatively, a modified SS-MC-MA scheme with flexible resource allocation can achieve a high throughput in the downlink [24].

SS-MC-MA can cope with a certain amount of asynchronism. It has been shown in [21] and [22] that it is possible to avoid any additional measures for uplink synchronization in cell radii up to several kilometers. The principle is to apply a synchronized downlink and each user transmits in the uplink directly after he has received its data without any additional time correction. A guard time shorter than the maximum time difference between the user signals is used, which increases the spectral efficiency of the system. Thus, SS-MC-MA can be achieved with a low-complex synchronization in the uplink.

Moreover, the SS-MC-MA scheme can be modified such that with not fully loaded systems, the additional available resources are used for more reliable transmission [6][7]. With a full load, these BER performance improvements can only be obtained by reducing the spectral efficiency of the system.

3.2.3 Interleaved FDMA (IFDMA)

The multiple access scheme IFDMA is based on the principle of FDMA where no multiple access interference occurs [34][35]. The signal is designed in such a way that the transmitted signal can be considered a multi-carrier signal where each user is exclusively assigned a sub-set of sub-carriers. The sub-carriers of the different users are interleaved. It is an inherent feature of the IFDMA signal that the sub-carriers of a user are equally spaced over the transmission bandwidth B, which guarantees a maximum exploitation of the available frequency diversity. The signal design of IFDMA is performed in the time domain and the resulting signal has the advantage of a low PAPR. However, IFDMA occupies a larger transmission bandwidth compared to the rectangular type spectrum with OFDM, which reduces the spectral efficiency.

The transmission of IFDMA signals in multipath channels results in ISI which requires more complex receivers than multi-carrier systems designed in the frequency domain. Compared to MC-CDMA, an IFDMA scheme is less flexible, since it does not support adaptive sub-carrier allocation and sub-carrier loading.

The IFDMA signal design is illustrated in Figure 3-7. A block of Q data symbols

$$\mathbf{d}^{(k)} = (d_0^{(k)}, d_1^{(k)}, \ldots, d_{Q-1}^{(k)})^T \tag{3.14}$$

assigned to user k is used for the construction of one IFDMA symbol. The duration of a data symbol is T and the duration of an IFDMA symbol is

$$T_s' = QT. \tag{3.15}$$

In order to limit the effect of ISI to one IFDMA symbol, a guard interval consisting of a cyclic extension of the symbol is included between adjacent IFDMA symbols, comparable

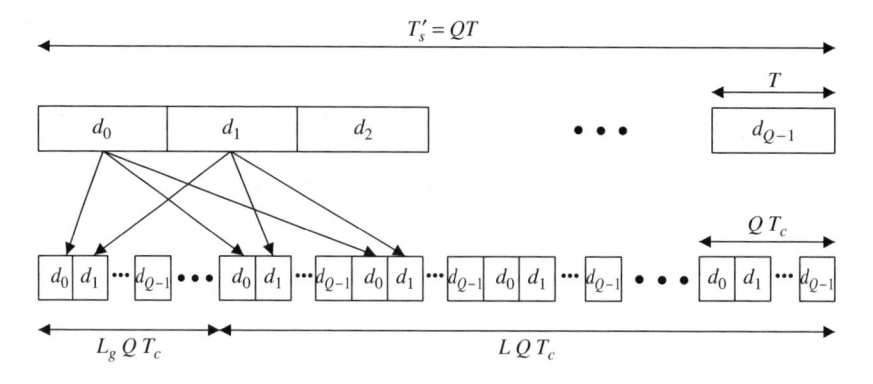

Figure 3-7 IFDMA signal design with guard interval

to the guard interval in multi-carrier systems. Each IFDMA symbol of duration T'_s includes the guard interval of duration

$$T_g = L_g Q T_c. \qquad (3.16)$$

An IFDMA symbol is obtained by compressing each of the Q symbols from symbol duration T to chip duration T_c, i.e.,

$$T_c = \frac{T}{L_g + L}, \qquad (3.17)$$

and repeating the resulting compressed block $(L_g + L)$ times. Thus, the transmission bandwidth is spread by the factor

$$P_G = L_g + L. \qquad (3.18)$$

The compressed vector of user k can be written as

$$\mathbf{s}^{(k)} = \frac{1}{L_g + L} \left(\underbrace{\mathbf{d}^{(k)^T}, \mathbf{d}^{(k)^T}, \ldots, \mathbf{d}^{(k)^T}}_{(L_g+L)\text{copies}} \right)^T. \qquad (3.19)$$

The transmission signal $\mathbf{x}^{(k)}$ is constructed by element-wise multiplication of the compressed vector $\mathbf{s}^{(k)}$ with a user-dependent phase vector $\mathbf{c}^{(k)}$ of length $(L_g + L)Q$ having the components

$$c_l^{(k)} = e^{-j2\pi l k/(QL)}, \quad l = 0, \ldots, (L_g + L)Q - 1. \qquad (3.20)$$

The element-wise multiplication of the two vectors $\mathbf{s}^{(k)}$ and $\mathbf{c}^{(k)}$ ensures that each user is assigned a set of sub-carriers orthogonal to the sub-carrier sets of all other users. Each sub-carrier set contains Q sub-carriers and the number of active users is restricted to

$$K \leqslant L. \qquad (3.21)$$

The IFDMA receiver has to perform an equalization to cope with the ISI which is present with IFDMA in multipath channels. For low numbers of Q, the optimum maximum likelihood sequence estimation can be applied with reasonable complexity whereas for higher numbers of Q, less complex suboptimum detection techniques such as linear equalization or decision feedback equalization are required to deal with the ISI.

Due to its low PAPR, a practical application of IFDMA can be an uplink where power-efficient terminal stations are required which benefit from the constant envelope and more complex receivers which have to cope with ISI are part of the base station.

3.3 Multi-Carrier TDMA

The combination of OFDM and TDMA is referred to as MC-TDMA or OFDM-TDMA. Due to its well understood TDMA component, MC-TDMA has achieved success and it is currently part of several high-rate wireless LAN standards, e.g., IEEE 802.11a/g/h, ETSI HIPERLAN/2, and MMAC, and is also part of the IEEE 802.16a and draft ETSI-HIPERMAN WLL standards [4][5][10][11] (see Chapter 5).

MC-TDMA transmission is done in a frame manner like in a TDMA system. One time frame within MC-TDMA has K time slots (or bursts), each allocated to one of the K terminal stations. One time slot/burst consists of one or several OFDM symbols. The allocation of time slots to the terminal stations is controlled by the base station medium access controller (MAC). Multiple access interference can be avoided when ISI between adjacent OFDM symbols can be prevented by using a sufficiently long guard interval or with a timing advance control mechanism.

Adaptive coding and modulation is usually applied in conjunction with MC-TDMA systems, where the coding and modulation can be easily adapted per transmitted burst.

The main advantages of MC-TDMA are in guaranteeing a high peak data rate, in its multiplexing gain (bursty transmission), in the absence of multiple access interference and in simple receiver structures that can be designed, for instance, by applying differential modulation in the frequency direction. In case of coherent demodulation a quite robust OFDM burst synchronization is needed, especially for the uplink. A frequency synchronous system where the terminal station transmitter is frequency-locked to the received signal in the downlink or spending a high amount of overhead transmitted per burst could remedy this problem.

Besides the complex synchronization mechanism required for an OFDM system, the other disadvantage of MC-TDMA is that diversity can only be exploited by using additional measures like channel coding or applying multiple transmit/receive antennas. As a TDMA system, the instantaneous transmitted power in the terminal station is high, which requires more powerful high power amplifiers than for FDMA systems. Furthermore, the MC-TDMA system as an OFDM system needs a high output power back-off.

As shown in Figure 3-8, the terminal station of an MC-TDMA system is synchronized to the base station in order to reduce the synchronization overhead. The transmitter of the terminal station extracts from the demodulated downlink data such as MAC messages burst allocation, power control and timing advance, and further clock and frequency synchronization information. In other words, the synchronization of the terminal station is achieved using the MAC control messages to perform time synchronization and using frequency information issued from the terminal station downlink demodulator (the recovered base station system clock). MAC control messages are processed by the MAC management block to instruct the terminal station modulator on the transmission resources assigned to it and to tune the access. Here, the pilot/reference symbols are inserted at the transmitter side to ease the burst synchronization and channel estimation tasks at the base station. At the base station, the received burst, issued by each terminal station, is detected and multi-carrier demodulated.

It should be emphasized that the transmitter and receiver structure of an MC-TDMA system is quite similar to an OFDM/OFDMA system. The same components, such as FFT, channel estimation, equalization and soft channel decoding, can be used for both, except that for an MC-TDMA system a burst synchronization is required, equivalent to a single-carrier TDMA system. Furthermore, a frequency synchronous system would simplify the MC-TDMA receiver synchronization tasks.

Combining OFDMA and MC-TDMA achieves a flexible multiuser system with high throughput [9].

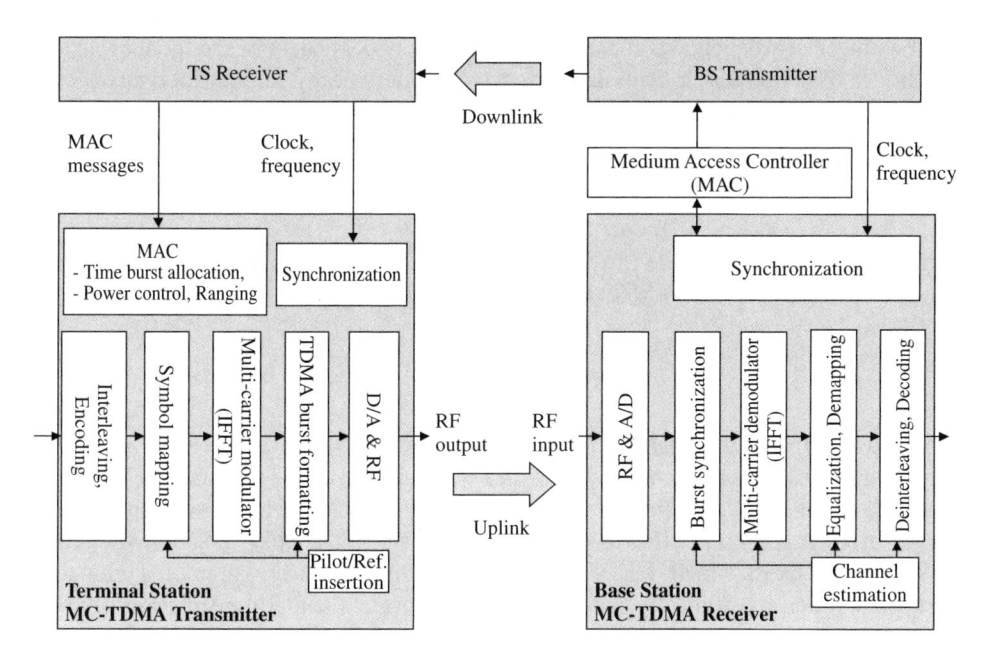

Figure 3-8 General MC-TDMA conceptual transceiver

3.4 Ultra Wide Band Systems

The technique for generating an ultra wide band (UWB) signal has existed for more than three decades [27], which is better known to the radar community as a *baseband carrier less short pulse* [1].

A classical way to generate an UWB signal is to spread the data with a code with a very large processing gain, i.e., 50 to 60 dB, resulting in a transmitted bandwidth of several GHz. Multiple access can be realized by classical CDMA where for each user a given spreading code is assigned. However, the main problem of such a technique is its implementation complexity.

As the power spectral density of the UWB signal is extremely low, the transmitted signal appears as a negligible white noise for other systems. In the increasingly crowded spectrum, the transmission of the data as a noise-like signal can be considered a main advantage for the UWB systems. However, its drawbacks are the small coverage and the low data rate for each user. Typically for short-range application (e.g., 100 m), the data rate assigned to each user can be about several kbit/s.

In [25] and [37] an alternative approach compared to classical CDMA is proposed for generating a UWB signal that does require sine-wave generation. It is based on time-hopping spread spectrum. The key advantages of this method are the ability to resolve multiple paths and the low complexity technology availability for its implementation.

3.4.1 Pseudo-Random PPM UWB Signal Generation

The idea of generating a UWB signal by transmitting ultra-short Gaussian monocycles with controlled pulse-to-pulse intervals can be found in [25]. The monocycle is a wideband signal with center frequency and bandwidth completely dependent of the monocycle duration. In the time domain, a Gaussian monocycle is derived by the first derivative of the Gaussian function given by

$$s(t) = 6a\sqrt{\frac{e\pi}{3}}\frac{t}{\tau}e^{-6\pi\left(\frac{t}{\tau}\right)^2}, \qquad (3.22)$$

where a is the peak amplitude of the monocycle and τ is the monocycle duration. In the frequency domain, the monocycle spectrum is given by

$$S(f) = -j\frac{2f\tau^2}{3}\sqrt{\frac{e\pi}{2}}e^{-\frac{\pi}{6}(f\tau)^2}, \qquad (3.23)$$

with center frequency and bandwidth approximately equal to $1/\tau$.

In Figure 3-9, a Gaussian monocycle with $\tau = 0.5$ ns duration is illustrated. This monocycle will result in a center frequency of 2 GHz with 3 dB bandwidth of approximately 2 GHz (from 1 GHz to 3.16 GHz). For data transmission, pulse position modulation (PPM) can be used, which varies the precise timing of transmission of a monocycle about the nominal position. By shifting each monocycle's actual transmission time over a large time frame in accordance with a specific PN code, i.e., performing time hopping (see Figure 3-10), this pseudo-random time modulation makes the UWB spectrum a pure white noise in the frequency domain. In the time domain each user will have a unique PN time-hopping code, hence resulting in a time-hopping multiple access.

A single data bit is generally spread over multiple monocycles, i.e., pulses. The duty cycle of each transmitted pulse is about 0.5–1%. Hence, the processing gain obtained by this technique is the sum of the duty cycle (ca. 20–23 dB) and the number of pulses used per data bit. As an example, if we consider a transmission with 10^6 pulses per second with a duty cycle of 0.5% and with a pulse duration of 0.5 ns ($B = 2$ GHz bandwidth), for 8 kbit/s transmitted data the resulting processing gain will be 54 dB, which is significantly high.

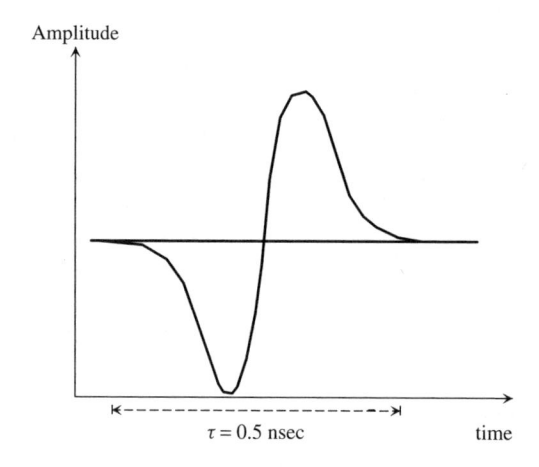

Figure 3-9 Gaussian monocycle with duration 0.5 ns

Ultra Wide Band Systems

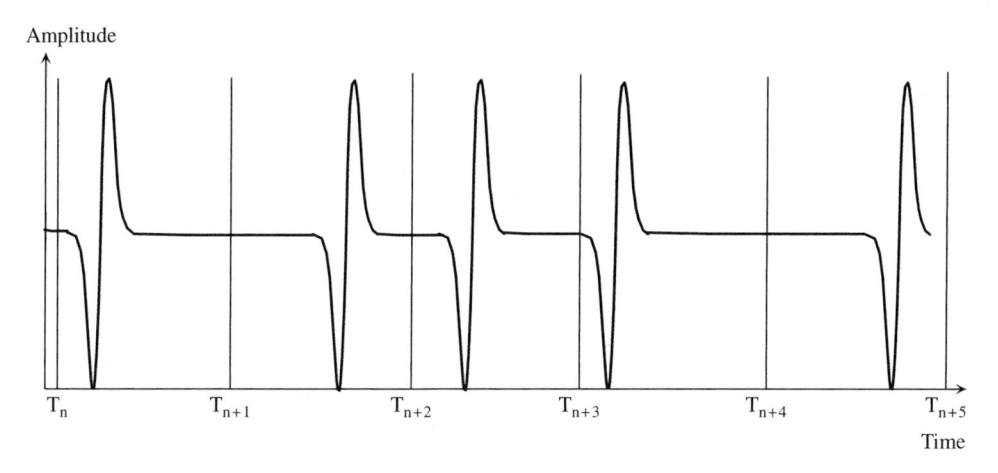

Figure 3-10 PN time modulation with 5 pulses

The ultra wide band signal generated above can be seen as a combination of spread spectrum with pulse position modulation.

3.4.2 UWB Transmission Schemes

A UWB transmission scheme for a multiuser environment is illustrated in Figure 3-11, where for each user a given time-hopping pattern, i.e., PN code, is assigned. The transmitter is quite simple. It does not include any amplifier or any IF generation. The signal of the transmitted data after pulse position modulation according to the user's PN code is emitted directly at the Tx antenna. A critical point of the transmitter is the antenna which may act as a filter.

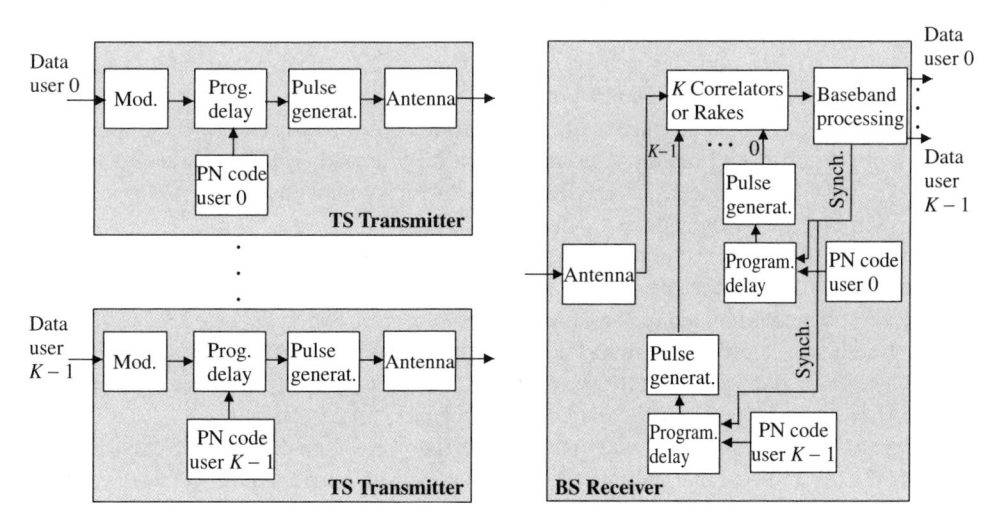

Figure 3-11 Multiuser UWB transmission scheme

The receiver components are similar to the transmitter. A rake receiver as in a conventional DS-CDMA system might be required to cope with multipath propagation. The baseband signal processing extracts the modulated signal and controls both signal acquisition and tracking.

The main application fields of UWB could be short range (e.g., indoor) multiuser communications, radar systems, and location determination/positioning. UWB may have a potential application in the automotive industry.

3.5 Comparison of Hybrid Multiple Access Schemes

A multitude of performance comparisons have been carried out between MC-CDMA and DS-CDMA as well as between the multi-carrier multiple access schemes MC-CDMA, MC-DS-CDMA, SS-MC-MA, OFDMA and MC-TDMA. It has been shown that MC-CDMA can significantly outperform DS-CDMA with respect to BER performance and bandwidth efficiency in the synchronous downlink [8][13][14]. The reason for better performance with MC-CDMA is that it can avoid ISI and ICI, allowing an efficient, simple user signal separation. The results of these comparisons are the motivation to consider MC-CDMA as a potential candidate for a future 4G mobile radio system which should outperform 3G systems based on DS-CDMA.

The design of a future air interface for broadband mobile communications requires a comprehensive comparison between the various multi-carrier based multiple access schemes. In Section 2.1.9, the performance of MC-CDMA, OFDMA, and MC-TDMA has been compared in a Rayleigh fading channel for scenarios with and without FEC channel coding, where different symbol mapping schemes have also been taken into account. It can generally be said that MC-CDMA outperforms the other multiple access schemes but requires additional complexity for signal spreading and detection. The reader is referred to Section 2.1.9 and to [15][17][23][26][29] to directly compare the performance of the various schemes.

In the following, we show a performance comparison between MC-CDMA and OFDMA for the downlink and between SS-MC-MA and OFDMA for the uplink. The transmission bandwidth is 2 MHz and the carrier frequency is 2 GHz. The guard interval exceeds the maximum delay of the channel. The mobile radio channels are chosen according to the COST 207 models. Simulations are carried out with a bad urban (BU) profile and a velocity of 3 km/h of the mobile user and with a hilly terrain (HT) profile and a velocity of 150 km/h of the mobile user. QPSK is chosen for symbol mapping. All systems are fully loaded and synchronized.

In Figure 3-12, the BER versus the SNR per bit for MC-CDMA and OFDMA systems with different channel code rates in the downlink is shown. The number of sub-carriers is 512. Perfect channel knowledge is assumed in the receiver. The results for MC-CDMA are obtained with soft interference cancellation [20] after the 1st iteration. It can be observed that MC-CDMA outperforms OFDMA. The SNR gain with MC-CDMA compared to OFDMA strongly depends on the propagation scenario and code rate.

Figure 3-13 shows the BER versus the SNR per bit of an SS-MC-MA system and an OFDMA system in the uplink. The number of sub-carriers is 256. Both systems apply one-dimensional channel estimation which requires an overhead on pilot symbols of 22.6%. The channel code rate is 2/3. The SS-MC-MA system applies maximum likelihood

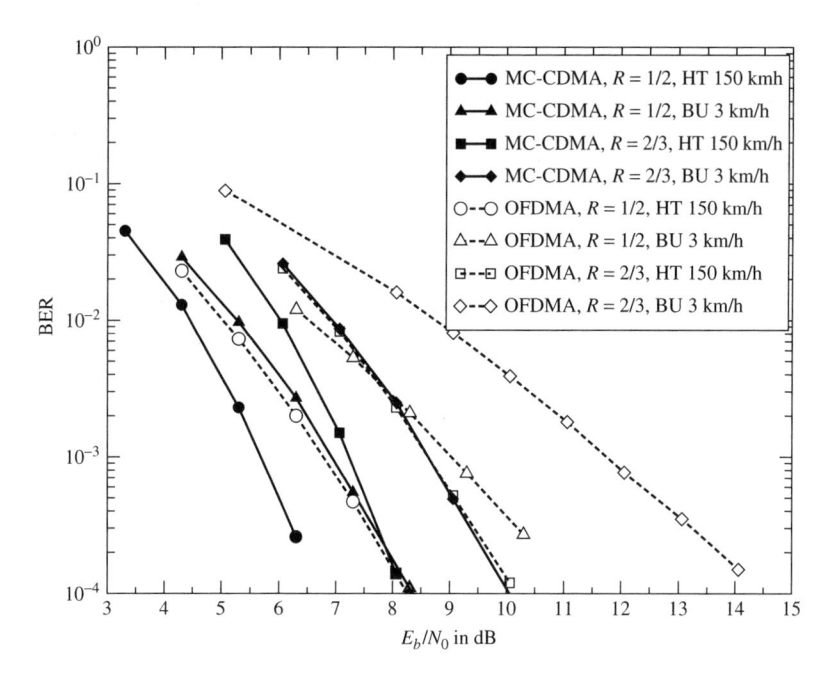

Figure 3-12 BER versus SNR of MC-CDMA and OFDMA in the downlink; QPSK; fully loaded system

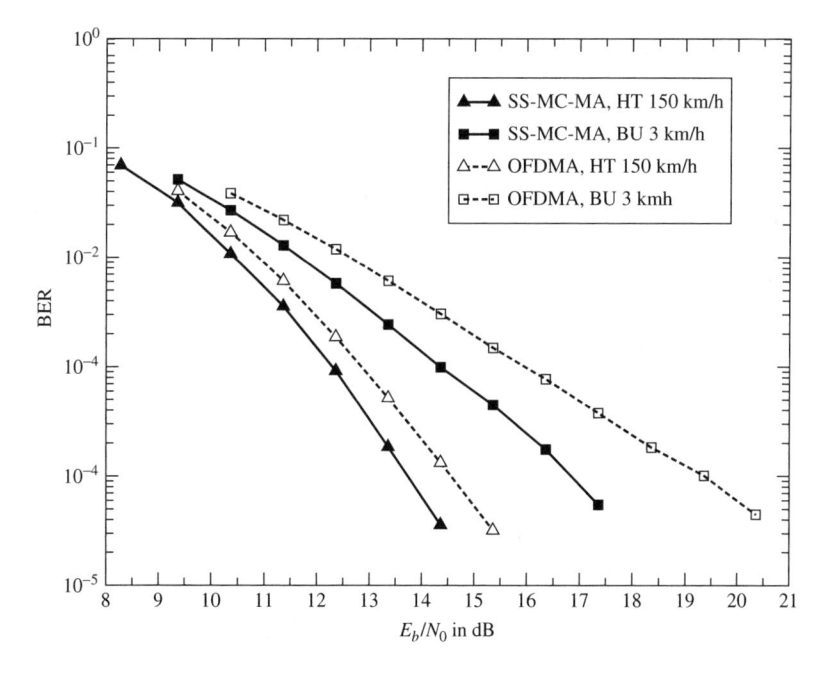

Figure 3-13 BER versus SNR of SS-MC-MA and OFDMA with one-dimensional pilot symbol aided channel estimation in the uplink; $R = 2/3$; QPSK; fully loaded system

detection. The performance of SS-MC-MA can be further improved by applying soft interference cancellation in the receiver. The SS-MC-MA system outperforms OFDMA in the uplink, however, it requires more complex receivers. The SS-MC-MA system and the OFDMA system would improve in performance by about 1 dB in the downlink due to reduced overheads with two-dimensional channel estimation.

3.6 References

[1] Bennett C.L. and Ross G.F., "Time-domain electromagnetics and its applications," *Proceedings IEEE*, vol. 66, pp. 299–318, March 1978.
[2] Chen Q., Sousa E.S. and Pasupathy S., "Multi-carrier CDMA with adaptive frequency hopping for mobile radio systems," *IEEE Journal on Selected Areas in Communications*, vol. 14, pp. 1852–1858, Dec. 1996.
[3] ETSI DVB-RCT (TS 301 958), "Interaction channel for digital terrestrial television (RCT) incorporating multiple access OFDM," Sophia Antipolis, France, March 2001.
[4] ETSI HIPERLAN (TS 101 475), "Broadband radio access networks HIPERLAN Type 2 functional specification – Part 1: Physical layer," Sophia Antipolis, France, Sept. 1999.
[5] ETSI HIPERMAN (Draft TS 102 177), "High performance metropolitan area network, Part 1: Physical layer," Sophia Antipolis, France, Feb. 2003.
[6] Giannakis G.B., Anghel P.A., Wang Z. and Scaglione A., "Generalized multi-carrier CDMA for MUI/ISI-resilient uplink transmissions irrespective of frequency-selective multipath," in *Proc. International Workshop on Multi-Carrier Spread-Spectrum & Related Topics (MC-SS'99)*, Oberpfaffenhofen, Germany, pp. 25–33, Sept. 1999.
[7] Giannakis G.B., Stamoulis A., Wang Z. and Anghel A., "Load-adaptive MUI/ISI-resilient generalized multi-carrier CDMA with linear and DF receivers," *European Transactions on Telecommunications (ETT)*, vol. 11, pp. 527–537, Nov./Dec. 2000.
[8] Hara S. and Prasad R., "Overview of multi-carrier CDMA," *IEEE Communications Magazine*, vol. 35, pp. 126–133, Dec. 1997.
[9] Ibars C. and Bar-Ness Y., "Rate-adaptive coded multi-user OFDM for downlink wireless systems," in *Proc. International Workshop on Multi-Carrier Spread-Spectrum & Related Topics (MC-SS 2001)*, Oberpfaffenhofen, Germany, pp. 199–207, Sept. 2001.
[10] IEEE-802.11 (P802.11a/D6.0), "LAN/MAN specific requirements – Part 2: Wireless MAC and PHY specifications – high speed physical layer in the 5 GHz band," IEEE 802.11, May 1999.
[11] IEEE 802.16ab-01/01, "Air interface for fixed broadband wireless access systems – Part A: Systems between 2 and 11 GHz," IEEE 802.16, June 2000.
[12] Jankiraman M. and Prasad R., "Wideband multimedia solution using hybrid CDMA/OFDM/SFH techniques," in *Proc. International Workshop on Multi-Carrier Spread-Spectrum & Related Topics (MC-SS'99)*, Oberpfaffenhofen, Germany, pp. 15–24, Sept. 1999.
[13] Kaiser S., "OFDM-CDMA versus DS-CDMA: Performance evaluation in fading channels," in *Proc. IEEE International Conference on Communications (ICC'95)*, Seattle, USA, pp. 1722–1726, June 1995.
[14] Kaiser S., "On the performance of different detection techniques for OFDM-CDMA in fading channels," in *Proc. IEEE Global Telecommunications Conference (GLOBECOM'95)*, Singapore, pp. 2059–2063, Nov. 1995.
[15] Kaiser S., "Trade-off between channel coding and spreading in multi-carrier CDMA systems," in *Proc. IEEE International Symposium on Spread Spectrum Techniques and Applications (ISSSTA'96)*, Mainz, Germany, pp. 1366–1370, Sept. 1996.
[16] Kaiser S., *Multi-Carrier CDMA Mobile Radio Systems – Analysis and Optimization of Detection, Decoding, and Channel Estimation*. Düsseldorf: VDI-Verlag, Fortschritt-Berichte VDI, series 10, no. 531, 1998, PhD thesis.
[17] Kaiser S., "MC-FDMA and MC-TDMA versus MC-CDMA and SS-MC-MA: Performance evaluation for fading channels," in *Proc. IEEE International Symposium on Spread Spectrum Techniques and Applications (ISSSTA'98)*, Sun City, South Africa, pp. 115–120, Sept. 1998.
[18] Kaiser S. and Fazel K., "A spread-spectrum multi-carrier multiple-access system for mobile communications," in *Proc. International Workshop on Multi-Carrier Spread-Spectrum (MC-SS'97)*, Oberpfaffenhofen, Germany, pp. 49–56, April 1997.

[19] Kaiser S. and Fazel K., "A flexible spread spectrum multi-carrier multiple-access system for multi-media applications," in *Proc. IEEE International Symposium on Personal, Indoor and Mobile Communications (PIMRC'97)*, Helsinki, Finland, pp. 100–104, Sept. 1997.
[20] Kaiser S. and Hagenauer J., "Multi-carrier CDMA with iterative decoding and soft-interference cancellation," in *Proc. IEEE Global Telecommunications Conference (GLOBECOM'97)*, Phoenix, USA, pp. 6–10, Nov. 1997.
[21] Kaiser S. and Krzymien W.A., "Performance effects of the uplink asynchronism in a spread spectrum multi-carrier multiple access system," *European Transactions on Telecommunications (ETT)*, vol. 10, pp. 399–406, July/Aug. 1999.
[22] Kaiser S., Krzymien W.A. and Fazel K., "SS-MC-MA systems with pilot symbol aided channel estimation in the asynchronous uplink," *European Transactions on Telecommunications (ETT)*, vol. 11, pp. 605–610, Nov./Dec. 2000.
[23] Lindner J., "On coding and spreading for MC-CDMA," in *Proc. International Workshop on Multi-Carrier Spread-Spectrum & Related Topics (MC-SS'99)*, Oberpfaffenhofen, Germany, pp. 89–98, Sept. 1999.
[24] Novak R. and Krzymien W.A., "A downlink SS-OFDM-F/TA packet data system employing multi-user diversity," in *Proc. International Workshop on Multi-Carrier Spread-Spectrum & Related Topics (MC-SS 2001)*, Oberpfaffenhofen, Germany, pp. 181–190, Sept. 2001.
[25] Petroff A. and Withington P., "Time modulated ultra-wideband (TM-UWB) overview," in *Proc. Wireless Symposium 2000*, San Jose, USA, Feb. 2000.
[26] Rohling H. and Grünheid R., "Performance comparison of different multiple access schemes for the downlink of an OFDM communication system," in *Proc. IEEE Vehicular Technology Conference (VTC'97)*, Phoenix, USA, pp. 1365–1369, May 1997.
[27] Ross G.F., "The transient analysis of certain TEM mode four-post networks," *IEEE Transactions on Microwave Theory and Techniques*, vol. 14, pp. 528–542, Nov. 1966
[28] Sari H., "Orthogonal frequency-division multiple access with frequency hopping and diversity," in *Proc. International Workshop on Multi-Carrier Spread-Spectrum (MC-SS'97)*, Oberpfaffenhofen, Germany, pp. 57–68, April 1997.
[29] Sari H., "A review of multi-carrier CDMA," in *Proc. International Workshop on Multi-Carrier Spread-Spectrum & Related Topics (MC-SS 2001)*, Oberpfaffenhofen, Germany, pp. 3–12, Sept. 2001.
[30] Sari H. and Karam G., "Orthogonal frequency-division multiple access and its application to CATV networks," *European Transactions on Telecommunications (ETT)*, vol. 9, pp. 507–516, Nov./Dec. 1998.
[31] Sari H., Levy Y. and Karam G., "Orthogonal frequency-division multiple access for the return channel on CATV networks," in *Proc. International Conference on Telecommunications (ICT'96)*, Istanbul, Turkey, pp. 52–57, April 1996.
[32] Sari H., Levy Y. and Karam G., "OFDMA – A new multiple access technique and its application to interactive CATV networks," in *Proc. European Conference on Multimedia Applications, Services and Techniques (ECMAST '96)*, Louvain-la-Neuve, Belgium, pp. 117–127, May 1996.
[33] Sari H., Vanhaverbeke F. and Moeneclaey M., "Some novel concepts in multiplexing and multiple access," in *Proc. International Workshop on Multi-Carrier Spread-Spectrum & Related Topics (MC-SS'99)*, Oberpfaffenhofen, Germany, pp. 3–12, Sept. 1999.
[34] Schnell M., De Broeck I. and Sorger U., "A promising new wideband multiple access scheme for future mobile communications systems," *European Transactions on Telecommunications (ETT)*, vol. 10, pp. 417–427, July/Aug. 1999.
[35] Sorger U., De Broeck I. and Schnell S., "Interleaved FDMA – a new spread-spectrum multiple access scheme," in *Proc. IEEE International Conference on Communications (ICC'98)*, Atlanta, USA, pp. 1013–1017, June 1998.
[36] Tomba L. and Krzymien W.A., "An OFDM/SFH-CDMA transmission scheme for the uplink," in *Proc. International Workshop on Multi-Carrier Spread-Spectrum (MC-SS'97)*, Oberpfaffenhofen, Germany, pp. 203–210, April 1997.
[37] Win M.Z. and Scholtz R.A., "Ultra-wideband bandwidth time-hopping spread-spectrum impulse radio for wireless multiple access communications," *IEEE Transactions on Communications*, vol. 48, pp. 679–691, April 2000.

4

Implementation Issues

A general block diagram of a multi-carrier transceiver employed in a cellular environment with a central base station (BS) and several terminal stations (TSs) in a point to multi-point topology is depicted in Figure 4-1.

For the downlink, transmission occurs in the base station and reception in the terminal station and for the uplink, transmission occurs in the terminal station and reception in the base station. Although very similar in concept, note that in general the base station equipment handles more than one terminal station, hence, its architecture is more complex.

The transmission operation starts with a stream of data symbols (bits, bytes or packets) sent from a higher protocol layer, i.e., the medium access control (MAC) layer. These data symbols are channel encoded, mapped into constellation symbols according to the designated symbol alphabet, spread (only in MC-SS) and optionally interleaved. The modulated symbols and the corresponding reference/pilot symbols are multiplexed to form a frame or a burst. The resulting symbols after framing or burst formatting are multiplexed and multi-carrier modulated by using OFDM and finally forwarded to the radio transmitter through a physical interface with digital-to-analog (D/A) conversion.

The reception operation starts with receiving an analog signal from the radio receiver. The analog-to-digital converter (A/D) converts the analog signal to the digital domain. After multi-carrier demodulation (IOFDM) and deframing, the extracted pilot symbols and reference symbols are used for channel estimation and synchronization. After optionally deinterleaving, despreading (only in the case of MC-SS) and demapping, the channel decoder corrects the channel errors to guarantee data integrity. Finally, the received data symbols (bits, bytes or a packet) are forwarded to the higher protocol layer for further processing.

Although the heart of an orthogonal multi-carrier transmission is the FFT/IFFT operation, synchronization and channel estimation process together with channel decoding play a major role. To ensure a low-cost receiver (low-cost local oscillator and RF components) and to guarantee a high spectral efficiency, robust digital synchronization and channel estimation mechanisms are needed. The throughput of an OFDM system not only depends on the used modulation constellation and FEC scheme but also on the amount of reference and pilot symbols spent to guarantee reliable synchronization and channel estimation.

Multi-Carrier and Spread Spectrum Systems K. Fazel and S. Kaiser
© 2003 John Wiley & Sons, Ltd ISBN: 0-470-84899-5

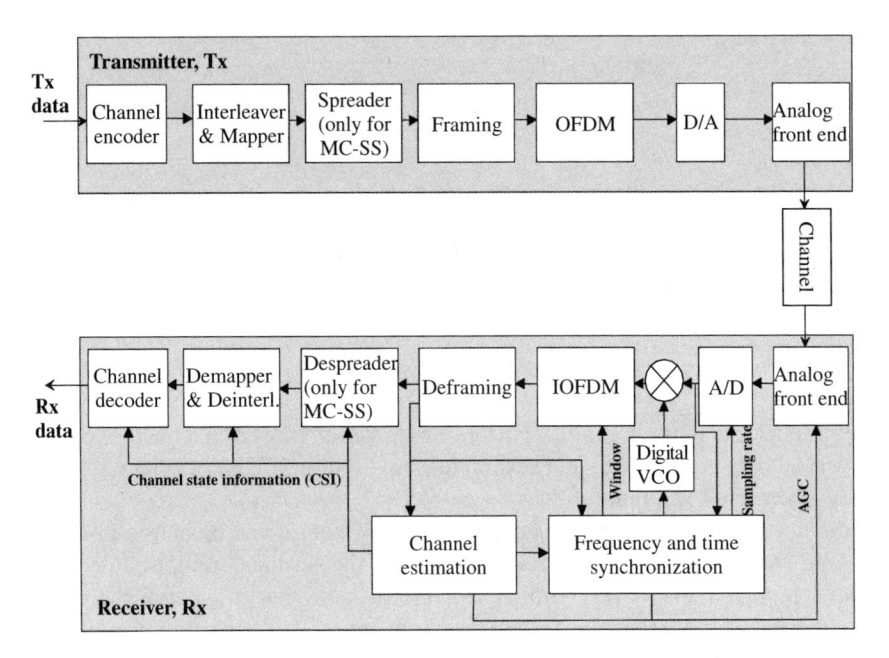

Figure 4-1 General block diagram of a multi-carrier transceiver

In Chapter 2 the different despreading and detection strategies for MC-SS systems were analysed. It was shown that with an appropriate detection strategy, especially in full load conditions (where all users are active) a high system capacity can be achieved. In the performance analysis in Chapter 2 we assumed that the modem is perfectly synchronized and the channel is perfectly known at the receiver.

The principal goal of this chapter is to describe in detail the remaining components of a multi-carrier transmission scheme with or without spreading. The focus is given to multi-carrier modulation/demodulation, digital I/Q generation, sampling, channel coding/decoding, framing/deframing, synchronization, and channel estimation mechanisms. Especially for synchronization and channel estimation units the effects of the transceiver imperfections (i.e., frequency drift, imperfect sampling time, phase noise) are highlighted. Finally, the effects of the amplifier non-linearity in multi-carrier transmission are analyzed.

4.1 Multi-Carrier Modulation and Demodulation

After symbol mapping (e.g., M-QAM) and spreading (in MC-SS), each block of N_c complex-valued symbols is serial-to-parallel (S/P) converted and submitted to the multi-carrier modulator, where the symbols are transmitted simultaneously on N_c parallel sub-carriers, each occupying a small fraction $(1/N_c)$ of the total available bandwidth B.

Figure 4-2 shows the block diagram of a multi-carrier transmitter. The transmitted baseband signal is given by

$$s(t) = \frac{1}{N_c} \sum_{i=-\infty}^{+\infty} \sum_{n=0}^{N_c-1} d_{n,i}\, g(t - iT_s)\, e^{j2\pi f_n t}, \qquad (4.1)$$

Multi-Carrier Modulation and Demodulation

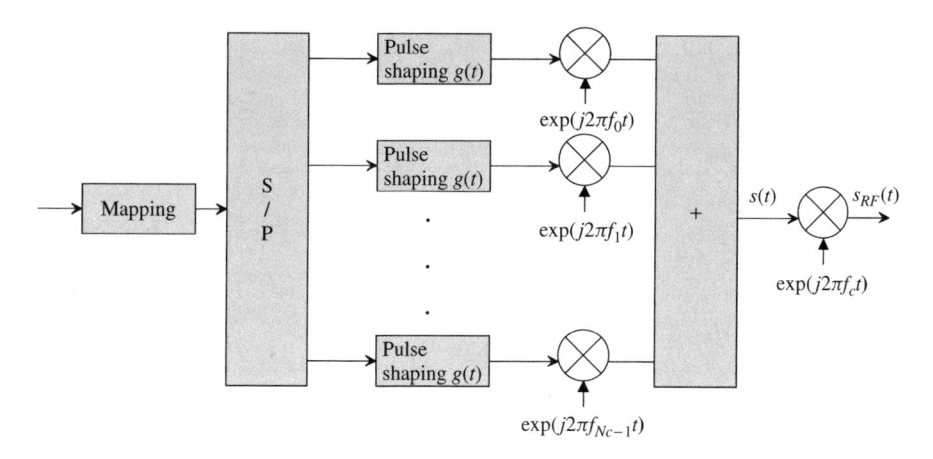

Figure 4-2 Block diagram of a multi-carrier transmitter

where N_c is the number of sub-carriers, $1/T_s$ is the symbol rate associated with each sub-carrier, $g(t)$ is the impulse response of the transmitter filters, $d_{n,i}$ is the complex constellation symbol, and f_n is the frequency of sub-carrier n. We assume that the sub-carriers are equally spaced, i.e.,

$$f_n = \frac{n}{T_s}, \quad n = 0, \ldots, N_c - 1. \tag{4.2}$$

The up-converted transmitted RF signal $s_{RF}(t)$ can be expressed by

$$s_{RF}(t) = \frac{1}{N_c} \mathrm{Re} \left\{ \sum_{i=-\infty}^{+\infty} \sum_{n=0}^{N_c-1} d_{n,i} g(t - iT_s) e^{j2\pi(f_n + f_c)t} \right\}$$

$$= \mathrm{Re}\{s(t) e^{j2\pi f_c t}\} \tag{4.3}$$

where f_c is the carrier frequency.

As shown in Figure 4-3, at the receiver side after down-conversion of the RF signal $r_{RF}(t)$, a bank of N_c matched filters is required to demodulate all sub-carriers. The received baseband signal after demodulation and filtering and before sampling at sub-carrier frequency f_m is given by

$$r_m(t) = [r(t) e^{-j2\pi f_m t}] \otimes h(t)$$

$$= \left[\sum_{i=-\infty}^{+\infty} \sum_{n=0}^{N_c-1} d_{n,i} g(t - iT_s) e^{j2\pi(f_n - f_m)t} \right] \otimes h(t), \tag{4.4}$$

where $h(t)$ is the impulse response of the receiver filter, which is matched to the transmitter filter (i.e., $h(t) = g^*(-t)$). The symbol $\otimes$ indicates the convolution operation. For simplicity, the received signal is given in the absence of fading and noise.

After sampling at optimum sampling time $t = lT_s$, the samples result in $r_m(lT_s) = d_{m,l}$, if the transmitter and the receiver of the multi-carrier transmission system fulfill both the ISI and ICI-free Nyquist conditions [65].

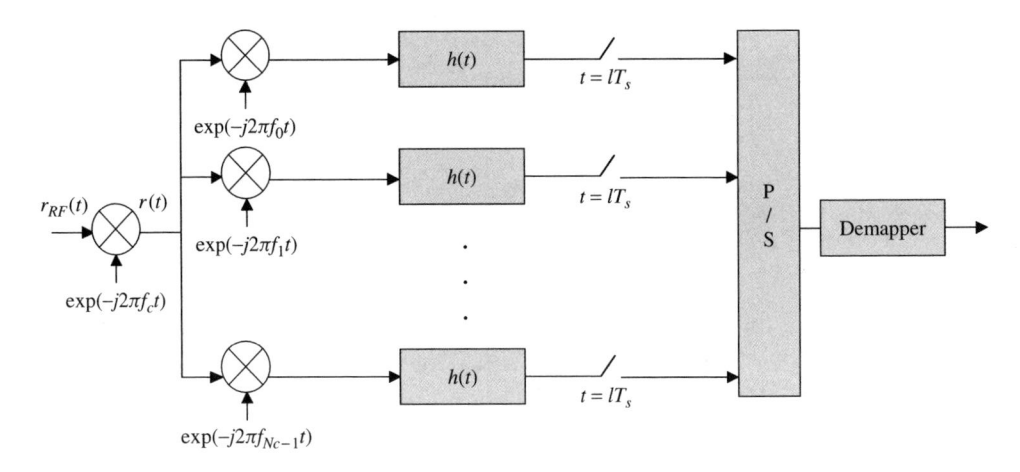

Figure 4-3 Block diagram of a multi-carrier receiver

To fulfill these conditions, different pulse shaping filtering can be used:

Rectangular band-limited system: Each sub-carrier has a rectangular band-limited transmission filter with impulse response

$$g(t) = \frac{\sin\left(\pi \frac{t}{T_s}\right)}{\pi \frac{t}{T_s}} = \text{sinc}\left(\pi \frac{t}{T_s}\right). \tag{4.5}$$

The spectral efficiency of the system is equal to the optimum value, i.e., normalized value of 1 bit/s/Hz.

Rectangular time-limited system: Each sub-carrier has a rectangular time-limited transmission filter with impulse response

$$g(t) = \text{rect}(t) = \begin{cases} 1 & 0 \leqslant t < T_s \\ 0 & \text{otherwise} \end{cases} \tag{4.6}$$

The spectral efficiency of the system is equal to normalized value $1/(1 + BT_s/N_c)$. For large N_c, it approaches the optimum normalized value of 1 bit/s/Hz.

Raised cosine filtering: Each sub-carrier is filtered by a time-limited ($t \in \{-kT'_s, kT'_s\}$) square root of a raised cosine filter with roll-off factor α and impulse response [65]

$$g(t) = \frac{\sin\left[\frac{\pi t}{T'_s}(1-\alpha)\right] + \frac{k\alpha t}{T'_s}\cos\left[\frac{\pi t}{T'_s}(1+\alpha)\right]}{\frac{\pi t}{T'_s}\left[1 - \left(\frac{k\alpha t}{T'_s}\right)^2\right]}, \tag{4.7}$$

where $T_s' = (1+\alpha)T_s$ and k is the maximum number of samples that the pulse shall not exceed. The spectral efficiency of the system is equal to $1/(1 + (1+\alpha)/N_c)$. For large N_c, it approaches the optimum normalized value of 1 bit/s/Hz.

4.1.1 Pulse Shaping in OFDM

OFDM employs a time-limited rectangular pulse shaping which leads to a simple digital implementation. OFDM without guard time is an optimum system, where for large numbers of sub-carriers its efficiency approaches the optimum normalized value of 1 bit/s/Hz.

The impulse response of the receiver filter is

$$h(t) = \begin{cases} 1 & \text{if } -T_s < t \leqslant 0 \\ 0 & \text{otherwise} \end{cases} \quad (4.8)$$

It can easily be shown that the condition of absence of ISI and ICI is fulfilled.

In case of inserting a guard time T_g, the spectral efficiency of OFDM will be reduced to $1 - T_g/(T_s + T_g)$ for large N_c.

4.1.2 Digital Implementation of OFDM

By omitting the time index i in (4.1), the transmitted OFDM baseband signal, i.e., one OFDM symbol with guard time, is given by

$$s(t) = \frac{1}{N_c} \sum_{n=0}^{N_c-1} d_n \, e^{j2\pi \frac{nt}{T_s}}, \quad -T_g \leqslant t < T_s, \quad (4.9)$$

where d_n is a complex-valued data symbol, T_s is the symbol duration and T_g is the guard time between two consecutive OFDM symbols in order to prevent ISI and ICI in a multipath channel. The sub-carriers are separated by $1/T_s$.

Note that for burst transmission, i.e., burst formatting, a pre-/postfix of duration T_a can be added to the original OFDM symbol of duration $T_s' = T_s + T_g$ so that the total OFDM symbol duration becomes

$$T' = T_s + T_g + T_a. \quad (4.10)$$

The pre-/postfix can be designed such that it has good correlation properties in order to perform channel estimation or synchronization. One possibility for the pre-/postfix is to extend the OFDM symbol by a specific PN sequence with good correlation properties. At the receiver, as guard time, the pre-/postfix is skipped and the OFDM symbol is rebuilt as described in Section 4.5.

From the above expression we note that the transmitted OFDM symbol can be performed by using an inverse complex FFT operation (IFFT), where the demultiplexing is done by an FFT operation. In the complex digital domain this operation leads to an IDFT operation with N_c points at the transmitter side and a DFT with N_c points at the receiver side (see Figure 4-4). Note that for guard time and pre-/postfix L_g samples are inserted after the IDFT operation at the transmitter side and removed before the DFT at the receiver side.

Highly repetitive structures based on elementary operations such as butterflies for the FFT operation can be applied if N_c is of the power of 2 [1]. Depending on the transmission media and the carrier frequency f_c, the actual OFDM transmission systems employ from

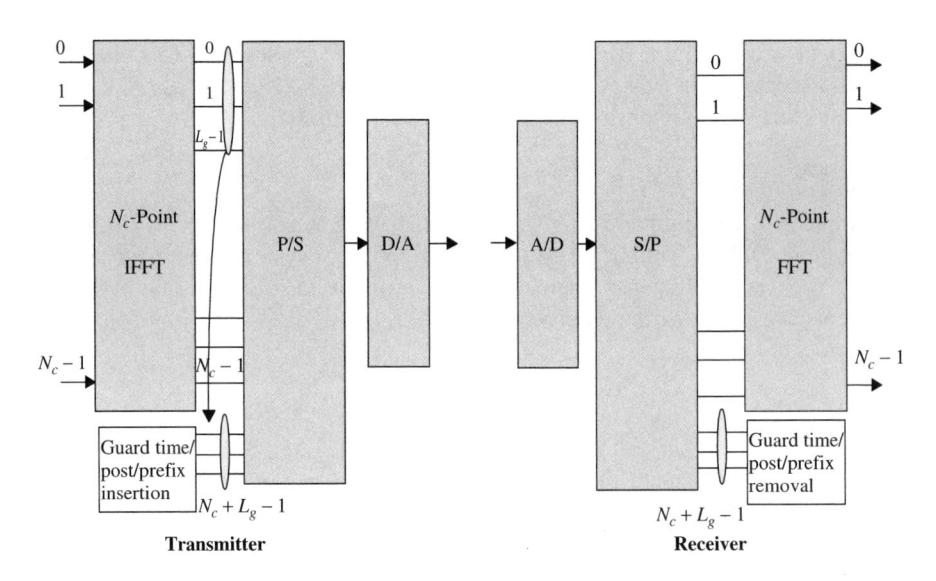

Figure 4-4 Digital implementation of OFDM

64 up to 2048 (2k) sub-carriers. In the DVB-T standard [16], up to 8192 (8k) sub-carriers are required to combat long echoes in a single frequency network operation.

The complexity of the FFT operation (multiplications and additions) depends on the number of FFT points N_c. It can be approximated by $(N_c/2) \log N_c$ operations [1]. Furthermore, large numbers of FFT points, resulting in long OFDM symbol durations T'_s, make the system more sensitive to the time variance of the channel (Doppler effect) and more vulnerable to the oscillator phase noise (technological limitation). However, on the other hand, a large symbol duration increases the spectral efficiency due to a decrease of the guard interval loss.

Therefore, for any OFDM realization a trade-off between the number of FFT points, the sensitivity to the Doppler and phase noise effects, and the loss due to the guard interval has to be found.

4.1.3 Virtual Sub-Carriers and DC Sub-Carrier

By employing large numbers of sub-carriers in OFDM transmission, a high frequency resolution in the channel bandwidth can be achieved. This enables a much easier implementation and design of the filters. If the number of FFT points is slightly higher than that required for data transmission, a simple filtering can be achieved by putting in both sides of the spectrum null sub-carriers (guard bands), called virtual sub-carriers (see Figure 4-5). Furthermore, in order to avoid the DC problem, a null sub-carrier can be put in the middle of the spectrum, i.e., the DC sub-carrier is not used.

4.1.4 D/A and A/D Conversion, I/Q Generation

The digital implementation of multi-carrier transmission at the transmitter and the receiver side requires digital-to-analog (D/A) and analog-to-digital (A/D) conversion and methods for modulating and demodulating a carrier with a complex OFDM time signal.

Multi-Carrier Modulation and Demodulation

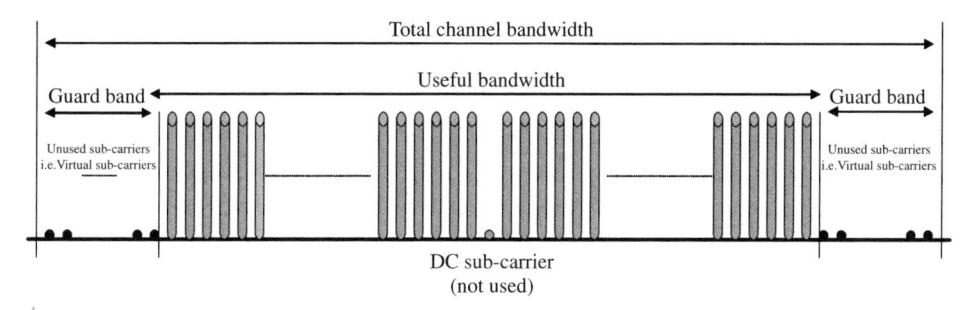

Figure 4-5 Virtual sub-carriers used for filtering

4.1.4.1 D/A and A/D Conversion and Sampling Rate

The main advantage of an OFDM transmission and reception is its digital implementation using digital FFT processing. Therefore, at the transmission side the digital signal after digital IFFT processing is converted to the analog domain with a D/A converter, ready for IF/RF up-conversion and vice versa at the receiver side.

The number of bits reserved for the D/A and A/D conversion depends on many parameters: i) accuracy needed for a given constellation, ii) required Tx/Rx dynamic ranges (e.g., difference between the maximum received power and the receiver sensitivity), and iii) used sampling rate, i.e., complexity. It should be noticed that at the receiver side, due to a higher disturbance, a more accurate converter is required. In practice, in order to achieve a good trade-off between complexity, performance, and implementation loss typically for a 64-QAM transmission, D/A converters with 8 bits or higher should be used, and 10 bits or higher are recommended for the receiver A/D converters. However, for low-order modulation, these constraints can be relaxed.

The sampling rate is a crucial parameter. To avoid any problem with aliasing, the sampling rate f_{samp} should be at least twice the maximum frequency of the signal. This requirement is theoretically satisfied by choosing the sampling rate [1]

$$f_{samp} = 1/T_{samp} = N_c/T_s = B. \qquad (4.11)$$

However, in order to provide a better channel selectivity in the receiver regarding adjacent channel interference, a higher sampling rate than the channel bandwidth might be used, i.e., $f_{samp} > N_c/T_s$.

4.1.4.2 I/Q Generation

At least two methods exist for modulating and demodulating a carrier (I and Q generation) with a complex OFDM time signal. These are described below.

Analog Quadrature Method

This is a conventional solution in which the in-phase carrier component I is fed by the real part of the modulating signal and the quadrature component Q is fed by the imaginary part of the modulating signal [65].

The receiver applies the inverse operations using the I/Q demodulator (see Figure 4-6). This method has two drawbacks for an OFDM transmission, especially for large numbers

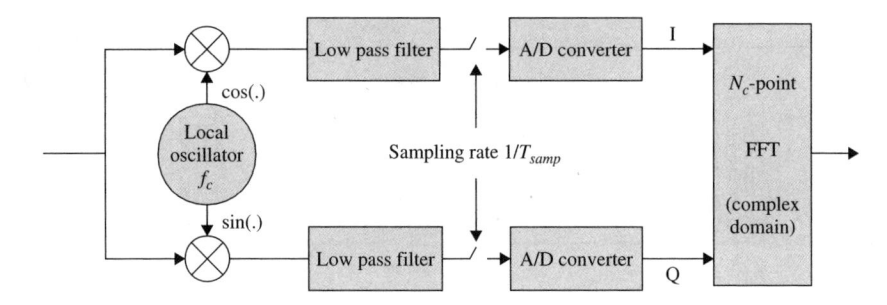

Figure 4-6 Conventional I/Q generation with two analog demodulators

of sub-carriers and high-order modulation (e.g., 64-QAM): i) due to imperfections in the RF components, it is difficult at moderate complexity to avoid a cross-talk between the I and Q signals and, hence, to maintain an accurate amplitude and phase matching between the I and Q components of the modulated carrier across the signal bandwidth. This imperfection may result in high received baseband signal degradation, i.e., interference, and ii) it requires two A/D converters.

A low cost front-end may result in I/Q mismatching, emanating from the gain mismatch between the I and Q signals and from non-perfect quadrature generation. These problems can be solved in the digital domain.

Digital FIR Filtering Method

The second approach is based on employing digital techniques in order to shift the complex time domain signal up in frequency and produce a signal with no imaginary components which is fed to a single modulator. Similarly, the receiver requires a single demodulator. However, the A/D converter has to work at double sampling frequency (see Figure 4-7).

The received analog signal can be written as

$$r(t) = I(t)\cos(\pi t/T_{samp}) + Q(t)\sin(\pi t/T_{samp}), \qquad (4.12)$$

where T_{samp} is the sampling period of each I and Q component. By doubling the sampling rate to $2/T_{samp}$ we get the sampled signal

$$r(l) = I(l)\cos(\pi l/2) + Q(l)\sin(\pi l/2). \qquad (4.13)$$

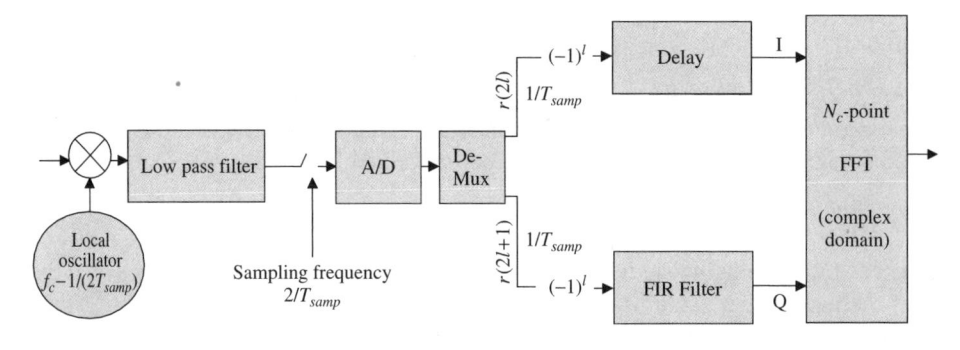

Figure 4-7 Digital I/Q generation using FIR filtering with single analog demodulator

This stream can be separated into two sub-streams with rate $1/T_{samp}$ by taking the even and odd samples

$$r(2l) = I(2l)\cos(\pi l) + Q(2l)\sin(\pi l)$$
$$r(2l+1) = I(2l+1)\cos(\pi(2l+1)/2) + Q(2l+1)\sin(\pi(2l+1)/2) \quad (4.14)$$

It is straightforward to show that the desired output I and Q components are related to $r(2l)$ and $r(2l+1)$ by

$$I(l) = (-1)^l r(2l) \quad (4.15)$$

and the $Q(l)$ outputs are obtained by delaying $(-1)^l r(2l+1)$ by $T_{samp}/2$, i.e., passing the $(-1)^l r(2l+1)$ samples through an interpolator filter (FIR). The $I(l)$ components have to be delayed as well to compensate the FIR filtering delay.

In other words, at the transmission side this method consists (at the output of the complex digital IFFT processing) of filtering the Q channel with an FIR interpolator filter to implement a 1/2 sample time shift. Both I and Q streams are then oversampled by a factor of 2. By taking the even and odd components of each stream, only one digital stream at twice the sampling frequency is formed. This digital signal is converted to analog and used to modulate the RF carrier. At the reception side, the inverse operation is applied. The incoming analog signal is down-converted and centered on a frequency $f_{samp}/2$, filtered and converted to digital by sampling at twice the sampling frequency (i.e., 2 f_{samp}). It is de-multiplexed into the 2 streams $r(2l)$ and $r(2l+1)$ at rate $f_{samp} = 1/T_{samp}$. The I and Q channels are multiplied by $(-1)^l$ to ensure transposition of the spectrum of the signal into baseband [1]. The Q channel is filtered using the same FIR interpolator filter as the transmitter while the I components are delayed by a corresponding amount so that the I and Q components can be delivered simultaneously to the digital FFT processing unit.

4.2 Synchronization

Reliable receiver synchronization is one of the most important issues in multi-carrier communication systems, and is especially demanding in fading channels when coherent detection of high-order modulation schemes is employed.

A general block diagram of a multi-carrier receiver synchronization unit is depicted in Figure 4-8. The incoming signal in the analog front end unit is first down-converted, performing the complex demodulation to baseband time domain digital I and Q signals of the received OFDM signal. The local oscillator(s) of the analog front end has/have to work with sufficient accuracy. Therefore, the local oscillator(s) is/are continuously adjusted by the frequency offset estimated in the synchronization unit. In addition, before the FFT operation a fine frequency offset correction signal might be required to reduce the ICI.

Furthermore, the sampling rate of the A/D clock needs to be controlled by the time synchronization unit as well, in order to prevent any frequency shift after the FFT operation that may result in an additional ICI. The correct positioning of the FFT window is another important task of the timing synchronization.

The remaining task of the OFDM synchronization unit is to estimate the phase and amplitude distortion of each sub-carrier, where this function is performed by the *channel estimation* core (see Section 4.3). These estimated channel state information (CSI) values

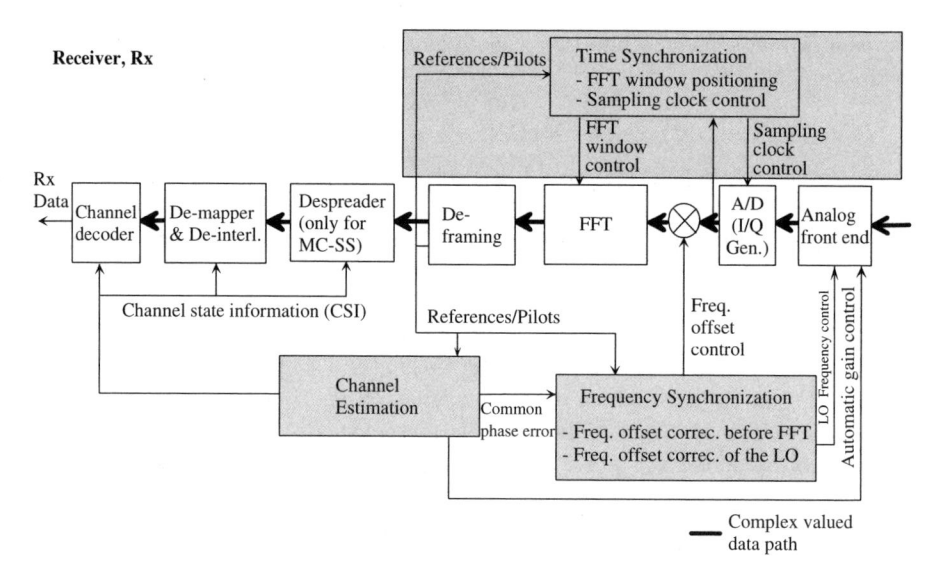

Figure 4-8 General block diagram of a multi-carrier synchronization unit

are used to derive for each demodulated symbol *reliability information* that is directly applied for despreading and/or for channel decoding.

An automatic gain control (AGC) of the incoming analog signal is also needed to adjust the gain of the received signal in its desired values.

The performance of any synchronization and channel estimation algorithm is determined by the following parameters:

— *Minimum SNR* under which the operation of synchronization is guaranteed,
— *Acquisition time* and *acquisition range* (e.g., maximum tolerable deviation range of timing offset, local oscillator frequency),
— *Overhead* in terms of reduced data rate or power excess,
— *Complexity*, regarding implementation aspects, and
— *Robustness* and *accuracy* in the presence of multipath and interference disturbances.

In a wireless cellular system with a point-to-multi-point topology, the base station acts as a central control of the available resources among several terminal stations. Signal transmission from the base station towards the terminal station in the downlink is often done in a continuous manner. However, the uplink transmission from the terminal station towards the base station might be different and can be performed in a bursty manner.

In case of a continuous downlink transmission, both acquisition and tracking algorithms for synchronization can be applied [22], where all fine adjustments to counteract time-dependent variations (e.g., local oscillator frequency offset, Doppler, timing drift, common phase error) are carried out in tracking mode. Furthermore, in case of a continuous transmission, non-pilot aided algorithms (blind synchronization) might be considered.

However, the situation is different for a bursty transmission. All synchronization parameters for each burst have to be derived with required accuracy within the limited time duration. Two ways exist to achieve simple and accurate burst synchronization:

— enough reference and pilot symbols are appended to each burst, or
— the terminal station is synchronized to the downlink, where the base station will continuously broadcast to all terminal stations synchronization signaling.

The first solution requires a significant amount of overhead, which leads to a considerable loss in uplink spectral efficiency. The second solution is widely adopted in burst transmission. Here all terminal stations synchronize their transmit frequency and clock to the received base station signal. The time-advance variation (moving vehicle) between the terminal station and the base station can be adjusted by transmitting ranging messages individually from the base station to each terminal station. Hence, the *burst receiver* at the base station does not need to regenerate the terminal station clock and carrier frequency; it only has to estimate the channel. Note that in FDD the uplink carrier frequency has only to be shifted.

In time- and frequency-synchronous multi-carrier transmission the receiver at the base station needs to detect the start position of an OFDM symbol or frame and to estimate the channel state information from some known pilot symbols inserted in each OFDM symbol. If the coherence time of the channel exceeds an OFDM symbol, the channel estimation can estimate the time variation as well. This strategy, which will be considered in the following, simplifies a burst receiver.

To summarize, in the next sections we make the following assumptions:

— the terminal stations are frequency/time-synchronized to the base station,
— the Doppler variation is slow enough to be considered constant during one OFDM symbol of duration T_s', and
— the guard interval duration T_g is larger than the channel impulse response.

4.2.1 General

The synchronization algorithms employed for multi-carrier demodulation are based either on the analysis of the received signal (*non-pilot aided*, i.e., blind synchronization) [10][11][35] or on the processing of special dedicated data time and/or frequency multiplexed with the transmitted data, i.e., *pilot-aided* synchronization [11][22][23][55][76]. For instance, in non-pilot aided synchronization some of these algorithms exploit the intrinsic redundancy present in the guard time (cyclic extension) of each OFDM symbol. Maximum likelihood estimation of parameters can also be applied, exploiting the guard-time redundancy [73] or using some dedicated transmitted reference symbols [55].

As shown in Figure 4-8, there are three main synchronization tasks around the FFT: i) timing recovery, ii) carrier frequency recovery and iii) carrier phase recovery. In this part, we concentrate on the first two items, since the carrier phase recovery is closely related to the channel estimation (see Section 4.3). Hence, the two main synchronization parameters that have to be estimated are: i) time-positioning of the FFT window including the sampling rate adjustment that can be controlled in a two-stage process, coarse- and fine-timing control and ii) the possible large frequency difference between the receiver and transmitter local oscillators that has to be corrected to a very high accuracy.

As known from DAB [14], DVB-T [16] and other standards, usually the transmission is performed in a *frame by frame* basis. An example of an OFDM frame is depicted in

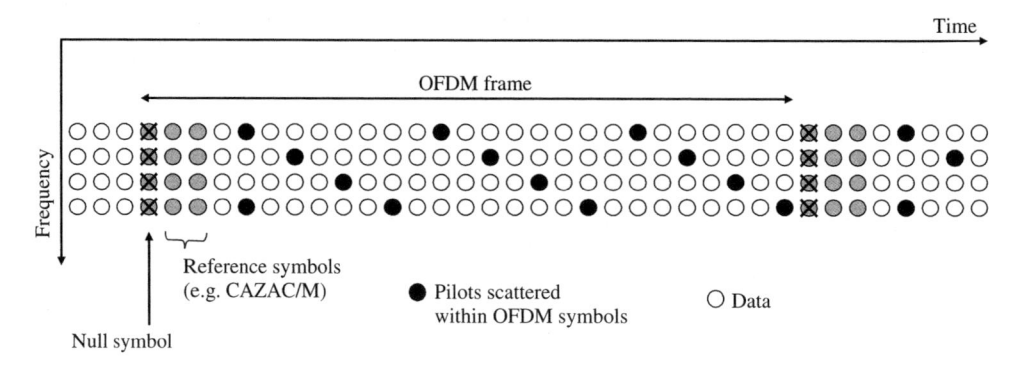

Figure 4-9 Example of an OFDM frame

Figure 4-9, where each frame consists of a so-called null symbol (without signal power) transmitted at the frame beginning, followed by some known reference symbols and data symbols. Furthermore, within data symbols some reference pilots are scattered in time and frequency. The null symbol may serve two important purposes: interference and noise estimation, and coarse timing control. The coarse timing control may use the null symbol as a mean of quickly establishing frame synchronization prior to fine time synchronization.

Fine timing control can be achieved by time [76] or frequency domain processing [12] using the reference symbols. These symbols have good partial autocorrelation properties. The resulting signal can either be used to directly control the fine positioning of the FFT window or to alter the sampling rate of the A/D converters. In addition, for time synchronization the properties of the guard time can be exploited [35][73].

If the frequency offset is smaller than half the sub-carrier spacing a maximum likelihood frequency estimation can be applied by exploiting the reference symbols [55] or the guard time redundancy [73]. In the case that the frequency offset exceeds several sub-carrier spacings, a frequency offset estimation technique using again the OFDM reference symbol as above for timing can be used [58][76]. These reference symbols allow coarse and fine adjustment of the local oscillator frequency in a two-step process. Here, frequency domain processing can be used. The more such special reference symbols are embedded into the OFDM frame, the faster the acquisition time and the higher the accuracy. Finally, a *common phase error (CPE)* estimation can be performed, that partially counters the effect of phase-noise of the local oscillator [69]. The common phase error estimation may exploit pilot symbols in each OFDM symbol (see Section 4.7.1.3) which can also be used for channel estimation.

In the following, after examining the effects of synchronization imperfections on multi-carrier transmission, we will detail the maximum likelihood estimation algorithms and other time and frequency synchronization techniques which are usually employed.

4.2.2 Effects of Synchronization Errors

Large timing and frequency errors in multi-carrier systems cause an increase of ISI and ICI, resulting in high performance degradations.

Let us assume that the receiver local oscillator frequency f_c (see Figure 4-3) is not perfectly locked to the transmitter frequency. The baseband received signal after down-conversion is

$$r(t) = s(t)\, e^{j2\pi f_{error} t} + n(t), \quad (4.16)$$

where f_{error} is the frequency error and $n(t)$ the complex-valued AWGN.

The above signal in the absence of fading after demodulation and filtering (i.e., convolution) at sub-carrier m can be written as [68]:

$$r_m(t) = [s(t)\, e^{j2\pi f_{error} t} + n(t)]\, e^{-j2\pi f_m t} \otimes h(t)$$

$$= \left[\sum_{i=-\infty}^{+\infty} \sum_{n=0}^{N_c-1} d_{n,i}\, g(t - iT_s)\, e^{-j2\pi\left(\frac{n-m}{T_s}\right)t}\, e^{j2\pi f_{error} t} \right] \otimes h(t) + n'(t) \quad (4.17)$$

where $h(t)$ is the impulse response of the receiver filter and $n'(t)$ is the filtered noise. Let us assume that the sampling clock has a static error τ_{error}. The sample at instant lT_s of the received signal at sub-carrier m is made up of four terms as follows

$$r_m(lT_s + \tau_{error}) = d_{m,l}\, A_m(\tau_{error})\, e^{j2\pi f_{error} lT_s} + ISI_{m,l} + ICI_{m,l} + n'(lT_s + \tau_{error}), \quad (4.18)$$

where the first term corresponds to the transmitted data $d_{m,i}$ which is attenuated and phase shifted. The second and third terms are the ISI and ICI given by

$$ISI_{m,l} = \sum_{\substack{i=-\infty \\ i \neq l}}^{+\infty} d_{m,i}\, A_m[(l-i)T_s + \tau_{error}]\, e^{j2\pi f_{error} iT_s} \quad (4.19)$$

$$ICI_{m,l} = \sum_{i=-\infty}^{+\infty} \sum_{\substack{n=0 \\ n \neq m}}^{N_c-1} d_{n,i}\, A_n[(l-i)T_s + \tau_{error}]\, e^{j2\pi f_{error} iT_s} \quad (4.20)$$

where

$$A_n(t) = \left(g(t)\, e^{j2\pi f_{error} t}\, e^{j2\pi \frac{(n-m)t}{T_s}} \right) \otimes h(t), \quad (4.21)$$

$g(t)$ is the impulse response of the transmitter filter and $A_n(lT_s)$ represents the sampled components of (4.21), i.e., samples after convolution.

4.2.2.1 Analysis of the SNR in the Presence of a Frequency Error

Here we consider only the effect of a frequency error, i.e., we put $\tau_{error} = 0$ in the above expressions. For simplicity, the guard time is omitted. Then (4.21) becomes [68]

$$A_n(t) = \begin{cases} e^{j\pi f_{error} t}\, e^{j\pi \frac{n-m}{T_s} t}\, \text{sinc}\left[\left(f_{error} + \frac{n-m}{T_s}\right)(T_s - t)\right]\left(1 - \frac{t}{T_s}\right) & 0 < t \leqslant T_s \\ 0 & \text{otherwise} \end{cases}$$

$$(4.22)$$

After sampling at instant lT_s, at sub-carrier $m = n$, $A_m(0) = \text{sinc}(f_{error}T_s)$, and $A_m(lT_s) = 0$. However, for $m \neq n$, $A_m(0) = \text{sinc}(f_{error}T_s + n - m)$ and $A_m(lT_s) = 0$.

Therefore, it can be shown that the received data after the FFT operation at time $t = 0$ and sub-carrier m can be written as [68]

$$r_m = d_m \text{sinc}(f_{error}T_s) e^{j2\pi f_{error}T_s} + ICI_m + n', \tag{4.23}$$

by omitting the time index. Note that the frequency error does not introduce any ISI.

Equation (4.23) shows that a frequency error creates besides the ICI a reduction of the received signal amplitude and a phase rotation of the symbol constellation on each sub-carrier. For large numbers of sub-carriers, the ICI can be modeled as AWGN. The resulting SNR can be written as

$$SNR_{ICI} \approx \frac{|d_m|^2 \text{sinc}^2(f_{error}T_s)}{\sum_{\substack{n=0 \\ n \neq m}}^{N_c-1} |d_n|^2 \text{sinc}^2[n - m + f_{error}T_s] + P_N}, \tag{4.24}$$

where P_N is the power of the noise n'. If E_s is the average received energy of the individual sub-carriers and $N_0/2$ is the noise power spectral density of the AWGN, then

$$\frac{E_s}{N_0} = \frac{|d_m|^2}{P_N} \tag{4.25}$$

and the SNR can be expressed as

$$SNR_{ICI} \approx \frac{E_s}{N_0} \frac{\text{sinc}^2(f_{error}T_s)}{1 + \frac{E_s}{N_0} \sum_{\substack{n=0 \\ n \neq m}}^{N_c-1} \text{sinc}^2[n - m + f_{error}T_s]}. \tag{4.26}$$

This equation shows that a frequency error can cause a significant loss in SNR. Furthermore, the SNR depends on the number of sub-carriers.

4.2.2.2 Analysis of the SNR in the Presence of a Clock Error

Here, we consider only the effect of a clock error, i.e., $f_{error} = 0$ in the above expressions. If the clock error is within the guard time, i.e., $|\tau_{error}| \leq T_g$ (i.e., early synchronization), the timing error is absorbed, hence, there is no ISI and no ICI. It results only in a phase shift at a given sub-carrier which can be compensated for by the channel estimation (see Section 4.3).

However, if the timing error exceeds the guard time, i.e., $|\tau_{error}| > T_g$ (i.e., late synchronization), both ISI and ICI appear. As Eqs (4.18) to (4.21) show, the clock error also introduces an amplitude reduction and a phase rotation which is proportional to the sub-carrier index. In a similar manner to above, the expression of the SNR can be derived as [68]

$$SNR_{ICI+ISI} \approx \frac{|d_m|^2(1 - \tau_{error}/T_s)^2}{(\tau_{error}/T_s)^2 \left(1 + 2 \sum_{\substack{n=0 \\ n \neq m}}^{N_c-1} |d_n|^2 \text{sinc}^2[(n-m)\tau_{error}/T_s]\right) + P_N}$$

$$\approx \frac{\frac{E_s}{N_o}(1 - \tau_{error}/T_s)^2}{\frac{E_s}{N_0}(\tau_{error}/T_s)^2 \left(1 + 2 \sum_{\substack{n=0 \\ n \neq m}}^{N_c-1} \text{sinc}^2[(n-m)\tau_{error}/T_s]\right) + 1} \quad (4.27)$$

It can be observed again that a clock error exceeding the guard time will introduce a reduction in SNR.

4.2.2.3 Requirements on OFDM Frequency and Clock Accuracy

Figures 4-10 and 4-11 show the simulated SNR degradation in dB for different bit error rates (BERs) versus the frequency error and timing error for QPSK, respectively. These diagrams show that an OFDM system is sensitive to frequency and to clock errors. In order to keep an acceptable performance degradation the error after frequency synchronization and time synchronization should not exceed the following limits [68]:

$$\tau_{error}/T_s < 0.01$$
$$f_{error}T_s < 0.02 \quad (4.28)$$

Thus, the error relative to the sampling period should fulfill $\tau_{error}/(N_c T_{samp}) < 0.01$ and the error relative to the sub-carrier spacing shall not be greater than 2% of the sub-carrier spacing, where the last is usually a quite difficult condition.

It should be noticed that for dimensioning the length of the guard time, the time synchronization inaccuracy should be taken into account. As long as the sum of the timing offset and the maximum multipath propagation delay is smaller than the guard time, the only effect is a phase rotation that can be estimated by the channel estimator (see Section 4.3) and compensated for by the channel equalizer (see Section 4.5).

4.2.3 Maximum Likelihood Parameter Estimation

Let us consider a frequency error f_{error} and a timing error τ_{error}. The joint maximum likelihood estimates $\hat{f}_{error}$ of the frequency error and $\hat{\tau}_{error}$ of the timing error are obtained by the maximization of the log-likelihood function (*LLF*) as follows [10][55][73],

$$LLF(f_{error}, \tau_{error}) = \log p(r|f_{error}, \tau_{error}), \quad (4.29)$$

where $p(r|f_{error}, \tau_{error})$ denotes the probability density function of observing the received signal r, given a frequency error f_{error} and timing error τ_{error}. In [73] it is shown that for

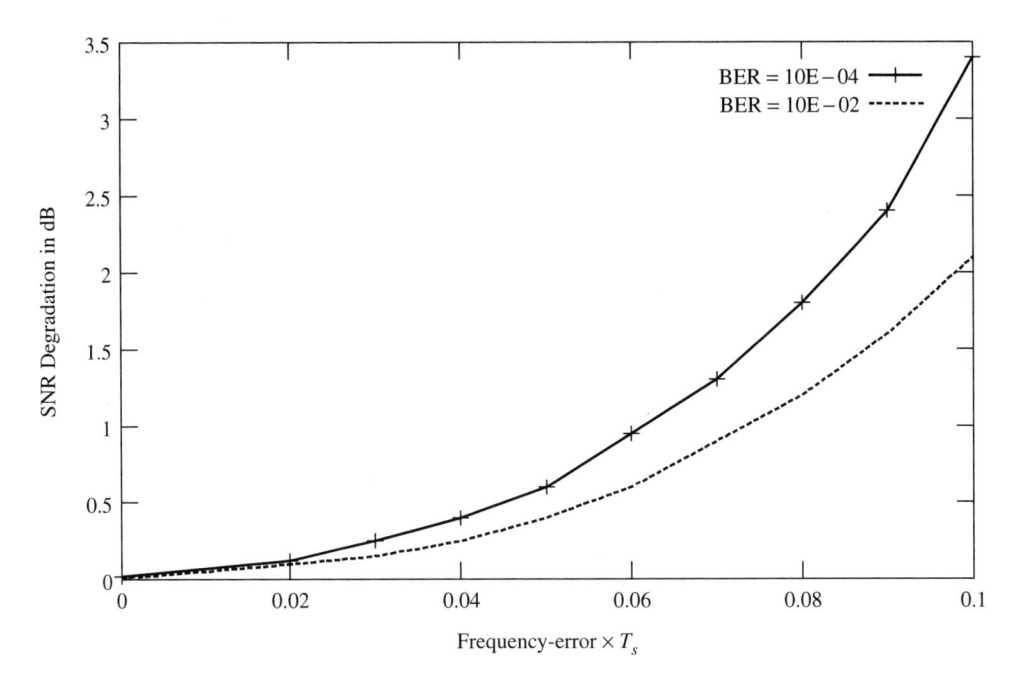

Figure 4-10 SNR degradation in dB versus the normalized frequency error $f_{error}T_s$; $N_c = 2048$ sub-carriers

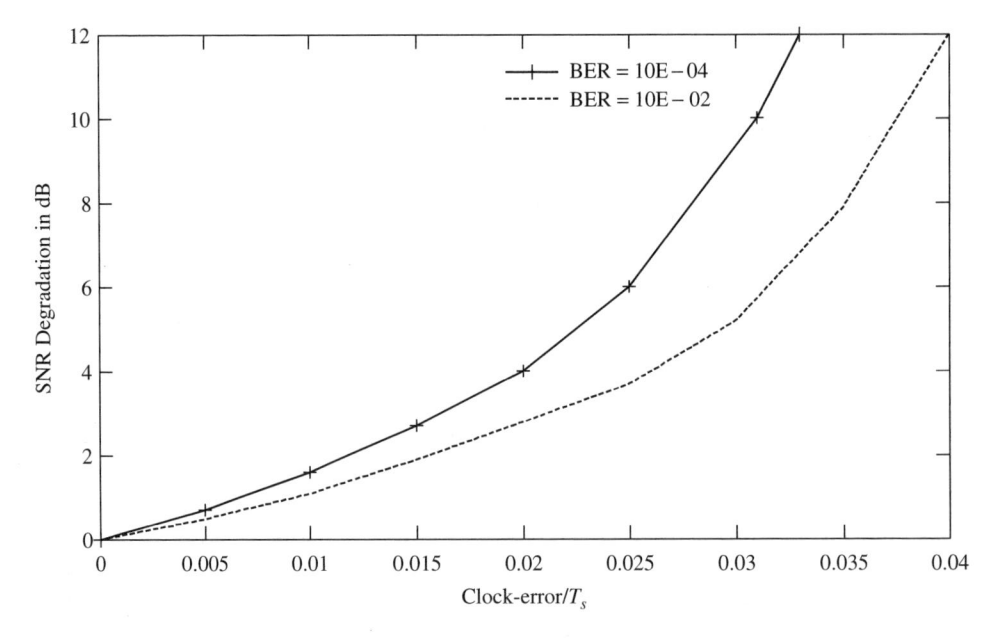

Figure 4-11 SNR degradation in dB versus the normalized timing error τ_{error}/T_s; $N_c = 2048$ sub-carriers

Synchronization

$N_c + M$ samples

$$LLF(f_{error}, \tau_{error}) = |\gamma(\tau_{error})|\cos(2\pi f_{error} + \angle\gamma(\tau_{error})) - \rho\Phi(\tau_{error}) \quad (4.30)$$

$$\gamma(m) = \sum_{k=m}^{m+M-1} r(k)r^*(k+N_c), \quad \Phi(m) = \sum_{k=m}^{m+M-1} |r(k)|^2 + |r(k+N_c)|^2 \quad (4.31)$$

where $\angle$ represents the argument of a complex number, ρ is a constant depending on the SNR which represents the magnitude of the correlation between the sequences $r(k)$ and $r(k+N_c)$, and m is the start of the correlation function of the received sequence (in case of no timing error one can start at $m=0$). Note that the first term in Eq. (4.30) is the weighted-magnitude of $\gamma(\tau_{error})$, which is the sum of M consecutive correlations.

These sequences $r(k)$ could be known in the receiver by transmitting, for instance, two consecutive reference symbols ($M = N_c$) as proposed by Moose [55] or one can exploit the presence of the guard time ($M = L_g$) [73].

The maximization of the above LLF can be done in two steps:

— a first maximization can be performed to find the frequency error estimate $\hat{f}_{error}$,
— then, the value of the given frequency error estimate is exploited for final maximization to find the timing error estimate $\hat{\tau}_{error}$.

The maximization of f_{error} is given by the partial derivation $\partial LLF(f_{error}, \tau_{error})/\partial f_{error} = 0$, which results in

$$\hat{f}_{error} = -\frac{1}{2\pi}\angle\gamma(\tau_{error}) + z = -\frac{1}{2\pi}\frac{\sum_{k=m}^{m+M-1}\operatorname{Im}[r(k)r^*(k+N_c)]}{\sum_{k=m}^{m+M-1}\operatorname{Re}[r(k)r^*(k+N_c)]} + z, \quad (4.32)$$

where z is an integer value.

By inserting $\hat{f}_{error}$ in Eq. (4.30), we obtain

$$LLF(\hat{f}_{error}, \tau_{error}) = |\gamma(\tau_{error})| - \rho\Phi(\tau_{error}) \quad (4.33)$$

and maximizing Eq. (4.33) gives us a joint estimate of $\hat{f}_{error}$ and $\hat{\tau}_{error}$

$$\hat{f}_{error} = -\frac{1}{2\pi}\angle\gamma(\hat{\tau}_{error}),$$

$$\hat{\tau}_{error} = \arg(\max_{\tau_{error}}\{|\gamma(\tau_{error})| - \rho\Phi(\tau_{error})\}). \quad (4.34)$$

Note that in case of $M = N_c$ (i.e, two reference symbols), $|\hat{f}_{error}| < 0.5$, $z = 0$, and no timing error $\tau_{error} = 0$ ($m = 0$), one obtains the same results as Moose [55] (see Figure 4-12).

The main drawback of the Moose maximum likelihood frequency detection is the small range of acquisition which is only half of the sub-carrier spacing F_s. When $f_{error} \to 0.5F_s$, the estimate $\hat{f}_{error}$ may, due to noise and the discontinuity of arctangent, jump to -0.5. When this happens, the estimate is no longer unbiased and in practice it becomes

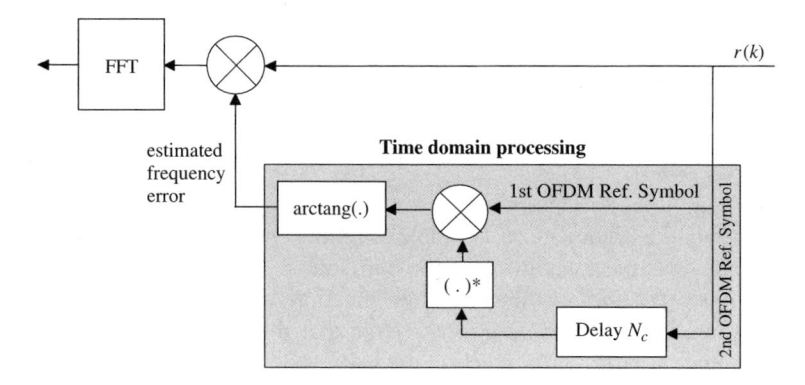

Figure 4-12 Moose maximum likelihood frequency estimator ($M = N_c$)

useless. Thus, for frequency errors exceeding one half of the sub-carrier spacing, an initial acquisition strategy, coarse frequency acquisition, should be applied. To enlarge the acquisition range of a maximum likelihood estimator, a modified version of this estimator was proposed in [10]. The basic idea is to modify the shape of the LLF.

The joint estimation of frequency and timing error using guard time may be sensitive in environments with several long echoes. In the following section, we will examine some approaches for time and frequency synchronization which are used in several implementations.

4.2.4 Time Synchronization

As we have explained before, the main objective of time synchronization for OFDM systems is to know when a received OFDM symbol starts. By using the guard time the timing requirements can be relaxed. A time offset, not exceeding the guard time, gives rise to a phase rotation of the sub-carriers. This phase rotation is larger on the edge of the frequency band. If a timing error is small enough to keep the channel impulse response within the guard time, the orthogonality is maintained and a symbol timing delay can be viewed as a phase shift introduced by the channel. This phase shift can be estimated by the channel estimator (see Section 4.3) and corrected by the channel equalizer (see Section 4.5). However, if a time shift is larger than the guard time, ISI and ICI occur and signal orthogonality is lost.

Basically the task of the time synchronization is to estimate the two main functions: FFT window positioning (OFDM symbol/frame synchronization) and sampling rate estimation for A/D conversion controlling.

The operation of time synchronization can be carried out in two steps: Coarse and fine symbol timing.

4.2.4.1 Coarse Symbol Timing

Different methods, depending on the transmission signal characteristics, can be used for coarse timing estimation [22][23][73].

Basically, the power at baseband can be monitored prior to FFT processing and for instance the dips resulting from null symbols (see Figure 4-9) might be used to control a 'flywheel'-type state transition algorithm as known from traditional frame synchronization [40].

Null Symbol Detection
A null symbol, containing no power, is transmitted for instance in DAB at the beginning of each OFDM frame (see Figure 4-13). By performing a simple power detection at the receiver side before the FFT operation, the beginning of the frame can be detected. That is, the receiver locates the null symbol by searching for a dip in the power of the received signal. This can be achieved, for instance, by using a flywheel algorithm to guard against occasional failures to detect the null symbol once in lock [40]. The basic function of this algorithm is that, when the receiver is out of lock, it searches continuously for the null symbols, whereas when in lock it searches for the symbol only at the expected null symbols. The null symbol detection gives only a coarse timing information.

Two Identical Half Reference Symbols
In [76] a timing synchronization is proposed that searches for a training symbol with two identical halves in the time domain, which can be sent at the beginning of an OFDM frame (see Figure 4-14). At the receiver side, these two identical time domain sequences may

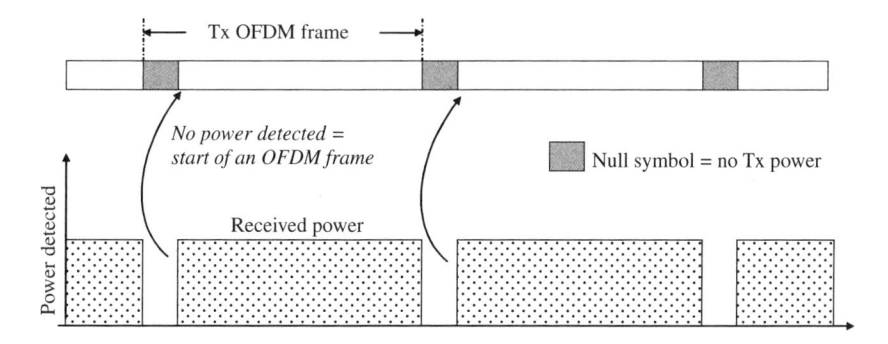

Figure 4-13 Coarse time synchronization based on null symbol detection

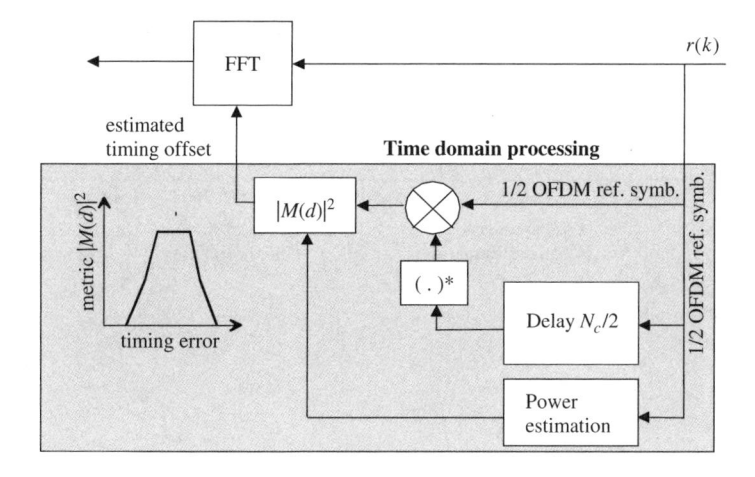

Figure 4-14 Time synchronization based on two identical half reference symbols

only be phase shifted $\phi = \pi T_s f_{error}$ due to the carrier frequency offset. The two halves of the training symbol are made identical by transmitting a PN sequence on the even frequencies, while zeros are used on the odd frequencies. Let there be M complex-valued samples in each half of the training symbol. The function for estimating the timing error d is defined as

$$M(d) = \frac{\sum_{m=0}^{M-1} r^*_{d+m} r_{d+m+M}}{\sum_{m=0}^{M-1} |r_{d+m+M}|^2}. \quad (4.35)$$

Finally, the estimate of the timing error is derived by taking the maximum quadratic value of the above function, i.e., max $|M(d)|^2$. The main drawback of this metric is its 'plateau' which may lead to some uncertainties.

Guard Time Exploitation

Each OFDM symbol is extended by a cyclic repetition of the transmitted data (see Figure 4-15). As the guard interval is just a duplication of a useful part of the OFDM symbol, a correlation of the part containing the cyclic extension (guard interval) with the given OFDM symbol enables a fast time synchronization [73]. The sampling rate can also be estimated based on this correlation method. The presence of strong noise or long echoes may prevent accurate symbol timing. However, the noise effect can be reduced by integration (filtering) on several peaks obtained from subsequent estimates. As far as echoes are concerned, if the guard time is chosen long enough to absorb all echoes, this technique can still be reliable.

4.2.4.2 Fine Symbol Timing

For fine time synchronization, several methods based on transmitted reference symbols can be used [12]. One straightforward solution applies the estimation of the channel impulse response. The received signal without noise $r(t) = s(t) \otimes h(t)$ is the convolution of the transmit signal $s(t)$ and the channel impulse response $h(t)$. In the frequency domain after FFT processing we obtain $R(f) = S(f)H(f)$. By transmitting special reference symbols

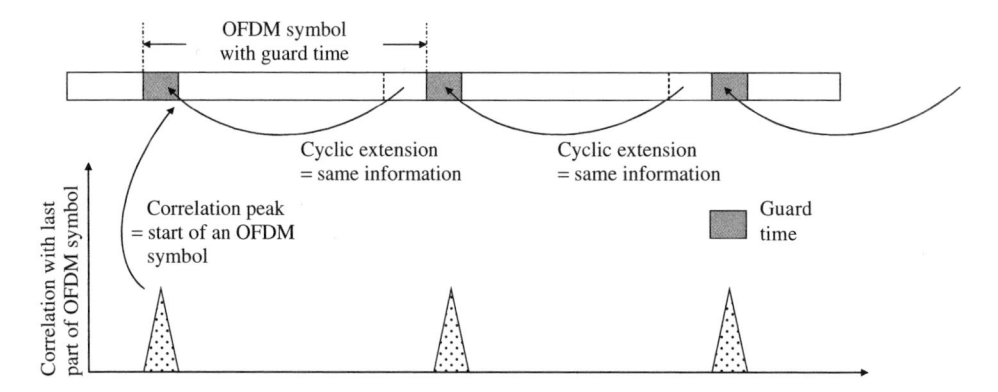

Figure 4-15 Time synchronization based on guard time correlation properties

(e.g., CAZAC sequences), $S(f)$ is *a priori* known by the receiver. Hence, after dividing $R(f)$ by $S(f)$ and IFFT processing, the channel impulse response $h(t)$ is obtained and an accurate timing information can be derived.

If the FFT window is not properly positioned, the received signal becomes

$$r(t) = s(t - t_0) \otimes h(t), \qquad (4.36)$$

which turns into

$$R(f) = S(f)H(f)e^{-j2\pi f t_0} \qquad (4.37)$$

after the FFT operation. After division of $R(f)$ by $S(f)$ and again performing an IFFT, the receiver obtains $h(t - t_0)$ and with that t_0. Finally, the fine time synchronization process consists of delaying the FFT window so that t_0 becomes quasi zero (see Figure 4-16).

In case of multipath propagation, the channel impulse response is made up of multiple Dirac pulses. Let C_p be the power of each constructive echo path and I_p be the power of a destructive path. An optimal time synchronization process is to maximize the C/I, the ratio of the total constructive path power to the total destructive path power. However, for ease of implementation a sub-optimal algorithm might be considered, where the FFT window positioning signal uses the first significant echo, i.e., the first echo above a fixed threshold. The threshold can be chosen from experience, but a reasonable starting value can be derived from the minimum carrier-to-noise ratio required.

4.2.4.3 Sampling Clock Adjustment

As we have seen, the received analog signal is first sampled at instants determined by the receiver clock before FFT operation. The effect of a clock frequency offset is twofold: the useful signal component is rotated and attenuated and, furthermore, ICI is introduced.

The sampling clock could be considered to be close to its theoretical value so it may have no effect on the result of the FFT. However, if the oscillator generating this clock is left free-running, the window opened for FFT may gently slide and will not match the useful interval of the symbols.

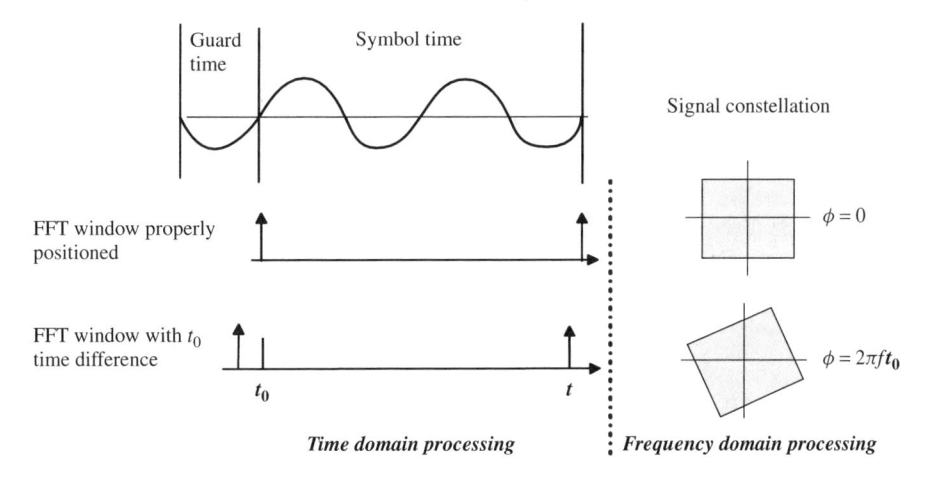

Figure 4-16 Fine time synchronization based on channel impulse response estimation

A first simple solution is to use the methods described above to evaluate the proper position of the window and to dynamically readjust it. However, this method generates a phase discontinuity between symbols where a readjustment of the FFT window occurs. This phase discontinuity requires additional filtering or interpolation after FFT operation.

A second method, although using a similar strategy, is to evaluate the shift of the FFT window that is proportional to the frequency offset of the clock oscillator. The shift can be used to control the oscillator with better accuracy. This method allows a fine adjustment of the FFT window without the drawback of phase discontinuity from one symbol to the other.

4.2.5 Frequency Synchronization

Another fundamental function of an OFDM receiver is the carrier frequency synchronization. Frequency offsets are introduced by differences in oscillator frequencies in the transmitter and receiver, Doppler shifts and phase noise. As we have seen earlier, the frequency offset leads to a reduction of the signal amplitude since the *sinc* functions are shifted and no longer sampled at the peak and to a loss of orthogonality between sub-carriers. This loss introduces ICI which results in a degradation of the global system performance [55][70][71].

In the previous sections we have seen that in order to avoid severe SNR degradation, the frequency synchronization accuracy should be better than 2%. Note that a multi-carrier system is much more sensitive to a frequency offset than a single carrier system [62].

As shown in Figure 4-8, the frequency error in an OFDM system is often corrected by a tracking loop with a frequency detector to estimate the frequency offset. Depending on the characteristics of the transmitted signal (pilot-based or not) several algorithms for frequency detection and synchronization can be applied:

— algorithms based on the analysis of special synchronization symbols embedded in the OFDM frame [7][50][55][58][76],
— algorithms based on the analysis of the received data at the output of the FFT (non-pilot aided) [10], and
— algorithms based on the analysis of guard time redundancy [11][35][73].

Like the time synchronization, the frequency synchronization can be performed in two steps: coarse and fine frequency synchronization.

4.2.5.1 Coarse Frequency Synchronization

We assume that the frequency offset is greater than half of the sub-carrier spacing, i.e.,

$$f_{error} = \frac{2z}{T_s} + \frac{\phi}{\pi T_s}, \qquad (4.38)$$

where the first term of the above equation represents the frequency offset which is a multiple of the sub-carrier spacing where z is an integer and the second term is the additional frequency offset being a fraction of the sub-carrier spacing, i.e., ϕ is smaller than π.

The aim of the coarse frequency estimation is mainly to estimate z. Depending on the transmitted OFDM signal, different approaches for coarse frequency synchronization can be used [10][11][12][58][73][76].

CAZAC/M Sequences

A general approach is to analyze the transmitted special reference symbols at the beginning of an OFDM frame; for instance, the CAZAC/M sequences [58] specified in the DVB-T standard [16]. As shown in Figure 4-17, CAZAC/M sequences are generated in the frequency domain and are embedded in I and R sequences. The CAZAC/M sequences are differentially modulated. The length of the M sequences is much larger than the length of the CAZAC sequences. The I and R sequences have the same length N_1, where in the I sequence (resp. R sequence) the imaginary (resp. real) components are 1 and the real (resp. imaginary) components are 0. The I and R sequences are used as start positions for the differential encoding/decoding of M sequences. A wide range coarse synchronization is achieved by correlating with the transmitted known M sequence reference data, shifted over $\pm N_1$ sub-carriers (e.g., $N_1 = 10$ to 20) from the expected center point [22][58]. The results from different sequences are averaged. The deviation of the correlation peak from the expected center point z with $-N_1 < z < +N_1$ is converted to an equivalent value used to correct the offset of the RF oscillator, or the baseband signal is corrected before the FFT operation. This process can be repeated until the deviation is less than $\pm N_2$ sub-carriers (e.g., $N_2 = 2$ to 5). For a fine-range estimation, in a similar manner the remaining CAZAC sequences can be applied that may reduce the frequency error to a few hertz.

The main advantage of this method is that it only uses one OFDM reference symbol. However, its drawback is the high amount of computation needed, which may not be adequate for burst transmission.

Schmidl and Cox

Similar to Moose [55], Schmidl and Cox [76] propose the use of two OFDM symbols for frequency synchronization (see Figure 4-18). However, these two OFDM symbols have a special construction which allows a frequency offset estimation greater than several sub-carrier spacings. The first OFDM training symbol in the time domain consists of two identical symbols generated in the frequency domain by a PN sequence on the even

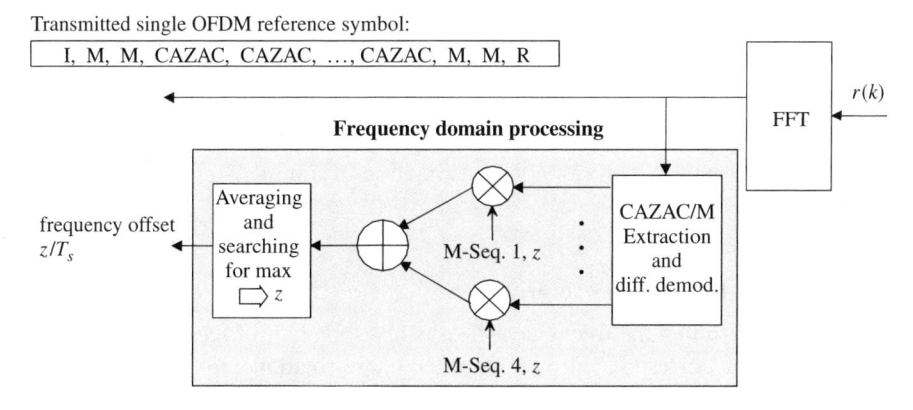

Figure 4-17 Coarse frequency offset estimation based on CAZAC/M sequences

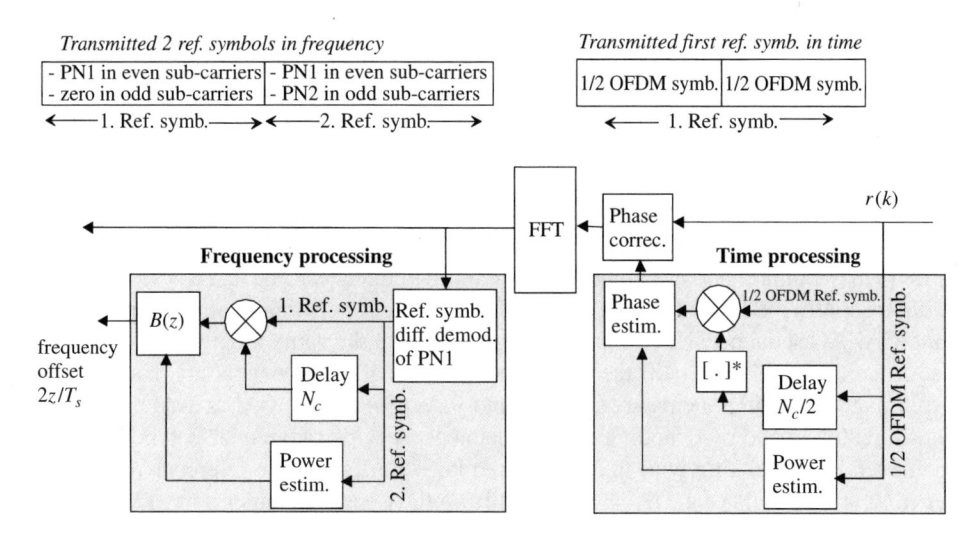

Figure 4-18 Schmidl and Cox frequency offset estimation using 2 OFDM symbols

sub-carriers and zeros on the odd sub-carriers. The second training symbol contains a differentially modulated PN sequence on the odd sub-carriers and another PN sequence on the even sub-carriers. Note that the selection of a particular PN sequence has little effect on the performance of the synchronization.

In Eq. (4.38), the second term can be estimated in a similar way to the Moose approach [55] by employing the two halves of the first training symbols, $\hat{\phi} = angle[M(d)]$ (see (4.35)). These two training symbols are frequency-corrected by $\hat{\phi}/(\pi T_s)$. Let their FFT be $x_{1,k}$ and $x_{2,k}$ and let the differentially modulated PN sequence on the even frequencies of the second training symbol be v_k and let X be the set of indices for the even sub-carriers. For the estimation of the integer sub-carrier offset given by z, the following metric is calculated,

$$B(z) = \frac{\left| \sum_{k \in X} x^*_{1,k+2z} \, v^*_k x_{2,k+2z} \right|^2}{2 \left(\sum_{k \in X} |x_{2,k}|^2 \right)^2}. \qquad (4.39)$$

The estimate of z is obtained by taking the maximum value of the above metric $B(z)$.

The main advantage of this method is its simplicity, which may be adequate for burst transmission. Furthermore, it allows a joint estimation of timing and frequency offset (see Section 4.2.4.1).

4.2.5.2 Fine Frequency Synchronization

Under the assumption that the frequency offset is less than half of the sub-carrier spacing, there is a one-to-one correspondence between the phase rotation and the frequency offset. The phase ambiguity limits the maximum frequency offset value. The phase offset can be estimated by using pilot/reference aided algorithms [76].

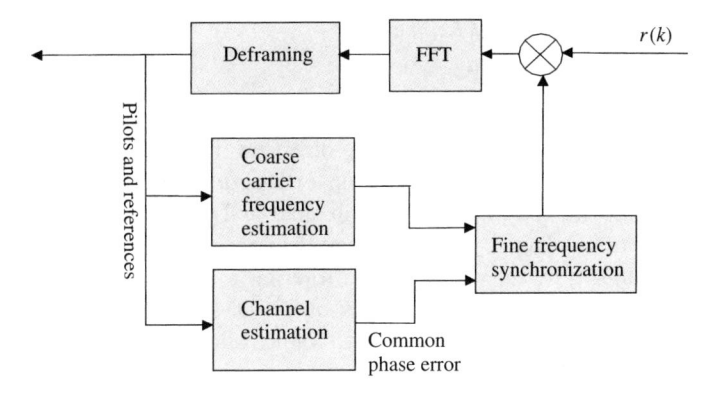

Figure 4-19 Frequency synchronization using reference symbols

Furthermore, as explained in Section 4.2.5.1, for fine frequency synchronization some other reference data (i.e., CAZAC sequences) can be used. Here, the correlation process in the frequency domain can be done over a limited number of sub-carrier frequencies (e.g., $\pm N_2$ sub-carriers).

As shown in Figure 4-19, channel estimation (see Section 4.3) can additionally deliver a common phase error estimation (see Section 4.7.1.3) which can be exploited for fine frequency synchronization.

4.2.6 Automatic Gain Control (AGC)

In order to maximize the input signal dynamic by avoiding saturation, the variation of the received signal field strength before FFT operation or before A/D conversion can be adjusted by an AGC function [12][76]. Two kinds of AGC can be implemented:

— *Controlling the time domain signal before A/D conversion*: First, in the digital domain, the average received power is computed by filtering. Then, the output signal is converted to analog (e.g., by a sigma-delta modulator) that controls the signal attenuation before the A/D conversion.
— *Controlling the time domain signal before FFT*: In the frequency domain the output of the FFT signal is analyzed and the result is used to control the signal before the FFT.

4.3 Channel Estimation

When applying receivers with coherent detection in fading channels, information about the channel state is required and has to be estimated by the receiver. The basic principle of pilot symbol aided channel estimation is to multiplex reference symbols, so-called pilot symbols, into the data stream. The receiver estimates the channel state information based on the received, known pilot symbols. The pilot symbols can be scattered in time and/or frequency direction in OFDM frames (see Figure 4-9). Special cases are either pilot tones which are sequences of pilot symbols in time direction on certain sub-carriers, or OFDM reference symbols which are OFDM symbols consisting completely of pilot symbols.

4.3.1 Two-Dimensional Channel Estimation

4.3.1.1 Two-Dimensional Filter

Multi-carrier systems allow channel estimation in two dimensions by inserting pilot symbols on several sub-carriers in the frequency direction in addition to the time direction with the intention to estimate the channel transfer function $H(f,t)$ [32][33][34][45]. By choosing the distances of the pilot symbols in time and frequency direction sufficiently small with respect to the channel coherence bandwidths, estimates of the channel transfer function can be obtained by interpolation and filtering.

The described channel estimation operates on OFDM frames where $H(f,t)$ is estimated separately for each transmitted OFDM frame, allowing burst transmission based on OFDM frames. The discrete frequency and time representation $H_{n,i}$ of the channel transfer function introduced in Section 1.1.6 is used here. The values $n = 0, \ldots, N_c - 1$ and $i = 0, \ldots, N_s - 1$ are the frequency and time indices of the fading process where N_c is the number of sub-carriers per OFDM symbol and N_s is the number of OFDM symbols per OFDM frame. The estimates of the discrete channel transfer function $H_{n,i}$ are denoted as $\hat{H}_{n,i}$. An OFDM frame consisting of 13 OFDM symbols, each with 11 sub-carriers, is shown as an example in Figure 4-20. The rectangular arrangement of the pilot symbols is referred to as a rectangular grid. The discrete distance in sub-carriers between two pilot symbols in frequency direction is N_f and in OFDM symbols in time direction is N_t. In the example given in Figure 4-20, N_f is equal to 5 and N_t is equal to 4.

The received symbols of an OFDM frame are given by

$$R_{n,i} = H_{n,i} S_{n,i} + N_{n,i}, \quad n = 0, \ldots, N_c - 1, \quad i = 0, \ldots, N_s - 1, \qquad (4.40)$$

where $S_{n,i}$ and $N_{n,i}$ are the transmitted symbols and the noise components, respectively.

The pilot symbols are written as $S_{n',i'}$, where the frequency and time indices at locations of pilot symbols are marked as n' and i'. Thus, for equally spaced pilot symbols we obtain

$$n' = pN_f, \quad p = 0, \ldots, \lceil N_c/N_f \rceil - 1 \qquad (4.41)$$

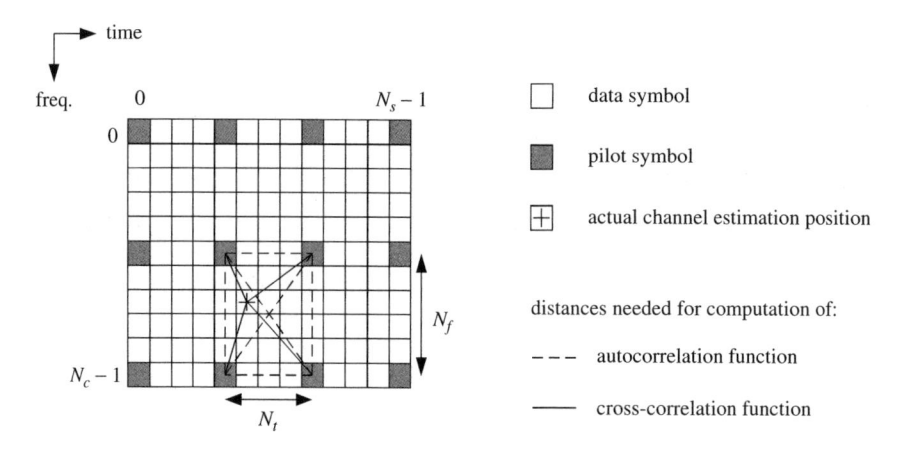

Figure 4-20 Pilot symbol grid for two-dimensional channel estimation

and
$$i' = qN_t, \quad q = 0, \ldots, \lceil N_s/N_t \rceil - 1, \qquad (4.42)$$

assuming that the first pilot symbol in the rectangular grid is located at the first sub-carrier of the first OFDM symbol in an OFDM frame. The number of pilot symbols in an OFDM frame results in

$$N_{grid} = \left\lceil \frac{N_c}{N_f} \right\rceil \left\lceil \frac{N_s}{N_t} \right\rceil. \qquad (4.43)$$

Pilot symbol aided channel estimation operates in two steps. In a first step, the initial estimate $\check{H}_{n',i'}$ of the channel transfer function at positions where pilot symbols are located is obtained by dividing the received pilot symbol $R_{n',i'}$ by the originally transmitted pilot symbol $S_{n',i'}$, i.e.,

$$\check{H}_{n',i'} = \frac{R_{n',i'}}{S_{n',i'}} = H_{n',i'} + \frac{N_{n',i'}}{S_{n',i'}}. \qquad (4.44)$$

In a second step, the final estimates of the complete channel transfer function belonging to the desired OFDM frame are obtained from the initial estimates $\check{H}_{n',i'}$ by two-dimensional interpolation or filtering. The two-dimensional filtering is given by

$$\hat{H}_{n,i} = \sum_{\{n',i'\} \in \Psi_{n,i}} \omega_{n',i',n,i} \check{H}_{n',i'}, \qquad (4.45)$$

where $\omega_{n',i',n,i}$ is the shift-variant two-dimensional impulse response of the filter. The subset $\Psi_{n,i}$ is the set of initial estimates $\check{H}_{n',i'}$ that is actually used for estimation of $\hat{H}_{n,i}$.

The number of filter coefficients is

$$N_{tap} = \|\Psi_{n,i}\| \leqslant N_{grid}. \qquad (4.46)$$

In the OFDM frame illustrated in Figure 4-20, N_{grid} is equal to 12 and N_{tap} is equal to 4.

Two-Dimensional Wiener Filter
The criterion for the evaluation of the channel estimator is the mean square value of the estimation error

$$\varepsilon_{n,i} = H_{n,i} - \hat{H}_{n,i}. \qquad (4.47)$$

The mean square error is given by

$$J_{n,i} = E\left\{|\varepsilon_{n,i}|^2\right\}. \qquad (4.48)$$

The optimal filter in the sense of minimizing $J_{n,i}$ with the minimum mean square error criterion is the two-dimensional Wiener filter. The filter coefficients of the two-dimensional Wiener filter are obtained by applying the orthogonality principle in linear mean square estimation,

$$E\left\{\varepsilon_{n,i} \check{H}^*_{n'',i''}\right\} = 0, \quad \forall \{n'', i''\} \in \Psi_{n,i}. \qquad (4.49)$$

The orthogonality principle states that the mean square error $J_{n,i}$ is minimum if the filter coefficients $\omega_{n',i',n,i}, \forall \{n', i'\} \in \Psi_{n,i}$ are selected such that the error $\varepsilon_{n,i}$ is orthogonal to all initial estimates $\check{H}^*_{n'',i''}, \forall \{n'', i''\} \in \Psi_{n,i}$. The orthogonality principle leads to the Wiener–Hopf equation, which states that

$$E\left\{H_{n,i} \check{H}^*_{n'',i''}\right\} = \sum_{\{n',i'\}\in\Psi_{n,i}} \omega_{n',i',n,i} E\left\{\check{H}_{n',i'} \check{H}^*_{n'',i''}\right\}, \quad \forall \{n'', i''\} \in \Psi_{n,i}. \quad (4.50)$$

With (4.44) and by assuming that $N_{n'',i''}$ has zero mean and is statistically independent from the pilot symbols $S_{n'',i''}$, the cross-correlation function $E\left\{H_{n,i} \check{H}^*_{n'',i''}\right\}$ is equal to the discrete time-frequency correlation function $E\left\{H_{n,i} H^*_{n'',i''}\right\}$, i.e., the cross-correlation function is given by

$$\theta_{n-n'',i-i''} = E\left\{H_{n,i} H^*_{n'',i''}\right\}. \quad (4.51)$$

The autocorrelation function in (4.50) is given by

$$\phi_{n'-n'',i'-i''} = E\left\{\check{H}_{n',i'} \check{H}^*_{n'',i''}\right\}. \quad (4.52)$$

When assuming that the mean energy of all symbols $S_{n,i}$ including pilot symbols is equal, the autocorrelation function can be written in the form

$$\phi_{n'-n'',i'-i''} = \theta_{n'-n'',i'-i''} + \sigma^2 \delta_{n'-n'',i'-i''}. \quad (4.53)$$

The cross-correlation function depends on the distances between the actual channel estimation position n, i and all pilot positions n'', i'', whereas the autocorrelation function depends only on the distances between the pilot positions and, hence, is independent of the actual channel estimation position n, i. Both relations are illustrated in Figure 4-20.

Inserting (4.51) and (4.52) into (4.53) yields, in vector notation,

$$\boldsymbol{\theta}^T_{n,i} = \boldsymbol{\omega}^T_{n,i} \boldsymbol{\Phi}, \quad (4.54)$$

where $\boldsymbol{\Phi}$ is the $N_{tap} \times N_{tap}$ autocorrelation matrix and $\boldsymbol{\theta}_{n,i}$ is the cross-correlation vector of length N_{tap}. The vector $\boldsymbol{\omega}_{n,i}$ of length N_{tap} represents the filter coefficients $\omega_{n',i',n,i}$ required to obtain the estimate $\hat{H}_{n,i}$. Hence, the filter coefficients of the optimum two-dimensional Wiener filter are

$$\boldsymbol{\omega}^T_{n,i} = \boldsymbol{\theta}^T_{n,i} \boldsymbol{\Phi}^{-1}, \quad (4.55)$$

when assuming that the auto- and cross-correlation functions are perfectly known and $N_{tap} = N_{grid}$. Since in practice the autocorrelation function $\boldsymbol{\Phi}$ and cross-correlation function $\boldsymbol{\theta}_{n,i}$ are not perfectly known in the receiver, estimates or assumptions about these correlation functions are necessary in the receiver.

4.3.1.2 Two Cascaded One-Dimensional Filters

Two-dimensional filters tend to have a large computational complexity. The choice of two cascaded one-dimensional filters working sequentially can give a good trade-off between performance and complexity. The principle of two cascaded one-dimensional filtering is depicted in Figure 4-21. Filtering in frequency direction on OFDM symbols containing pilot symbols, followed by filtering in time direction on all sub-carriers is shown. This ordering is chosen to enable filtering in frequency direction directly after receiving a pilot symbol

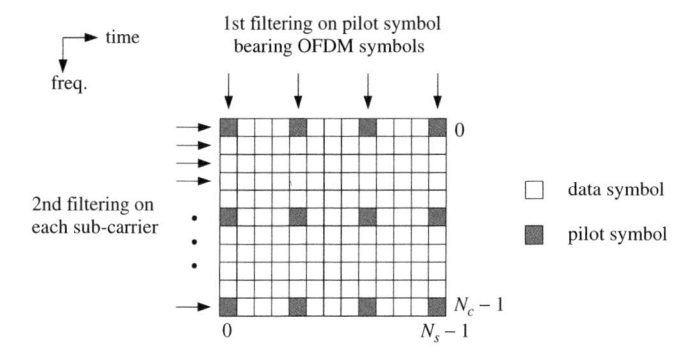

Figure 4-21 Two cascaded one-dimensional filter approach

bearing OFDM symbol and, thus, to reduce the overall filtering delay. However, the opposite ordering would achieve the same performance due to the linearity of the filters.

The mean square error of the two cascaded one-dimensional filters working sequentially is obtained in two steps. Values and functions related to the first filtering are marked with the index [1] and values and functions related to the second filtering are marked with the index [2]. The estimates delivered by the first one-dimensional filter are

$$\hat{H}_{n,i'}^{[1]} = \sum_{\{n',i'\} \in \Psi_{n,i'}} \omega_{n',n}^{[1]} \check{H}_{n',i'}. \qquad (4.56)$$

The filter coefficients $\omega_{n',n}^{[1]}$ only depend on the frequency index n. This operation is performed in all $\lceil N_s/N_t \rceil$ pilot symbol bearing OFDM symbols. The estimates delivered by the second one-dimensional filter are

$$\hat{H}_{n,i} = \hat{H}_{n,i}^{[2]} = \sum_{\{n,i'\} \in \Psi_{n,i}} \omega_{i',i}^{[2]} \hat{H}_{n,i'}^{[1]}. \qquad (4.57)$$

The filter coefficients $\omega_{i',i}^{[2]}$ only depend on the time index i. The estimates $\hat{H}_{n,i'}^{[1]}$ obtained from the first filtering are used as pilot symbols for the second filtering on sub-carrier n. The second filtering is performed on all N_c sub-carriers.

4.3.2 One-Dimensional Channel Estimation

One-dimensional channel estimation can be considered a special case of two-dimensional channel estimation, where the second dimension is omitted. These schemes require a higher overhead on pilot symbols, since the correlation of the fading in the second dimension is not exploited in the filtering process. The overhead on pilot symbols with one-dimensional channel estimation Λ_{1D} compared to two-dimensional channel estimation Λ_{2D} is

$$\Lambda_{1D} = N_f \Lambda_{2D} \qquad (4.58)$$

with one-dimensional filtering in time direction or

$$\Lambda_{1D} = N_t \Lambda_{2D} \qquad (4.59)$$

with one-dimensional filtering in frequency direction.

4.3.3 Filter Design

4.3.3.1 Adaptive Design

A filter is designed by determining the filter coefficients $\omega_{n,i}$. In the following, two-dimensional filtering is considered. The filter coefficients for one-dimensional filters are obtained from the two-dimensional filter coefficients by omitting the dimension which is not required in the corresponding one-dimensional filter. The two-dimensional filter coefficients can be calculated, given the discrete time-frequency correlation function of the channel $\theta_{n-n'',i-i''}$ and the variance of the noise σ^2. In the mobile radio channel, it can be assumed that the delay power density spectrum $\rho(\tau)$ and the Doppler power density spectrum $S(f_D)$ are statistically independent. Thus, the time-frequency correlation function $\theta_{n-n'',i-i''}$ can be separated in the frequency correlation function $\theta_{n-n''}$ and the time correlation function $\theta_{i-i''}$. Hence, the optimum filter has to adapt the filter coefficients to the actual power density spectra $\rho(\tau)$ and $S(f_D)$ of the channel. The resulting channel estimation error can be minimized with this approach since the filter mismatch can be minimized. Investigations with adaptive filters show significant performance improvements with adaptive filters [56][60]. Of importance for the adaptive filter scheme is that the actual power density spectra of the channel should be estimated with high accuracy, low delay and reasonable effort.

4.3.3.2 Robust Non-Adaptive Design

A low-complex selection of the filter coefficients is to choose a fixed set of filter coefficients which is designed such that a great variety of power density spectra with different shapes and maximum values is covered [34][45]. No further adaptation to the time-variant channel statistics is performed during the estimation process. A reasonable approach is to adapt the filters to uniform power density spectra. By choosing the filter parameter τ_{filter} equal to the maximum expected delay of the channel τ_{max}, the normalized delay power density spectrum used for the filter design is given by

$$\rho_{filter}(\tau) = \begin{cases} \dfrac{1}{\tau_{filter}} & |\tau| < \dfrac{\tau_{filter}}{2} \\ 0 & \text{otherwise} \end{cases}. \tag{4.60}$$

Furthermore, by choosing the filter parameter $f_{D,filter}$ equal to the maximum expected Doppler frequency of the channel $f_{D,max}$, the normalized Doppler power density spectrum used for the filter design is

$$S_{filter}(f_D) = \begin{cases} \dfrac{1}{2f_{D,filter}} & |f_D| < f_{D,filter} \\ 0 & \text{otherwise} \end{cases} \tag{4.61}$$

With the selection of uniform power density spectra, the discrete frequency correlation functions results in

$$\theta_{n-n''} = \frac{\sin(\pi \tau_{filter}(n - n'')F_s)}{\pi \tau_{filter}(n - n'')F_s} = \operatorname{sinc}(\pi \tau_{filter}(n - n'')F_s) \tag{4.62}$$

Channel Estimation

and the discrete time correlation function yields

$$\theta_{i-i''} = \frac{\sin(2\pi f_{D,filter}(i-i'')T_s')}{2\pi f_{D,filter}(i-i'')T_s'} = \text{sinc}(2\pi f_{D,filter}(i-i'')T_s'). \quad (4.63)$$

The autocorrelation function is obtained according to (4.53).

4.3.4 Implementation Issues
4.3.4.1 Pilot Distances

In this section, the arrangement of the pilot symbols in an OFDM frame is shown by applying the two-dimensional sampling theorem. The choice of a rectangular grid is motivated by the results presented in [34] where channel estimation performance with rectangular, diagonal, and random grids is investigated. Channel estimation either with rectangular or diagonal grid shows similar performance but outperforms channel estimation with a random grid.

Given the normalized filter bandwidths $\tau_{filter} F_s$ and $f_{D,filter} T_s'$, the sampling theorem requires that the distance of the pilot symbols in the frequency direction is

$$N_f \leq \frac{1}{\tau_{filter} F_s}, \quad (4.64)$$

and in the time direction is

$$N_t \leq \frac{1}{2 f_{D,filter} T_s'}. \quad (4.65)$$

An optimum sampling of the channel transfer function is given by a balanced design which guarantees that the channel is sampled in time and in frequency with the same sampling rate. A balanced design is defined as [33]

$$N_f \tau_{filter} F_s \approx 2 N_t f_{D,filter} T_s'. \quad (4.66)$$

A practically proven value of the sampling rate is the selection of approximately two-times oversampling to achieve a reasonably low complexity with respect to the filter length and performance. A practical hint concerning the performance of the channel estimation is to design the pilot grid such that the first and the last OFDM symbol and sub-carrier, respectively, in an OFDM frame contain pilot symbols (see Figure 4-20). This avoids the channel estimation having to perform channel prediction, which is more unreliable than interpolation. In the special case that the downlink can be considered as a broadcasting scenario with continuous transmission, it is possible to continuously use pilot symbols in the time direction without requiring additional pilots at the end of an OFDM frame.

4.3.4.2 Overhead

Besides the mean square error of a channel estimation, one criterion for the efficiency of a channel estimation is the overhead and the loss in SNR due to pilot symbols. The overhead due to pilot symbols is given by

$$\Lambda = \frac{N_{\text{grid}}}{N_c N_s} \quad (4.67)$$

and the SNR loss in dB is defined as

$$V_{\text{pilot}} = 10 \log_{10}\left(\frac{1}{1-\Lambda}\right). \tag{4.68}$$

4.3.4.3 Signal-to-Pilot Power Ratio

The pilot symbols $S_{n',i'}$ can be transmitted with higher average energy than the data-bearing symbols $S_{n,i}$, $n \neq n'$, $i \neq i'$. Pilot symbols with increased energy are called *boosted pilot symbols* [16]. The boosting of pilot symbols is specified in the European DVB-T standard and achieves better estimates of the channel but reduces the average SNR of the data symbols. The choice of an appropriate boosting level for the pilot symbols is investigated in [34].

4.3.4.4 Complexity and Performance Aspects

A simple alternative to optimum Wiener filtering is a DFT-based channel estimator illustrated in Figure 4-22. The channel is first estimated in the frequency domain on the N_{pilot} sub-carriers where pilots have been transmitted on. In the next step, this N_{pilot} estimates are transformed with an N_{pilot} point IDFT in the time domain and the resulting time sequence can be weighted before it is transformed back in the frequency domain with an N_c point DFT. As long as $N_{\text{pilot}} < N_c$, the DFT-based channel estimation performs interpolation. The DFT-based channel estimation can also be applied in two dimensions where the time direction is processed in the same way as described above for the frequency direction. An appropriate weighting function between IDFT and DFT applies the minimum mean square error criterion [84].

The application of singular value decomposition within channel estimation is an approach to reduce the complexity of the channel estimator as long as an irreducible error floor can be tolerated in the system design [13].

The performance of pilot symbol based channel estimation concepts can be further improved by iterative channel estimation and decoding, where reliable decisions obtained from the decoding are exploited for channel estimation. A two-dimensional implementation is possible with estimation in the time and frequency directions [74].

When assumptions about the channel can be made in advance, the channel estimator performance can be improved. An approach is based on a parametric channel model where the channel frequency response is estimated by using an N_p path channel model [85]. The

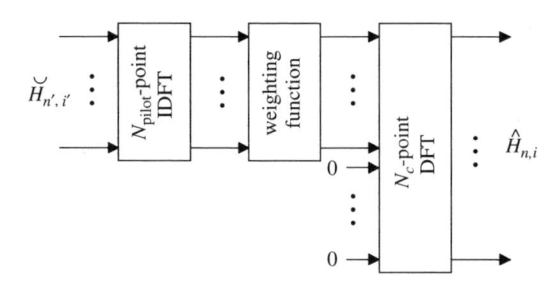

Figure 4-22 Low-complexity DFT based channel estimation

Channel Estimation

parametric channel modeling requires additional complexity to estimate the multipath time delays.

4.3.5 Performance Analysis

In this section, the mean square error performance of pilot symbol aided channel estimation in multi-carrier systems is shown. To evaluate and optimize the channel estimation, a multi-carrier reference scenario typical for mobile radio systems is defined. The frequency band has a bandwidth of $B = 2$ MHz and is located at a carrier frequency of $f_c = 1.8$ GHz. The OFDM operation and its inverse are achieved with an IFFT and FFT, respectively, of size 512. The considered multi-carrier transmission scheme processes one OFDM frame per estimation cycle. An OFDM frame consists of $N_s = 24$ OFDM symbols. The OFDM frame duration results in $T_{fr} = 6.6$ ms. The filter parameters are chosen as $\tau_{filter} = 20$ µs and $f_{D,filter} = 333.3$ Hz, where $f_{D,filter}$ corresponds to a velocity of $v = 200$ km/h. A balanced design with approximately two-times oversampling is given with a pilot symbol spacing of $N_f = 6$ in the frequency direction and of $N_t = 3$ in the time direction. The resulting system parameters are summarized in Table 4-1. The chosen OFDM frame structure with pilot symbol aided channel estimation in two dimensions is illustrated in Figure 4-23.

In time direction, the last OFDM symbol with pilot symbols of the previous OFDM frame can be used for filtering. With a pilot spacing of 6 in the frequency direction, while starting and ending with a sub-carrier containing pilot symbols, a number of 511 used sub-carriers per OFDM symbol is obtained and is considered in the following. The resulting overhead Λ due to pilot symbols is 5.6%. With pilot symbols and data symbols having the same average energy, the loss in SNR V_{pilot} due to pilot symbols is only 0.3 dB in the defined multi-carrier scenario.

The performance criterion for the evaluation of the channel estimation is the mean square error which is averaged over an OFDM frame. Thus, edge effects are also taken

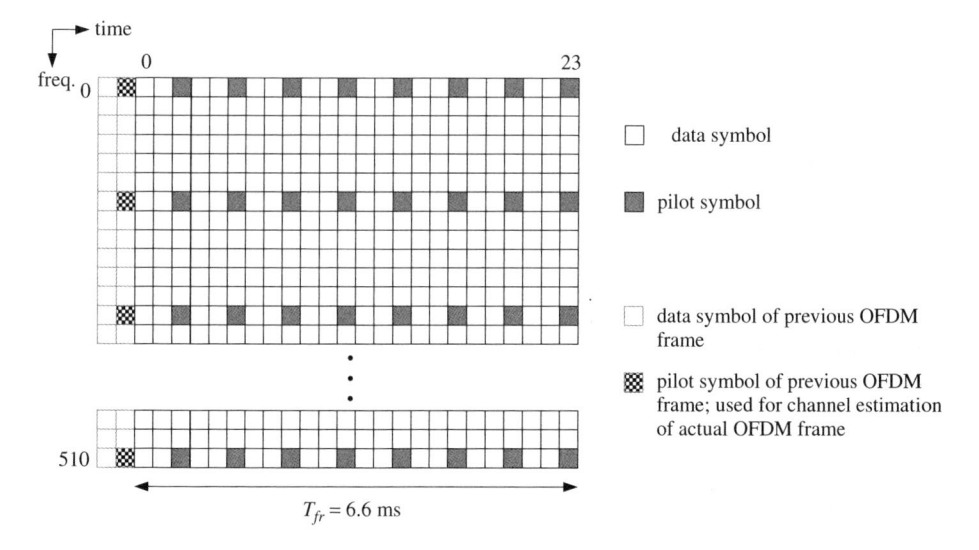

Figure 4-23 OFDM frame with pilot grid for channel estimation in two dimensions

Table 4-1 Parameters for pilot symbol aided channel estimation in two dimensions

Parameter	Value
Bandwidth	$B = 2$ MHz
Carrier frequency	$f_c = 1.8$ GHz
OFDM frame duration	$T_{fr} = 6.6$ ms
OFDM symbols per OFDM frame	$N_s = 24$
FFT size	512
OFDM symbol duration	$T_s = 256$ µs
Cyclic prefix duration	$T_g = 20$ µs
Sub-carrier spacing	$F_s = 3.9$ kHz
Number of used sub-carriers	511
Pilot symbol distance in frequency direction	$N_f = 6$
Pilot symbol distance in time direction	$N_t = 3$
Delay filter bandwidth	$\tau_{filter} = 20$ µs
Doppler filter bandwidth	$f_{D,filter} = 333.3$ Hz
Filter characteristic	Wiener filter with and without model mismatch

into account. The information about the variance σ^2 required for the calculation of the autocorrelation matrix $\boldsymbol{\Phi}$ is assumed to be known perfectly at the receiver unless otherwise stated. In practice, the variance σ^2 can be estimated by transmitting a null symbol without signal energy at the beginning of each OFDM frame (see Section 4.2.1). Alternatively, the autocorrelation function can be optimized for the highest σ^2 at which successful data transmission should be possible.

The mean square error of two-dimensional (2-D) filtering without model mismatch and two cascaded one-dimensional (2×1-D) filtering applied in a multi-carrier system is presented and compared in the following. The mobile radio channel used has a uniform delay power spectrum with $\tau_{max} = 20$ µs and a uniform Doppler power density spectrum with $f_{D,max} = 333.3$ Hz.

In Figure 4-24, the mean square error (MSE) versus the SNR for 2-D filtering without model mismatch with different numbers of filter taps is shown. The corresponding results for 2×1-D filtering are presented in Figure 4-25. It can be observed in both figures that the mean square error decreases with increasing numbers of taps.

Channel Estimation

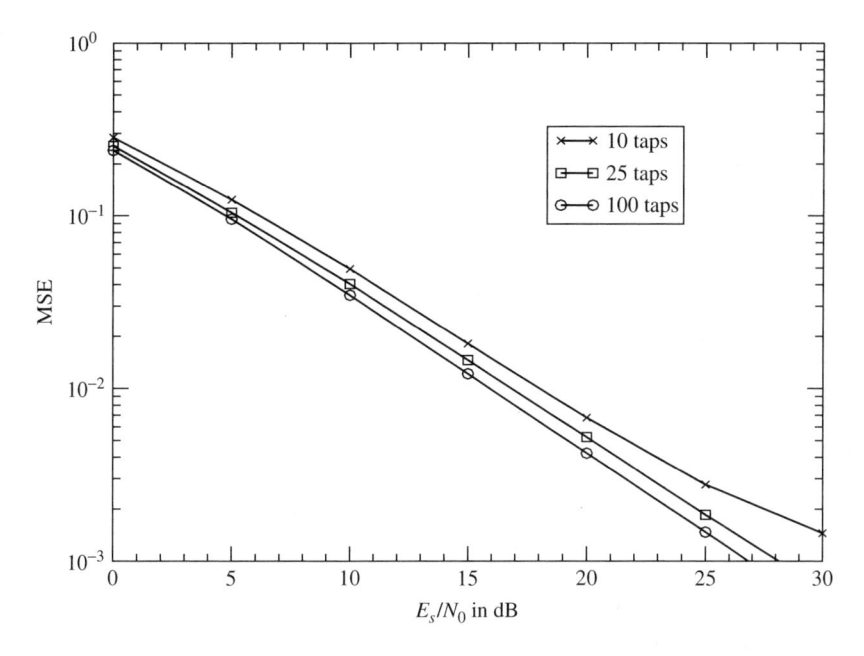

Figure 4-24 MSE for 2-D channel estimation with different numbers of filter taps; no model mismatch

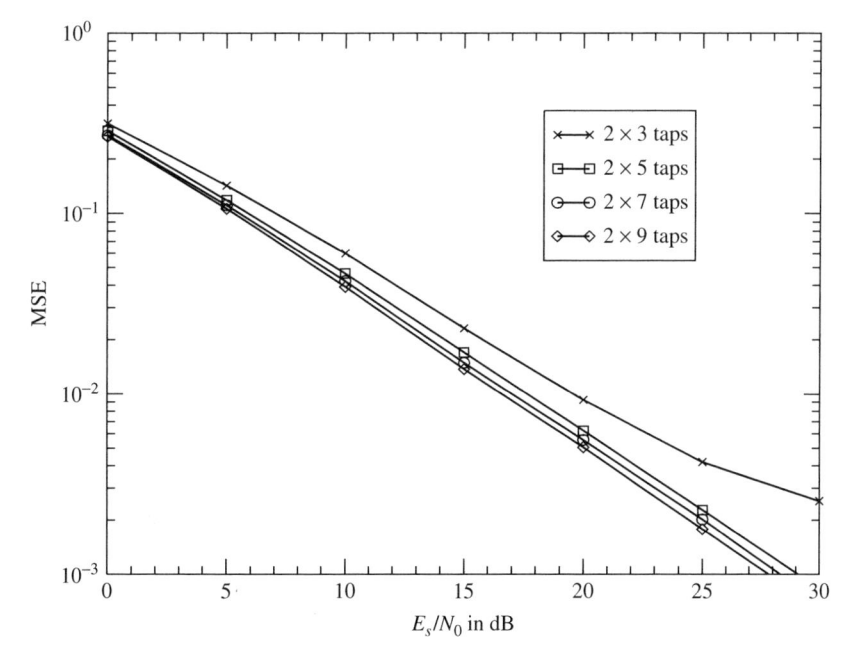

Figure 4-25 MSE for 2 × 1D channel estimation with different numbers of filter taps; no model mismatch

The mean square error presented with 2-D filtering using 100 taps can be considered as a lower bound. In the case of 2-D filtering, 25 taps seems to be reasonable with respect to mean square error performance and complexity. In the case of $2 \times$ 1-D filtering, 2×5 taps is a reasonable choice. A further increase of the number of taps only reduces the mean square error slightly. Moreover, with 2×5 taps, the performance with $2 \times$ 1-D filtering is similar to the performance with 2-D filtering with 25 taps. Based on these results, $2 \times$ 1-D filtering with 2×5 taps is chosen for channel estimation in this section.

In the following, the focus is on degradation due to model mismatch in the filter design. In a first step, the mean square error of the defined channel estimation in different COST 207 channel models versus the SNR is shown in Figure 4-26 for $2 \times$ 1-D filtering with 2×5 taps. The velocity of the terminal station is equal to 3 km/h. As a reference, the mean square error curve without model mismatch is given. The presented channel estimation provides a better performance in channels with large delay spread than in channels with low delay spread. The reason for this effect is that a channel with a delay power density spectrum which matches more closely with that chosen for the filter design also matches more closely with performance without model mismatch and, thus, shows a better mean square error performance.

In the second step, the influence of different Doppler power density spectra is considered. Furthermore, the effect when the condition of two-times oversampling is not fulfilled in mobile radio channels with Doppler frequencies larger than 333.3 Hz is investigated. Figure 4-27 shows the mean square error of the presented channel estimation with $2 \times$ 1-D filtering and 2×5 taps for different velocities v of the terminal station in the

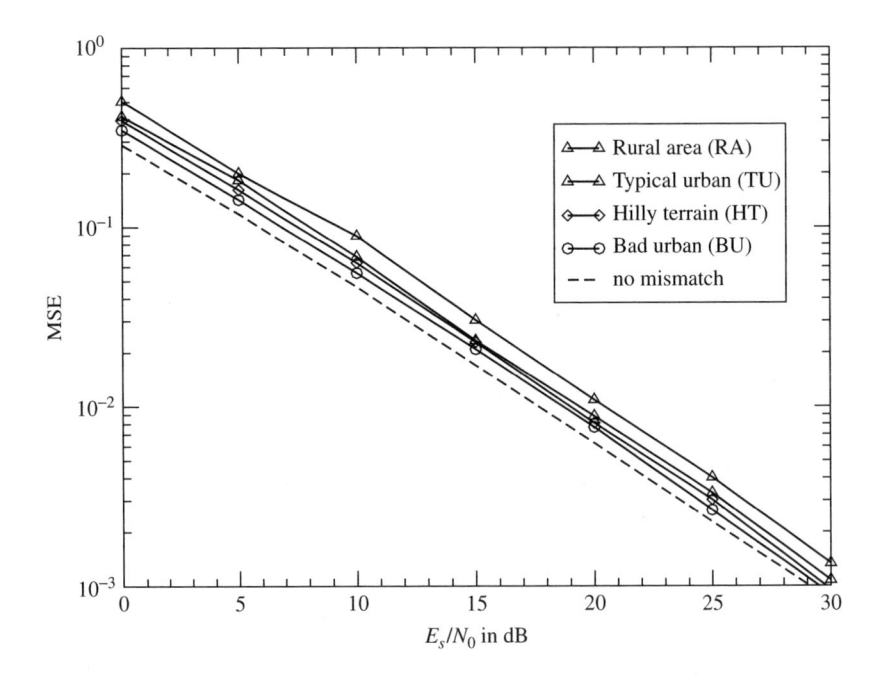

Figure 4-26 MSE for $2 \times$ 1D channel estimation with model mismatch; $v = 3$ km/h

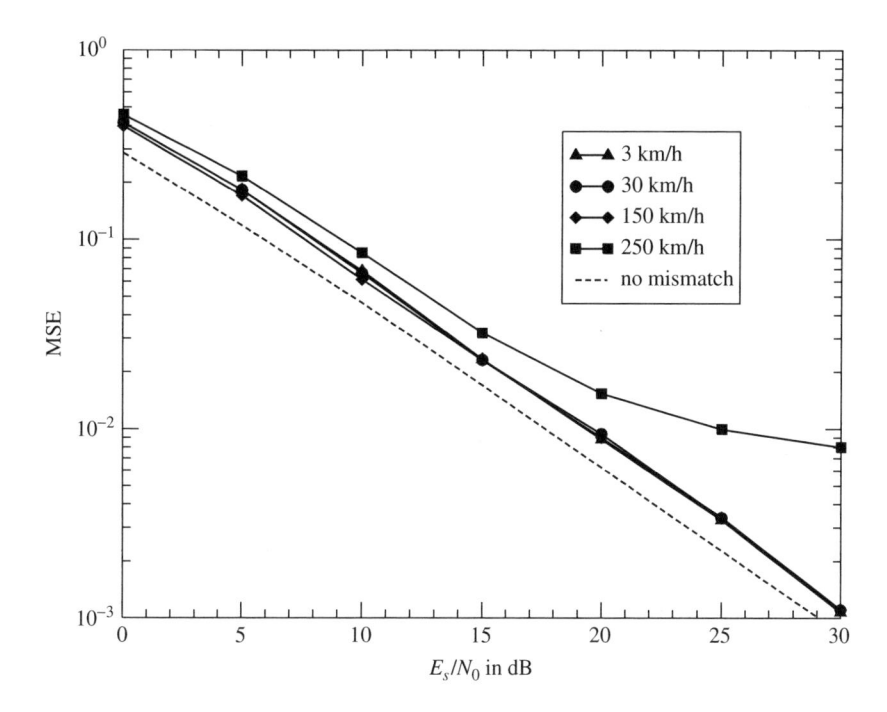

Figure 4-27 MSE for 2×1D channel estimation with model mismatch; COST 207 channel model HT

COST 207 channel model HT versus the SNR. As a reference, the mean square error curve without model mismatch is given. It can be observed that there is approximately no change in the mean square error with different velocities v of the mobile station as long as two-times oversampling of the fading process is guaranteed. For the defined multi-carrier system, two-times oversampling is given for a velocity of 200 km/h. As soon as the rule of two-times oversampling is not fulfilled, shown for velocity of $v = 250$ km/h, the mean square error considerably increases.

Figure 4-28 shows the mean square error obtained with the proposed channel estimation using 2×1-D filtering and 2×5 taps when the autocorrelation matrix $\mathbf{\Phi}$ is optimized for an SNR of 10 dB. During the channel estimation no information about the actual variance σ^2 is used to optimally adapt the filter coefficients. It can be seen that accurate results are obtained in a span of about 10 dB with mean at 10 dB. For higher SNRs, the mean square error obtained with the simplified channel estimation flattens out. Thus, the application of the simplified channel estimation which requires no estimation of σ^2 depends on whether the error floor is acceptable or not. Otherwise, it depends on the SNR dynamic range at the input of the receiver whether the simplified channel estimation is applicable or not.

4.3.6 Time Domain Channel Estimation

Channel estimation for OFDM systems can be applied in the time domain, as in single-carrier systems. The basic principle is to periodically insert PN sequences in the time

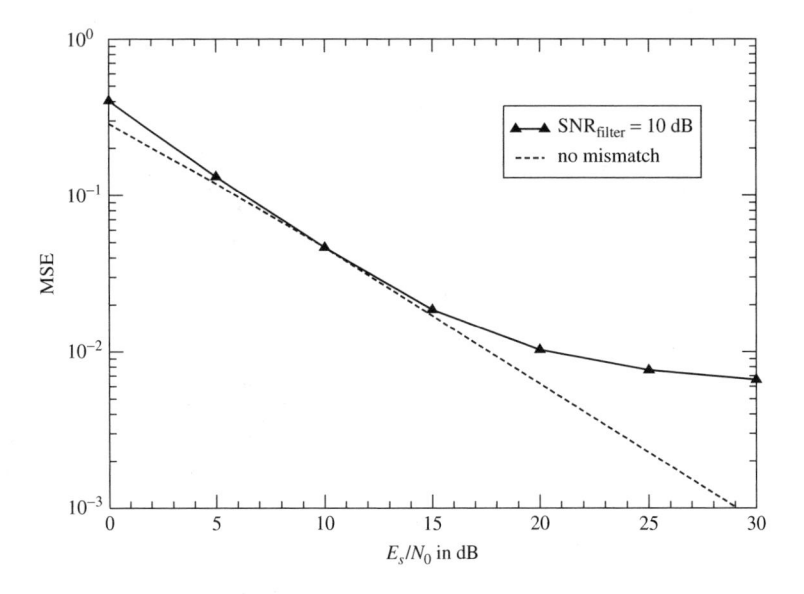

Figure 4-28 MSE for $2 \times 1D$ channel estimation with σ^2 mismatch

domain between OFDM symbols where the channel impulse response is obtained by correlation in the time domain [86][87]. In OFDM systems, the signal used for time domain channel estimation can be generated in the transmitter in the frequency domain. The insertion of equal spaced pilot symbols in the frequency domain results in an OFDM symbol which can be considered as split into several time slots, each with a pilot. Assuming that the impulse response of each time slot is identical, the channel response in the time domain can be obtained by averaging the impulse responses of the time slots [53][86]. The cost of time domain channel estimation compared to frequency domain channel estimation is its higher complexity. Time domain channel estimation can be combined with time domain channel equalization [8].

4.3.7 Decision Directed Channel Estimation

Under the assumption that the channel is quasi-stationary over two OFDM symbols, the decisions from the previous OFDM symbol can be used for data detection in the current OFDM symbol. These schemes can outperform differentially modulated schemes and are efficient if FEC coding is included in the reliable reconstruction of the transmitted sequence [51]. The principle of decision directed channel estimation including FEC decoding is illustrated in Figure 4-29.

In the first estimation step (start), reference symbols have to be used for the initial channel estimation. These reference symbols are the symbols $S_{n,0}$, $n = 0, \ldots, N_c - 1$, of the initial OFDM symbol. The initial estimation of the channel transfer function is given by

$$\hat{H}_{n,0} = \frac{R_{n,0}}{S_{n,0}} = H_{n,0} + \frac{N_{n,0}}{S_{n,0}}. \tag{4.69}$$

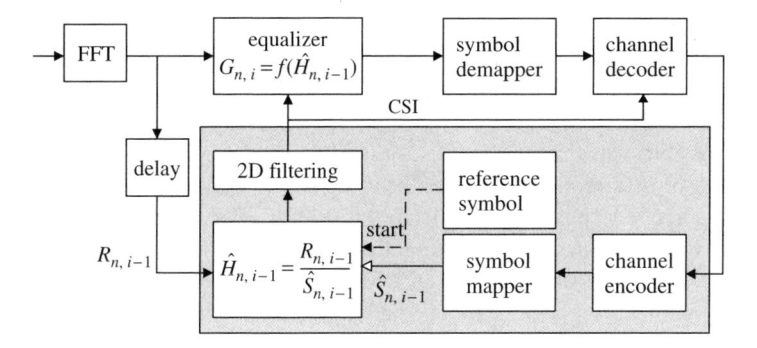

Figure 4-29 Decision directed channel estimation

The channel can be oversampled in the frequency and time direction with respect to filtering, since all re-encoded data symbols can be used as pilot symbols. Thus, the channel transfer function can be low-pass filtered in the frequency and/or time direction in order to reduce the noise. Without filtering in either frequency or time domain, the BER performance of differential modulation is obtained. Based on the obtained channel estimate $\hat{H}_{n,i-1}$, the received symbols $R_{n,i}$ are equalized according to

$$U_{n,i} = G_{n,i} R_{n,i}, \qquad (4.70)$$

where

$$G_{n,i} = f(\hat{H}_{n,i-1}) \qquad (4.71)$$

is an equalization coefficient depending on the previous $\hat{H}_{n,i-1}$. In order to achieve a reliable estimation of the transmitted sequence, the error correction capability of the channel coding is used. The bit sequence at the output of the channel decoder is re-encoded and remodulated to obtain reliable estimates $\hat{S}_{n,i-1}$ of the transmitted sequence $S_{n,i-1}$. These estimates are fed back to obtain a channel estimate $\hat{H}_{n,i-1}$, which is used for the detection of the following symbol $S_{n,i}$. With decision directed channel estimation the channel estimates are updated symbol-by-symbol.

Decision directed channel estimation can reduce the amount of reference data required to only an initial OFDM reference symbol. This can significantly reduce the overhead due to pilot symbols at the expense of additional complexity [26]. The achievable estimation accuracy with decision directed channel estimation including FEC decoding is comparable to pilot symbol aided channel estimation and with respect to BER outperforms classical differential demodulation schemes [51]. Classical differential modulation schemes can also benefit from decision directed channel estimation where the correlations in time and frequency of previously received symbols are filtered and used for estimation of the actual $\hat{H}_{n,i}$.

4.3.8 Blind and Semi-Blind Channel Estimation

The cyclic extension of an OFDM symbol can be used as an inherent reference signal within the data, enabling channel estimation based on the cyclic extension [31][59][83]. Since no additional pilot symbols are required with this method in OFDM schemes, this can be considered blind channel estimation. The advantage of blind channel estimation based on the cyclic prefix is that this channel estimation concept is standard-compliant and can be applied to all commonly used OFDM systems which use a cyclic prefix.

A further approach of blind detection without the necessity of pilot symbols for coherent detection is possible when joint equalization and detection is applied. This is possible by trellis decoding of differentially encoded PSK signals [49] where the trellis decoding can efficiently be achieved applying the Viterbi algorithm. The differential encoding can be performed in the frequency or time direction, while the detector exploits correlations between adjacent sub-carriers and/or OFDM symbols. These blind detection schemes require a low number of pilot symbols and outperform classical differential detection schemes. The complexity of a blind scheme with joint equalization and detection is higher than that of differential or coherent receivers due to the additional implementation of the Viterbi algorithm.

Statistical methods for blind channel estimations have also been proposed [31] which, however, require several OFDM symbols to be able to estimate the channel and might fail in mobile radio channels with fast fading. Blind channel estimation concepts exploiting the feature that the transmitted data are confined to a finite alphabet set can perform channel estimation from a single OFDM symbol [88]. Modifications of the finite alphabet approach attempt to reduce the enormous computational effort of these schemes [41].

Semi-blind channel estimation takes advantage of pilot symbols which are included in the data stream. E.g., most existing OFDM systems have pilot symbols multiplexed in the data streams such that these symbols in combination with blind algorithms result in semi-blind schemes with improved estimation accuracy.

4.3.9 *Channel Estimation in MC-SS Systems*
4.3.9.1 Downlink
MC-CDMA

The synchronous downlink of MC-CDMA systems is a broadcast scenario where all K users can exploit the same pilot symbols within an OFDM frame [39]. The power of the common pilot symbols has to be adjusted such that the terminal station with the most critical channel conditions is able to estimate the channel. So, it is possible to realize a pilot symbol scheme with adaptive power adjustment.

The performance of an MC-CDMA mobile radio system with two-dimensional channel estimation in the downlink is shown in Figure 4-30. The transmission bandwidth is 2 MHz and the carrier frequency is located at 1.8 GHz. The number of sub-carriers is 512. MMSE equalization is chosen as a low-complex detection technique in the terminal station and the system is fully loaded. The system parameters are given in Table 4-1 and the pilot symbol grid corresponds to the structure shown in Figure 4-23. The COST 207 channel models have been chosen as propagation models. It can be observed that two-dimensional channel estimation can handle different propagation scenarios with high velocities up to 250 km/h. For the chosen scenario, the SNR degradation compared to perfect channel estimation is in the order of 2 dB.

4.3.9.2 Uplink
Channel estimation in the uplink requires separate sets of pilot symbols for each user for the estimation of the user specific channel state information. This leads to a much

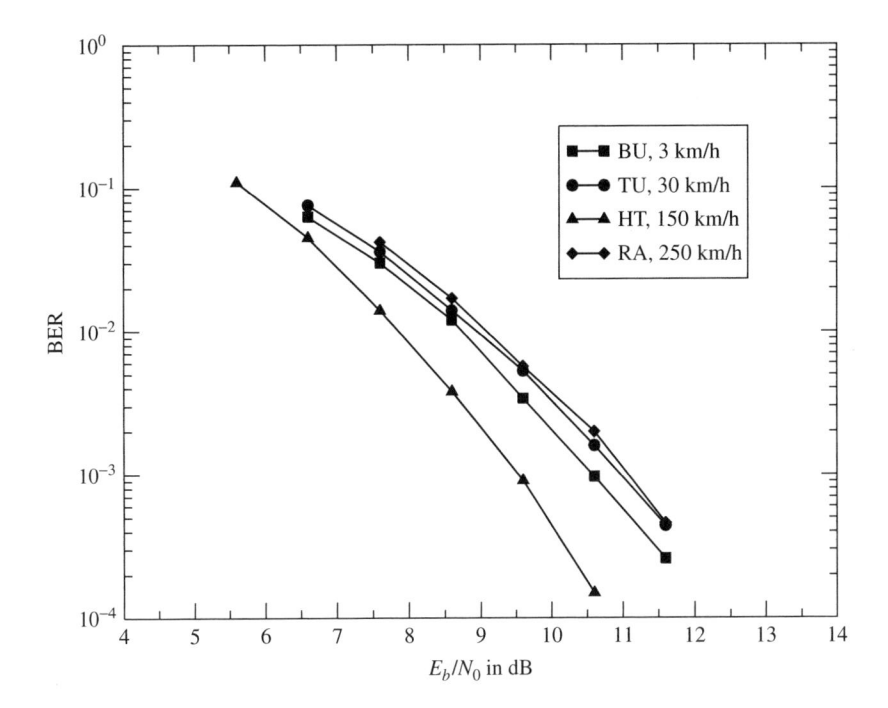

Figure 4-30 BER versus SNR of MC-CDMA with pilot symbol aided channel estimation in the downlink

higher overhead in pilot symbols in the uplink compared to the downlink. The increase of the overhead in the uplink is proportional to the number of users K. Alternatives to this approach are systems with pre-equalization in the uplink (see Section 2.1.6).

MC-CDMA

In synchronous uplinks of MC-CDMA systems, the channel impulse responses of the different uplink channels can be estimated by assigning each user an exclusive set of pilot symbol positions which the user can exploit for channel estimation. This can either be a one- or two-dimensional channel estimation per user. The channel estimation concept per user is identical to the concept in the downlink used for all users. In mobile radio channels with high time and frequency selectivity, the possible number of active users can be quite small, due to the high number of required pilot symbols.

Alternatively, a pilot symbol design in the time domain is possible by using for each user a different fraction of an OFDM symbol [77]. These methods have restrictions on the maximum length of the channel impulse response τ_{max}. When fulfilling the condition

$$\tau_{max} \leq \frac{T_s}{K}, \qquad (4.72)$$

the system is able to estimate K different channel impulse responses within one OFDM symbol. Since the OFDM system design typically results in an OFDM symbol duration T_s much larger than the guard interval T_g and with that than τ_{max}, several users can be estimated in one OFDM symbol. The principle of time domain channel estimation in

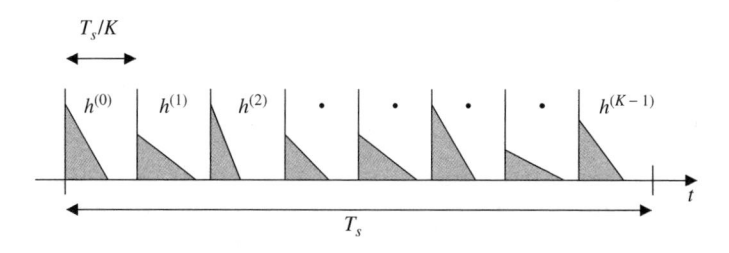

Figure 4-31 MC-CDMA uplink channel estimation

the uplink of an MC-CDMA system with 8 users is illustratively shown in Figure 4-31, where $h^{(k)}$ is the channel impulse response $h^{(k)}(\tau,t)$ of user k.

The structure for channel estimation shown in Figure 4-31 is obtained by inserting pilot symbols in an OFDM symbol and rotating the phase of the complex pilot symbols according to the user index. This results in a delayed version of the time reference signal for each user.

Since whole OFDM symbols are used for this type of channel estimation, the overhead in pilot symbols is comparable to that of scattering pilots in the frequency domain, as described at the beginning of this section.

Additionally, filtering in the time direction can be introduced, which reduces the overhead on pilot symbols. The distance between OFDM symbols with reference information is N_t OFDM symbols. Depending on the system complexity, the filter in the time direction can be designed as described in Section 4.3.3.

Alternatively, in a synchronous uplink, a single-carrier modulated midamble can be inserted between OFDM symbols of an OFDM frame, such that classical single-carrier multiuser channel estimation concepts can be applied [3].

Another approach is to use differential modulation in an MC-CDMA uplink to overcome the problem with channel estimation. Using chip- and symbol-level differential encoding for the uplink has been investigated [80]. The results show that the capacity is limited to moderate system loads and that the performance of this approach strongly depends on the propagation environment.

SS-MC-MA

In the following, the performance of an SS-MC-MA mobile radio system with pilot symbol aided channel estimation in the uplink is presented. The SS-MC-MA system transmits on 256 sub-carriers. The transmission bandwidth is 2 MHz and the carrier frequency is located at 1.8 GHz. The channel estimation in the considered SS-MC-MA system is based on filtering in the time direction. An OFDM frame consists of 31 OFDM symbols as illustrated in Figure 4-32.

Each user exclusively transmits on a subset of 8 sub-carriers. The spreading is performed with Walsh–Hadamard codes of length $L = 8$. The receiver applies maximum likelihood detection. Convolutional codes of rate 1/2 and memory 6 are used for channel coding.

The BER versus the SNR for an SS-MC-MA system in the uplink is shown in Figure 4-33. The COST 207 channel models, each with different velocity of the terminal station, are considered. It should be noted that the performance of the SS-MC-MA system is independent of the number of active users due to the avoidance of multiple access interference, i.e., the performance presented is valid for any system load, assuming that ICI

Channel Estimation

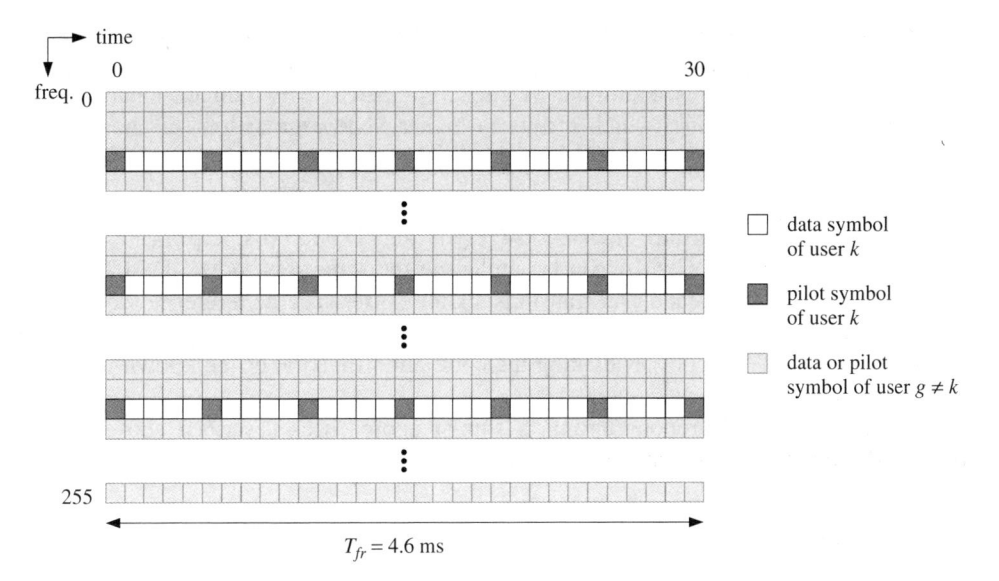

Figure 4-32 Exemplary OFDM frame of an SS-MC-MA system in the uplink

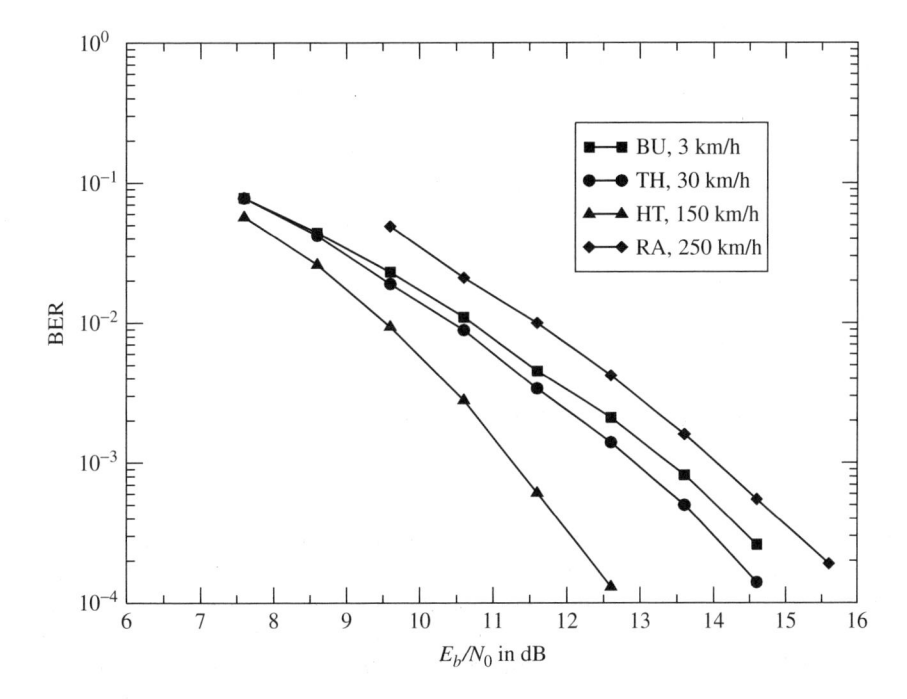

Figure 4-33 BER versus SNR of SS-MC-MA with pilot symbol aided channel estimation

can be neglected. It can be observed that one-dimensional channel estimation can handle different propagation scenarios with high velocities of 250 km/h. One-dimensional uplink channel estimation requires additional overhead compared to two-dimensional channel estimation for the downlink, which results in higher SNR degradation (of about 1 dB) due to pilot symbols.

MC-DS-CDMA

Due to the close relationship of MC-DS-CDMA and DS-CDMA, channel estimation techniques design for DS-CDMA can be applied in the same way to MC-DS-CDMA. A common approach is that each user transmits a known user-specific reference sequence at a certain period of time. These user-specific sequences are used to estimate the channel impulse response $h^{(k)}(\tau, t)$ by evaluating the correlation function of the known reference sequences with the received sequences.

In the case of a synchronous uplink, a similar approach as described above for the MC-CDMA uplink can be applied to MC-DS-CDMA.

4.3.10 Channel Estimation in MIMO-OFDM Systems

The basic concept of applying OFDM in *multiple input multiple output* (MIMO) systems, i.e., employing multiple transmit and receive antennas, is described in Chapter 6.

Regarding channel estimation, pilot based or decision directed channel estimation with multiple antennas can simultaneously estimate multiple channel transfer functions if the channel has moderate delay spread, such that the correlations between adjacent sub-carriers are high. Based on the assumption that fading on adjacent sub-carriers is equal, it is possible to decouple the channel transfer functions corresponding to the different transmit antennas as long as transmitted data on the different antennas are known at the receiver by either pilot symbols or the decoded data can be used to generate the reference data [44][46][54]. Making use of the fact that the channel delay profiles of the various channels in a MIMO scheme should have similar delay profiles, this knowledge allows the accuracy of the channel estimator to be further improved [47].

Blind channel estimation in MIMO-OFDM systems can be achieved using statistical methods and separating signals from the different transmit antennas by performing a periodic pre-coding of the individual data streams prior to transmission [5]. Each transmit antenna is assigned a different pre-coding sequence.

4.4 Channel Coding and Decoding

Channel coding is an inherent part of any multi-carrier system. By using channel state information (CSI) in a maximum likelihood type FEC decoding process a high diversity, hence, high coding gain can be achieved, especially in fading channels [1][22]. Therefore, it is crucial to choose the encoder in such a way that it enables the exploitation of soft information for decoding. Furthermore, flexibility on the coding scheme to derive different code rates (e.g., for unequal error protection) from the same mother code is always preferred. This flexibility may allow one to adapt the transmission scheme to different transmission conditions.

The following channel coding schemes have been proposed for multi-carrier transmission [1][14][15][16][17][18][22]:

- punctured convolutional coding,
- concatenated coding (e.g., inner convolutional and outer block code, i.e., Reed Solomon code), and
- Turbo coding (block or convolutional).

4.4.1 Punctured Convolutional Coding

A punctured convolutional code that provides from the mother code rate 1/2, memory v (e.g., $v = 6$ resulting in 64 states), a wide range of higher inner code rates R (e.g., $R = 2/3, 3/4, 5/6$ and $7/8$) is usually applied, for instance with a generator polynomial $G_1 = 171_{oct}, G_2 = 133_{oct}$.

The puncturing patterns of a convolutional code with 64 states for different inner code rates R are given in Table 4-2. In this table "0" means that the coded bit is not transmitted (i.e., punctured or masked) and "1" means that the coded bit is transmitted. It should be noticed that each matrix has two rows and several columns, where the puncturing vector for each row corresponds to the outputs of the encoder X and Y, respectively (see Figures 4-34 and 4-35). For decoding the received data a soft input maximum likelihood sequence estimator efficiently realized with the Viterbi algorithm can be employed [65]. Deriving the soft values by taking the channel state information gives a high diversity for decoding, resulting in high performance. The number of bits that could be used for soft values is typically 3–4 bits.

Table 4-3 shows the performance of punctured convolutional coding for different modulation schemes in AWGN, Rayleigh, and Ricean (10 dB Rice factor) fading channels with perfect channel estimation.

4.4.2 Concatenated Convolutional and Reed–Solomon Coding

Compared to a single code, the main advantage of concatenated coding schemes is to obtain much higher coding gains at low BERs with reduced complexity.

Table 4-2 Puncturing patterns of an 64-state convolutional code (X_1 is sent first)

Inner code rate R	Puncturing patterns	Min. distance d_{min}	Transmitted sequences (after P/S conversion)
1/2	X: 1 Y: 1	10	$X_1 Y_1$
2/3	X: 1 0 Y: 1 1	6	$X_1 Y_1 Y_2$
3/4	X: 1 0 1 Y: 1 1 0	5	$X_1 Y_1 Y_2 X_3$
5/6	X: 1 0 1 0 1 Y: 1 1 0 1 0	4	$X_1 Y_1 Y_2 X_3 Y_4 X_5$
7/8	X: 1 0 0 0 1 0 1 Y: 1 1 1 1 0 1 0	3	$X_1 Y_1 Y_2 Y_3 Y_4 X_5 Y_6 X_7$

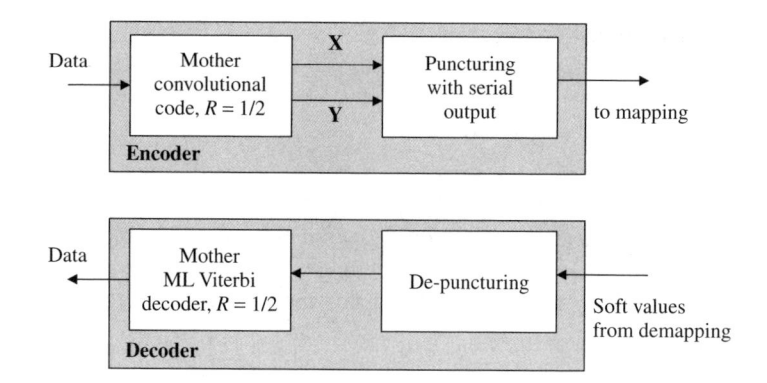

Figure 4-34 Punctured convolutional coding

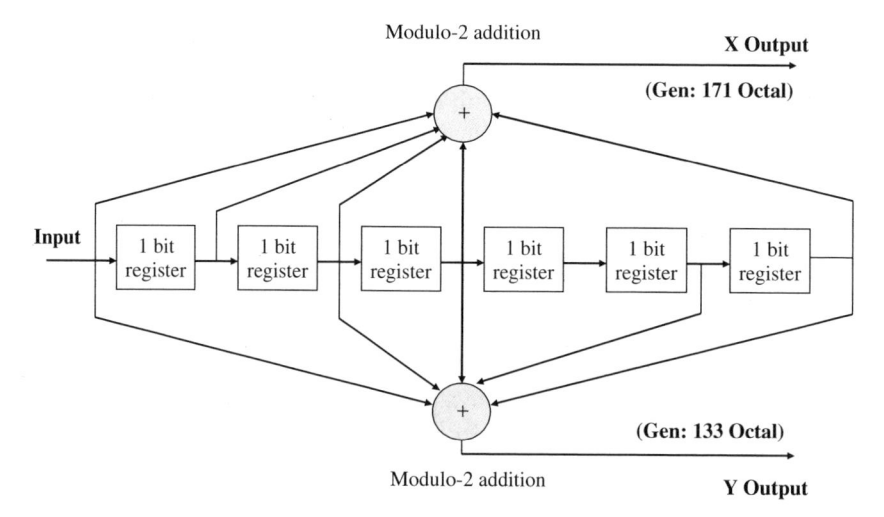

Figure 4-35 Example of inner mother convolutional code of rate 1/2 with memory 6

For concatenated coding, usually as outer code, a shortened Reed–Solomon code and as inner code a punctured convolutional code are used (see Figures 4-36 and 4-37). An optional interleaving between these codes can be inserted. The role of this byte interleaving is to scatter the bursty errors at the output of the inner decoder, i.e, the Viterbi decoder [65]. This type of coding scheme is quite flexible, providing different outer and inner code rates.

The outer shortened Reed–Solomon RS$(K + 2t, K, t)$ code can transmit up to K bytes. This code can be derived, for instance, from the original systematic Reed–Solomon RS(255, 239, $t = 8$) code, able to correct up to $t = 8$ byte errors. As field generator polynomial for Reed–Solomon codes over Galois filed GF(256) the generator polynomial $P(x) = x^8 + x^4 + x^3 + x^2 + 1$ can be employed. As inner code, the same convolutional coding as described in Section 4.4.1 can be used.

Shortened RS codes together with convolutional coding can easily be adapted for the downlink and uplink packet transmission:

Channel Coding and Decoding 161

Table 4-3 E_s/N_0 for punctured convolutional codes with perfect CSI for M-QAM modulation in AWGN, independent Rayleigh and Rician fading channels for BER $= 2 \cdot 10^{-4}$

Modulation	CC rate R	AWGN	Ricean fading (Rice factor 10 dB)	Rayleigh fading
QPSK	1/2	3.1 dB	3.6 dB	5.4 dB
	2/3	4.9 dB	5.7 dB	8.4 dB
	3/4	5.9 dB	6.8 dB	10.7 dB
	5/6	6.9 dB	8.0 dB	13.1 dB
	7/8	7.7 dB	8.7 dB	16.3 dB
16-QAM	1/2	8.8 dB	9.6 dB	11.2 dB
	2/3	11.1 dB	11.6 dB	14.2 dB
	3/4	12.5 dB	13.0 dB	16.7 dB
	5/6	13.5 dB	14.4 dB	19.3 dB
	7/8	13.9 dB	15.0 dB	22.8 dB
64-QAM	1/2	14.4 dB	14.7 dB	16.0 dB
	2/3	16.5 dB	17.1 dB	19.3 dB
	3/4	18.0 dB	18.6 dB	21.7 dB
	5/6	19.3 dB	20.0 dB	25.3 dB
	7/8	20.1 dB	21.0 dB	27.9 dB

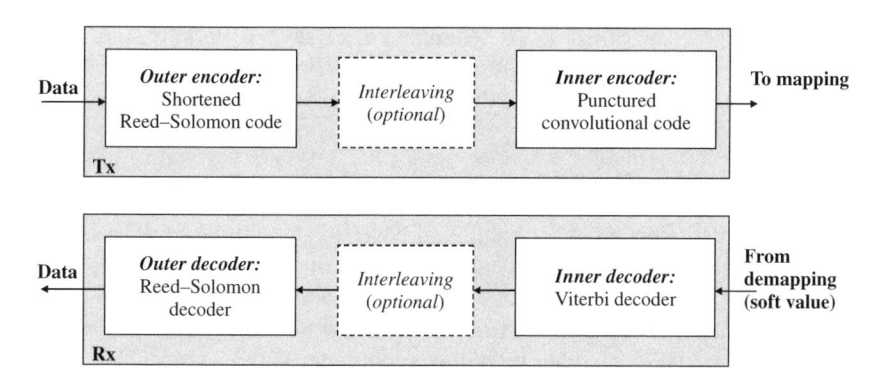

Figure 4-36 Concatenated Reed–Solomon and convolutional code

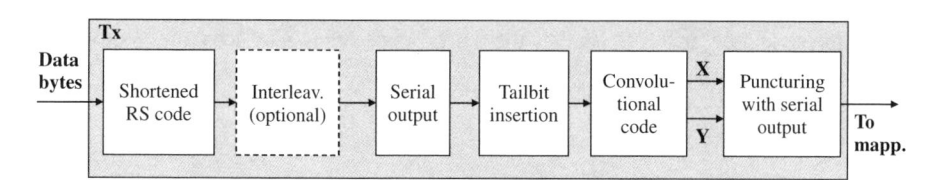

Figure 4-37 Coding procedure for packet transmission

— One or several MAC packets can be mapped to the information part (K bytes of the outer code).
— If the total number of K bytes is smaller than 239 bytes, the remaining $239 - K$ bytes are filled by zero bytes. Systematic RS(255, 239, $t = 8$) coding is applied. After RS

coding, the systematic structure of the RS code allows one to shorten the code, i.e., remove the filled $239 - K$ zero bytes before transmission. Then, each coded packet of length $K + 16$ bytes will be serial bit converted.
— At the end of the each packet, tailbits (e.g., 6 bits for memory 6) can be inserted for inner code trellis termination purposes.
— A block consisting of $[(K + 16) \times 8 + 6]$ bits is encoded by the inner convolutional mother binary code of rate 1/2. After convolutional coding, the puncturing operation is applied following the used inner code rate R for the given packet. This results in a total of $[(K + 16) \times 8 + 6]/R$ bits. Finally, the punctured bits are serial-to-parallel converted and submitted to the symbol mapper.

If the BER before RS decoding is guaranteed to be about $2 \cdot 10^{-4}$, then with sufficient interleaving (e.g., 8 RS code words) for the same SNR values given in Table 4-3, a quasi error-free (i.e., BER $< 10^{-12}$) transmission after RS decoding is guaranteed. However, if no interleaving is employed, depending on the inner coding rate, a loss of about 1.5–2.5 dB has to be considered to achieve a quasi error-free transmission [20].

4.4.3 Turbo Coding

Recently, interest has focused on iterative decoding of parallel or serial concatenated codes using *soft-in/soft-out* (SISO) decoders with simple code components in an interleaved scheme [4][28][29][30][66]. These codes, after several iterations, provide near-Shannon performance [29][30]. We will consider here two classes of codes with iterative decoding: convolutional and block Turbo codes. These codes are already adopted in several standards.

4.4.3.1 Convolutional Turbo Coding

By applying systematic recursive convolutional codes in an iterative scheme and by introducing an interleaver between the two parallel encoders, impressive results can be obtained with so-called convolutional Turbo codes [4]. Convolutional Turbo codes are currently of great interest because of their good performance at low SNRs.

Figure 4-38 shows the block diagram of a convolutional Turbo encoder. The code structure consists of two parallel recursive systematic punctured convolutional codes. A block of encoded bits consists of three parts. The two parity bit parts and the systematic part which is the same in both code bit streams and, hence, has to be transmitted only once. The code bit sequence at the output of the Turbo encoder is given by the vector $\mathbf{b}^{(k)}$.

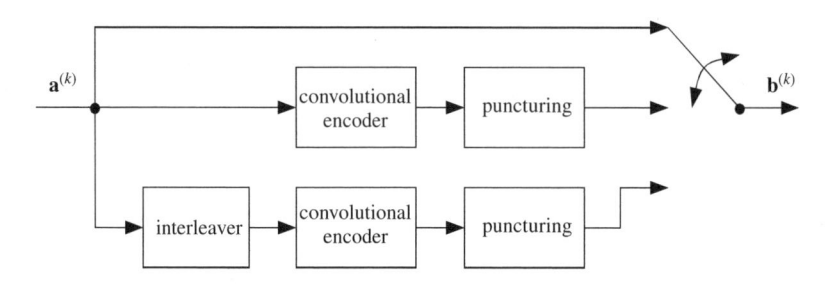

Figure 4-38 Convolutional Turbo encoder

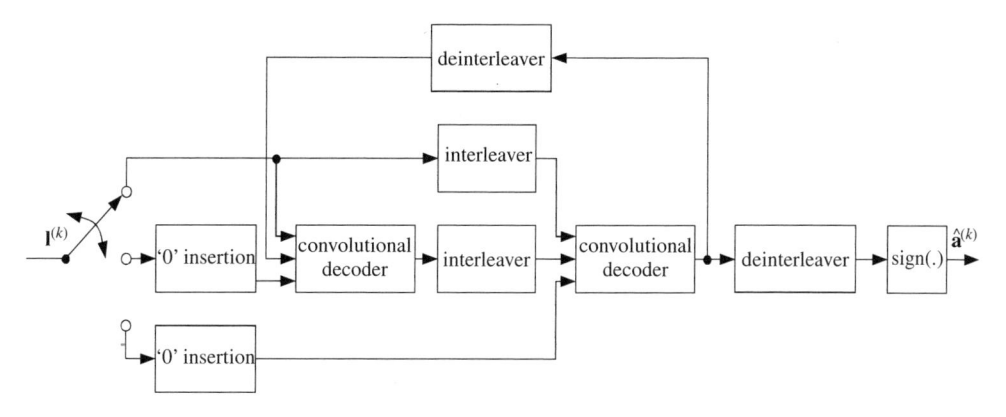

Figure 4-39 Convolutional Turbo decoder

In the receiver, the decoding is performed iteratively. Figure 4-39 shows the block diagram of the convolutional Turbo decoder. The component decoders are soft output decoders providing log-likelihood ratios (LLRs) of the decoded bits (see Section 2.1.7). The basic idea of iterative decoding is to feed forward/backward the soft decoder output in the form of LLRs, improving the next decoding step. In the initial stage, the non-interleaved part of the coded bits $\mathbf{b}^{(k)}$ is decoded. Only the LLRs given by the vector $\mathbf{l}^{(k)}$ at the input of the Turbo decoder are used. In the second stage, the interleaved part is decoded. In addition to the LLRs given by $\mathbf{l}^{(k)}$, the decoder uses the output of the first decoding step as *a priori* information about the coded bits. This is possible due to the separation of the two codes by the interleaver. In the next iteration cycle, this procedure is repeated, but now the non-interleaved part can be decoded using an *a priori* information delivered from the last decoding step. Hence, this decoding run has a better performance than the first one and the decoding improves. Since in each individual decoding step the decoder combines soft information from different sources, the representation of the soft information is crucial.

It is shown in [29] and [30] that the soft value at the decoder input should be a LLR to guarantee that after combining the soft information at the input of the decoder LLRs are again available. The size of the Turbo code interleaver and the number of iterations essentially determine the performance of the Turbo coding scheme.

The performance of Turbo codes as channel codes in different multi-carrier multiple access schemes is analyzed for the following Turbo coding scheme. The component codes of the Turbo code are recursive systematic punctured convolutional codes, each of rate 2/3, resulting in an overall Turbo code rate of $R = 1/2$. Since performance with Turbo codes in fading channels cannot be improved with a memory greater than 2 for a BER of 10^{-3} [30], we consider a convolutional Turbo code with memory 2 in order to minimize the computational complexity. The component decoders exploit the soft output Viterbi algorithm (SOVA) [28]. The Turbo code interleaver is implemented as a random interleaver. Iterative Turbo decoding in the channel decoder uses 10 iterations. The SNR gain with Turbo codes relative to convolutional codes with $R = 1/2$ and memory 6 versus the Turbo code interleaver size I_{TC} is given in Figure 4-40 for the BER of 10^{-3}.

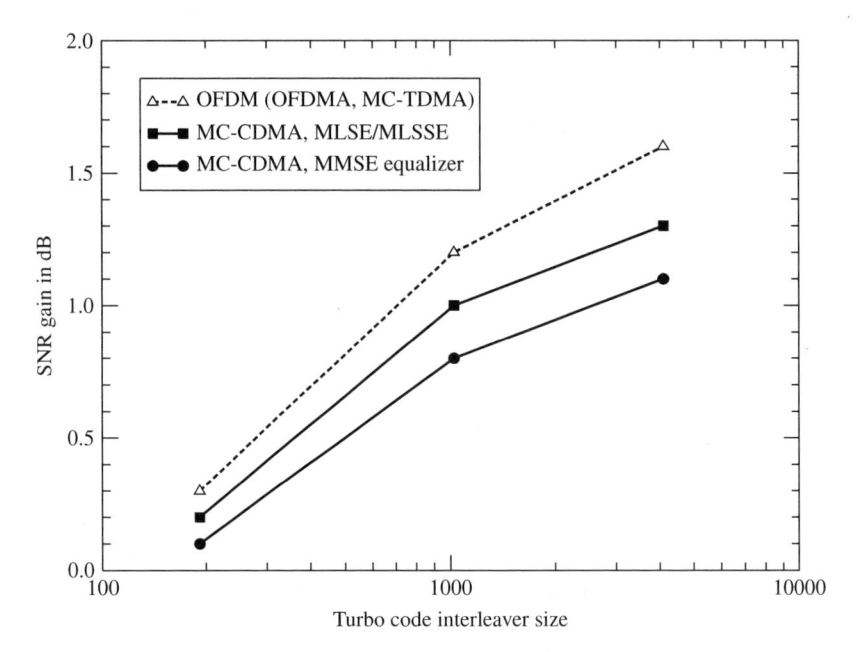

Figure 4-40 SNR gain with Turbo codes relative to convolutional codes versus interleaver size I_{TC}

The results show that OFDMA and MC-TDMA systems benefit more from the application of Turbo codes than MC-CDMA systems. It can be observed that the improvements with Turbo codes at interleaver sizes smaller than 1000 are small.

Due to the large interleaver sizes required for convolutional Turbo codes, they are of special interest for non-real time applications.

4.4.3.2 Block Turbo Coding

The idea of product block or block Turbo coding is to use the well-known product codes with block codes as components for two-dimensional coding (or three dimensions) [66]. The two-dimensional code is depicted in Figure 4-41. The k_r information bits in the rows are encoded into n_r bits, by using a binary block code $C_r(n_r, k_r)$. The redundancy of the code is $r_r = n_r - k_r$ and d_r the minimum distance. After encoding the rows, the columns are encoded using another block code $C_c(n_c, k_c)$, where the check bits of the first code are also encoded.

The two-dimensional code has the following characteristics

— overall block size $n = n_r \cdot n_c$,
— number of information bits $k_r \cdot k_c$,
— code rate $R = R_r \cdot R_c$, where $R_i = k_i/n_i$, $i = c, r$, and
— minimum distance $d_{\min} = d_r \cdot d_c$.

The binary block codes employed for rows and columns could be systematic BCH (Bose–Chaudhuri–Hocquenghem) or Hamming codes [48]. Furthermore, the constituent

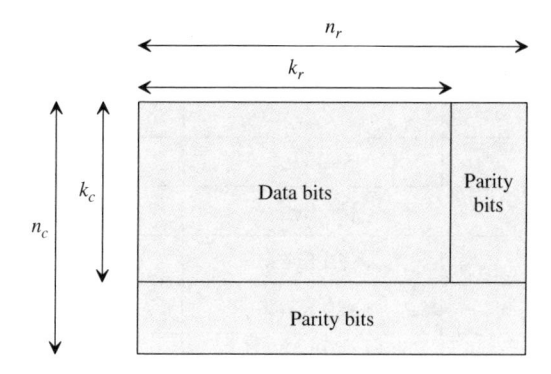

Figure 4-41 Two-dimensional product code matrix

Table 4-4 Generator polynomials of Hamming codes as block Turbo code components

n_i	k_i	Generator
7	4	$x^3 + x + 1$
15	11	$x^4 + x + 1$
31	26	$x^5 + x^2 + 1$
63	57	$x^6 + x + 1$
127	120	$x^7 + x^3 + 1$

codes of rows or columns can be extended with an extra parity bit to obtain *extended* BCH or Hamming codes. Table 4-4 gives the generator polynomials of the Hamming codes used in block Turbo codes.

The main advantage of block Turbo codes is in their application for packet transmission, where the interleaver as it is used in convolutional Turbo coding is not necessary. Furthermore, as block codes, block Turbo codes are efficient for high code rates.

To match packet sizes, a product code can be shortened by removing symbols. In the two-dimensional case, either rows or columns can be removed until the appropriate size is reached. Unlike one-dimensional codes (such as Reed–Solomon codes), parity bits are removed as part of the shortening process, helping to keep the code rate high.

As with convolutional Turbo codes, the decoding of block Turbo codes is done in an iterative way [66]. First, all the horizontal blocks are decoded then all the vertical received blocks are decoded (or vice versa). The decoding procedure is iterated several times to maximize the decoder performance. The core of the decoding process is the soft-in/soft-out (SISO) constituent code decoder. High-performance iterative decoding requires the constituent code decoders to not only determine a transmitted sequence, but also to yield a soft decision metric (i.e., LLR) which is a measure of the likelihood or confidence of each bit in that sequence. Since most algebraic block decoders do not operate with soft

Table 4-5 Performance of block Turbo codes in AWGN channel after three iterations

BTC constituent codes	Coded packet size	Code rate	E_b/N_0 at BER = 10^{-9}
(23,17)(31,25)	53 bytes	0.596	4.5 dB
(16,15)(64,57)	106 bytes	0.834	6.2 dB
(56,49)(32,26)	159 bytes	0.711	3.8 dB
(49,42)(32,31)	159 bytes	0.827	6.5 dB
(43,42)(32,31)	159 bytes	0.945	8.5 dB

inputs or generate soft outputs, such block decoders have primarily been used using the soft output Viterbi algorithm (SOVA) [28] or a soft output variant of the modified Chase algorithm(s) [6]. However, other SISO block decoding algorithms can also be used for deriving the LLR.

The decoding structure of block Turbo codes is similar to that of Figure 4-39, where instead of convolutional decoders, the row and column decoders are applied. Note that here the interleaving is simply a read/write mechanism of rows and columns of the code matrix. The performance of block Turbo codes with three iterations for different packet sizes in an AWGN channel is given in Table 4-5.

4.4.4 OFDM with Code Division Multiplexing: OFDM-CDM

OFDM-CDM is a multiplexing scheme which can better exploit diversity than conventional OFDM systems. Each data symbol is spread over several sub-carriers and/or several OFDM symbols, exploiting additional frequency- and/or time-diversity [36][37]. By using orthogonal spreading codes, self-interference between data symbols can be minimized. Nevertheless, self-interference occurs in fading channels due to a loss of orthogonality between the spreading codes. To reduce this degradation, an efficient data detection and decoding technique is required. The principle of OFDM-CDM is shown in Figure 4-42.

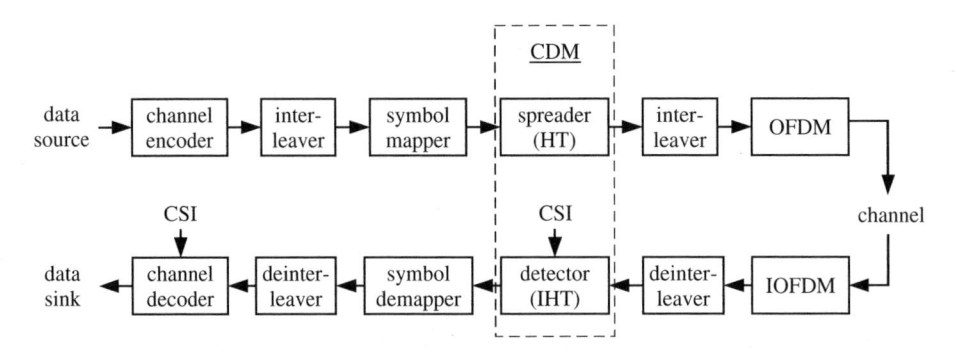

Figure 4-42 OFDM-CDM transmitter and receiver

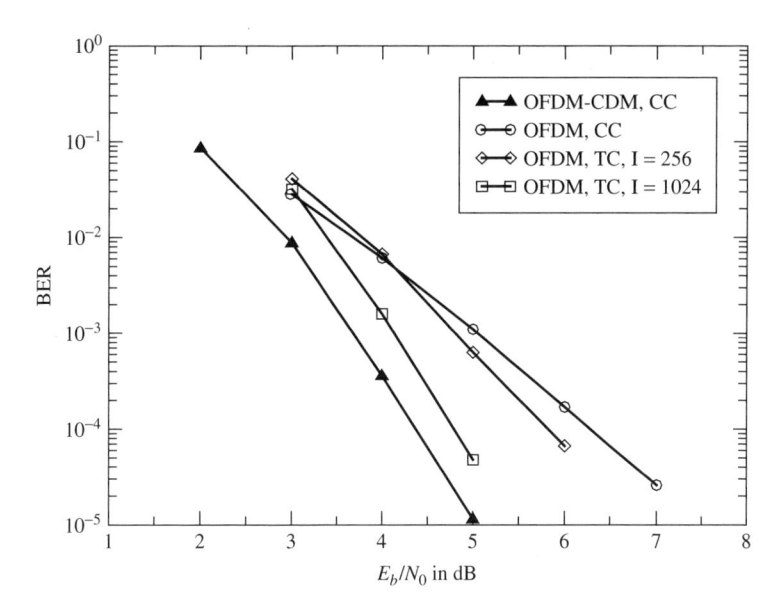

Figure 4-43 Performance of OFDM-CDM with classical convolutional codes versus OFDM with Turbo codes and interleaver size 256 and 1024, respectively

MC-CDMA and SS-MC-MA can be considered special cases of OFDM-CDM. In MC-CDMA, CDM is applied for user separation and in SS-MC-MA different CDM blocks of spread symbols are assigned to different users.

The OFDM-CDM receiver applies single-symbol detection or more complex multi-symbol detection techniques which correspond to single-user or multiuser detection techniques, respectively, in the case of MC-CDMA. The reader is referred to Section 2.1.5 for a description of the different detection techniques.

In Figure 4-43, the performance of OFDM-CDM using classical convolutional codes is compared with the performance of OFDM using Turbo codes. The BER versus the SNR for code rate 1/2 and QPSK symbol mapping is shown. Results are given for OFDM-CDM with soft IC after the 1st iteration and for OFDM using Turbo codes with interleaver sizes $I = 256$ and $I = 1024$ and iterative decoding with 10 iterations. As reference, the performance of OFDM with classical convolutional codes is given. It can be observed that OFDM-CDM with soft IC and classical convolutional codes can outperform OFDM with Turbo codes.

4.5 Signal Constellation, Mapping, Demapping, and Equalization

4.5.1 Signal Constellation and Mapping

The modulation employed in multi-carrier systems is usually based on quadrature amplitude modulation (QAM) with 2^m constellation points, where m is the number of bits transmitted per modulated symbol, and $M = 2^m$ is the number of constellation points. The general principle of modulation schemes is illustrated in Figure 4-44, which is valid for both uplink and downlink.

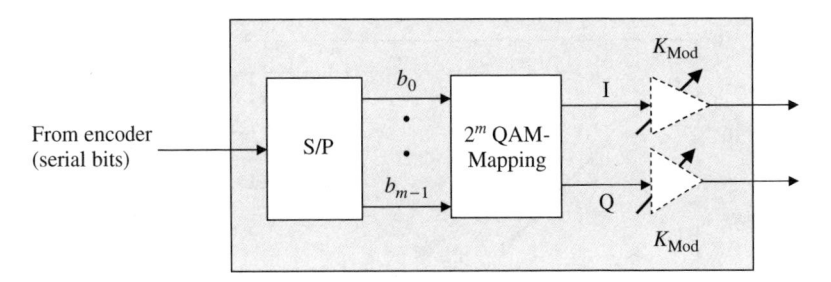

Figure 4-44 Signal mapping block diagram

Table 4-6 Bit mapping with 4-QAM

b_0	b_1	I	Q
0	0	1	1
0	1	1	−1
1	0	−1	1
1	1	−1	−1

For the downlink, high-order modulation can be used such as 4-QAM ($m = 2$) up to 64-QAM ($m = 6$). For the uplink, more robust 4-QAM is preferred. Typically, constellation mappings are based on Gray mapping, where adjacent constellation points differ by only one bit.

Table 4-6 defines the constellation for 4-QAM modulation. In this table, $b_l, l = 0, \ldots, m - 1$, denotes the modulation bit order after serial to parallel conversion.

The complex modulated symbol takes the value $I + jQ$ from the 2^m point constellation (see Figure 4-45). In the case of transmission of mixed constellations in the downlink frame, i.e., adaptive modulation (from 4-QAM up to 64-QAM), a constant RMS should be guaranteed. Unlike the uplink transmission, this would provide the advantage that the downlink interference from all base stations has a quasi-constant behavior. Therefore, the output complex values are formed by multiplying the resulting $I + jQ$ value by a normalization factor K_{MOD} as shown in Figure 4-44. The normalization K_{MOD} depends on the modulation as prescribed in Table 4-7.

Symbol mapping can also be performed differentially as with D-QPSK applied in the DAB standard [14]. Differential modulation avoids the necessity of estimating the carrier phase. Instead, the received signal is compared to the phase of the preceding symbol [65]. However, since one wrong decision results in 2 decision errors, differential modulation performs worse than non-differential modulation with accurate knowledge of the channel in the receiver. Differential demodulation can be improved by applying a two-dimensional demodulation, where the correlation of the channel in time and frequency direction is taken into account in the demodulation [27].

Signal Constellation, Mapping, Demapping, and Equalization 169

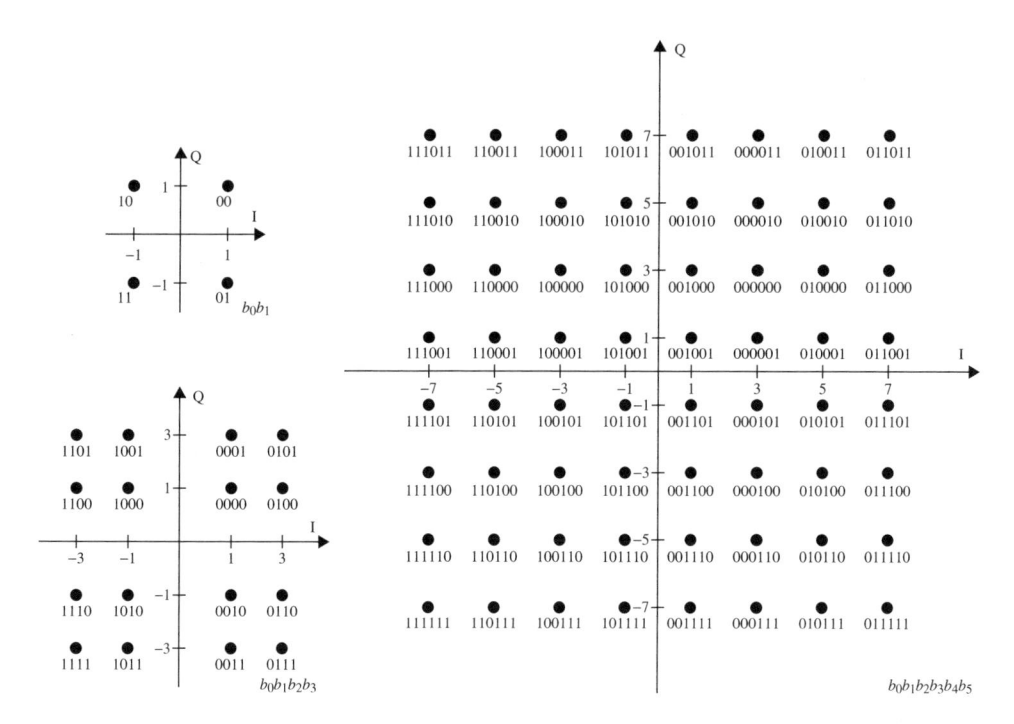

Figure 4-45 M-QAM signal constellation

Table 4-7 Modulation dependent normalization factor K_{MOD}

Modulation	K_{MOD}
4-QAM	1
16-QAM	$1/\sqrt{5}$
64-QAM	$1/\sqrt{21}$

4.5.2 Equalization and Demapping

The channel estimation unit in the receiver provides for each sub-carrier n an estimate of the channel transfer function $H_n = a_n e^{j\varphi_n}$. In mobile communications, each sub-carrier is attenuated (or amplified) by a Rayleigh or Ricean distributed variable $a_n = |H_n|$ and phase distorted by φ_n. Therefore, after FFT operation a correction of the amplitude and the phase of each sub-carrier is required. This can be done by a simple channel inversion, i.e., multiplying each sub-carrier by $1/H_n$. Since each sub-carrier suffers also from noise, this channel correction for small values of a_n leads to a noise amplification. To counteract this effect, the SNR value $\gamma_n = 4|H_n|^2/\sigma_n^2$ of each sub-carrier (where σ_n^2 is the noise variance at sub-carrier n) should be considered for soft metric estimation. Moreover, in order to provide soft information for the channel decoder, i.e., Viterbi decoder, the received channel corrected data (after FFT

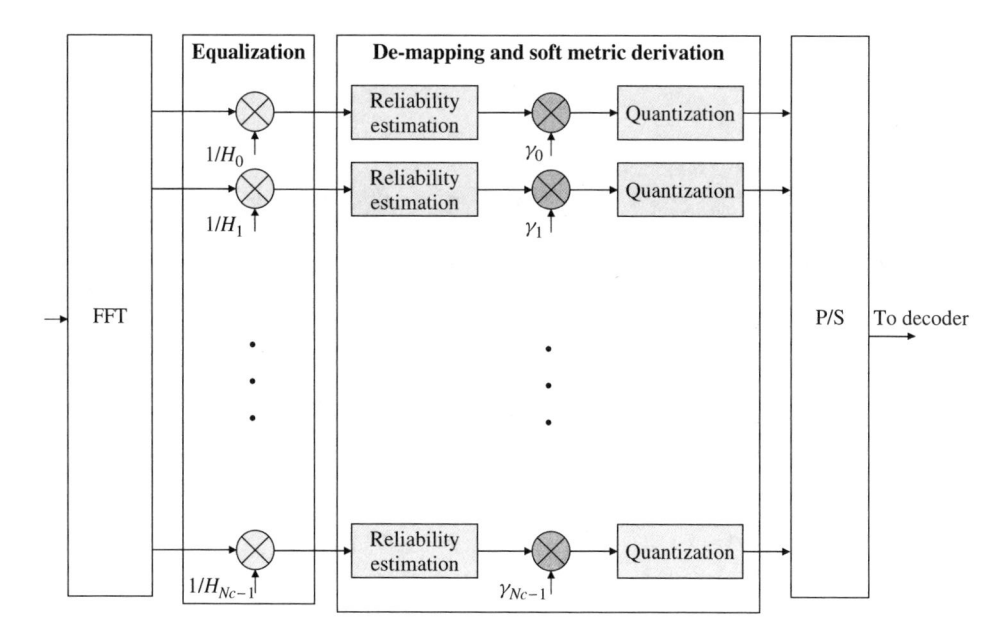

Figure 4-46 Channel equalization and soft metric derivation

and equalization with CSI coefficients) should be optimally converted to soft metric information. Thus, the channel-corrected data have to be combined with the reliability information exploitary channel state information for each sub-carrier, so that each encoded bit has an associated soft metric value and a hard decision that are provided to the Viterbi decoder.

As shown in Figure 4-46, after channel correction, i.e., equalization, for each mapped bit of the constellation a reliability information is provided. This reliability information corresponds to the minimum distance from the nearest decision boundary that affects the decision of the current bit. This metric corresponds to LLR values (see Section 2.1.7) [65] after it is multiplied with the corresponding value of the SNR of each sub-carrier $\gamma_n = 4|H_n|^2/\sigma_n^2$. Finally after quantization (typically 3–4 bits for amplitude and 1 bit for the sign), these soft values are submitted to the channel decoder.

4.6 Adaptive Techniques in Multi-Carrier Transmission

As shown in Chapter 1, the radio channel suffers especially from time and frequency selectivity. Co-channel and adjacent channel interference (CCI and ACI) are further impairments that are present in cellular environments due to the high frequency reuse. Each terminal station may have different channel conditions. For instance, the terminal stations located near the base station receive the highest power which results in a high carrier-to-noise and -interference power ratio $C/(N + I)$. However, the terminal station at the cell border has a lower $C/(N + I)$.

In order to exploit the channel characteristics and to use the spectrum in an efficient way, several adaptive techniques can be applied, namely adaptive FEC, adaptive modulation, and adaptive power leveling. Note that the criteria for these adaptive techniques can be based on the measured $C/(N+I)$ or the received average power per symbol or per sub-carrier. These measured data have to be communicated to the transmitter via a return channel, which may be seen as a disadvantage for any adaptive techniques.

In TDD systems this disadvantage can be reduced, since the channel coefficients are typically highly correlated between successive uplink and downlink slots and, thus, are also available at the transmitter. Only if significant interference occurs at the receiver, this has to be communicated to the transmitter via a return channel.

4.6.1 Nulling of Weak Sub-Carriers

The most straightforward solution for reducing the effect of noise amplification during equalization is the technique of nulling weak sub-carriers which can be applied in an adaptive way. Sub-carriers with the weakest received power are discarded at the transmission side. However, by using strong channel coding or long spreading codes, the gain obtained by nulling weak sub-carriers is reduced.

4.6.2 Adaptive Channel Coding and Modulation

Adaptive coding and modulation in conjunction with multi-carrier transmission can be applied in several ways. The most commonly used method is to adapt channel coding and modulation during each transmit OFDM frame/burst, assigned to a given terminal station [16][17][18]. The most efficient coding and modulation will be used for the terminal station having the highest $C/(N+I)$, where the most robust one will be applied for the terminal station having the worst $C/(N+I)$ (see Figure 4-47). The spectral efficiency in a cellular environment is doubled using this adaptive technique [20].

An alternative technique that can be used in multi-carrier transmission is to apply the most efficient modulation for sub-carriers with the highest received power, where the most robust modulation is applied for sub-carriers suffering from multipath fading (see Figure 4-48).

Furthermore, this technique can be applied in combination with power control to reduce *out of band* emission, where for sub-carriers located at the channel bandwidth border low-order modulation with low transmit power and for sub-carriers in the middle of the bandwidth higher order modulation with higher power can be used.

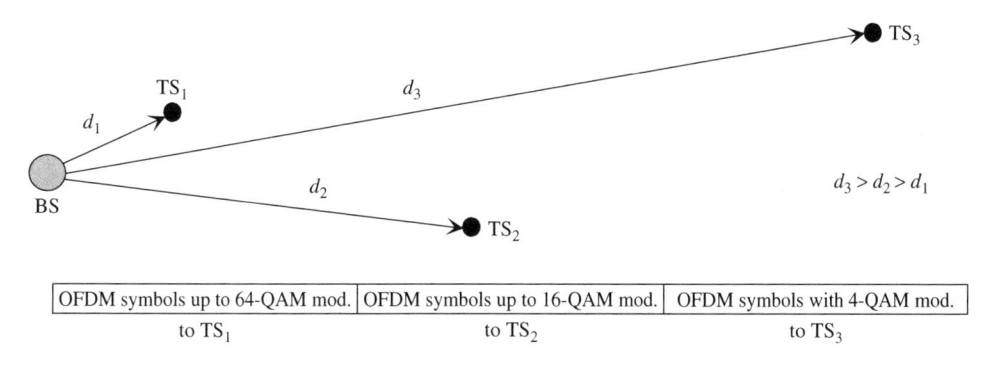

OFDM symbols up to 64-QAM mod. to TS_1	OFDM symbols up to 16-QAM mod. to TS_2	OFDM symbols with 4-QAM mod. to TS_3

Figure 4-47 Adaptive channel coding and modulation per OFDM symbol

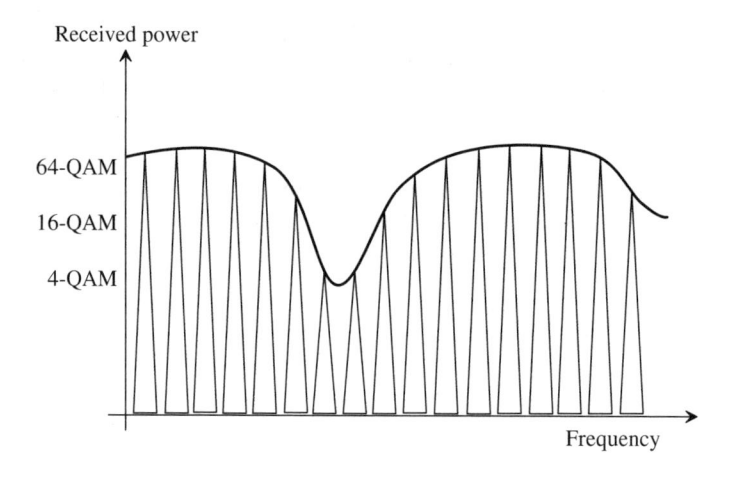

Figure 4-48 Adaptive channel coding and modulation per sub-carrier

4.6.3 Adaptive Power Control

Beside the adaptation of coding and modulation, the transmit power of each OFDM symbol or each sub-carrier can be adjusted to counteract, for instance, the near–far problem or shadowing. A combination of adaptive coding and modulation (the first approach) with power adjustment per OFDM symbol is usually adopted [17][18].

4.7 RF Issues

A simplified OFDM transmitter front-end is illustrated in Figure 4-49. The transmitter comprises an I/Q generator with a local oscillator with carrier frequency f_c, low-pass filters, a mixer, channel pass-band filters, and a power amplifier. After power amplification and filtering, the RF analog signal is submitted to the transmit antenna. The receiver front-end comprises similar components.

Especially in cellular environments due to employing low gain antennas, i.e., non-directive antennas, high-power amplifiers are needed to guarantee a given coverage and

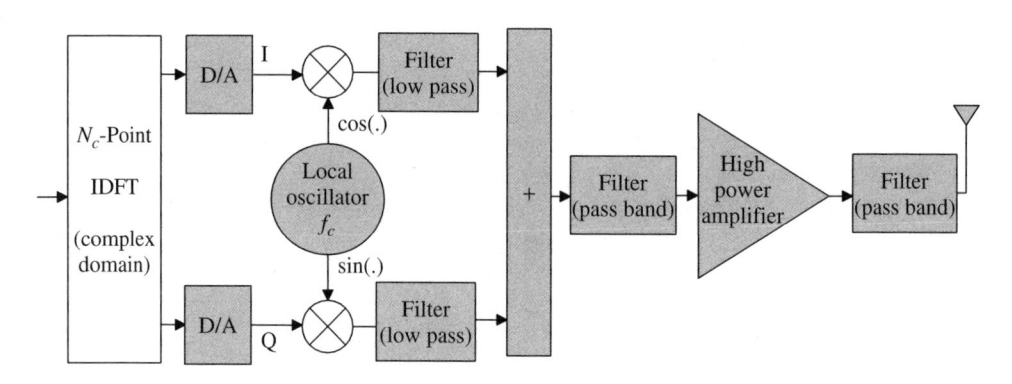

Figure 4-49 Simplified OFDM transmitter front end

hence reduce, for instance, infrastructure costs by installing fewer base stations. Unfortunately, high power amplifiers are non-linear devices, where the maximum efficiency is achieved at saturation point.

Furthermore, at high carrier frequencies (e.g., HIPERLAN/2 at 5 GHz) low cost RF transmit and receive oscillators can be applied at the expense of higher phase noise.

The main objective of this section is to analyze the performance of multi-carrier and multi-carrier CDMA transmission with a high number of sub-carriers in the presence of low cost oscillators with phase noise and HPAs with both AM/AM and AM/PM non-linear conversions. First, a commonly accepted *phase noise model* is described. After analyzing its effects in multi-carrier transmission with high order modulation, measures in the digital domain based on *common phase error (CPE)* correction are discussed. The effects of two classes of non-linear power amplifiers are presented, namely traveling wave tube amplifiers (TWTAs) and solid state power amplifiers (SSPAs). Two techniques based on pre-distortion and spreading code selection are discussed. Finally, in order to estimate the required transmit RF power for a given coverage area, a link budget analysis is carried out.

4.7.1 Phase Noise

The performance of multi-carrier synchronization tracking loops depends strongly on the RF oscillator phase noise characteristics. Phase noise instabilities can be expressed and measured in the time and/or frequency domain.

4.7.1.1 Phase Noise Modeling

Various phase noise models exist for the analysis of phase noise effects. Two often used phase noise models which assume instability of the phase only are described in the following.

Lorenzian Power Density Spectrum
The phase noise generated by the oscillators can be modeled by a Wiener–Lèvy process [63], i.e.,

$$\theta(t) = 2\pi \int_0^t \mu(\tau)\, d\tau, \qquad (4.73)$$

where $\mu(\tau)$ represents white Gaussian frequency noise with power spectral density N_0. The resulting power density spectrum is Lorenzian, i.e.,

$$H(f) = \frac{2}{\left(\pi\beta + \left(\frac{2\pi f}{\beta}\right)^2\right)}. \qquad (4.74)$$

The two-sided 3 dB bandwidth of the Lorenzian power density spectrum is given by β, also referred to as the line-width of the oscillator.

Measurement Based Power Density Spectrum

An approach used within the standardization of DVB-T is the application of the power density spectrum defined by [69]

$$H(f) = 10^{-c} + \begin{cases} 10^{-a} & |f| \leq f_1 \\ 10^{b(f_1-f)/(f_2-f_1)-a} & f > f_1 \\ 10^{b(f_1+f)/(f_2-f_1)-a} & f < -f_1 \end{cases} \quad (4.75)$$

The parameters a and f_1 characterize the phase lock loop (PLL) and the parameter c the noise floor. The steepness of the linear slope is given by b and the frequency f_2 indicates where the noise floor becomes dominant. A plot of the power density spectrum with typical parameters ($a = 6.5$, $b = 4$, $c = 10.5$, $f_1 = 1$ kHz, and $f_2 = 10$ kHz) is shown in Figure 4-50.

This phase noise process can be modeled using two white Gaussian noise processes as shown in Figure 4-51. The first noise term is filtered by an analog filter with a transfer

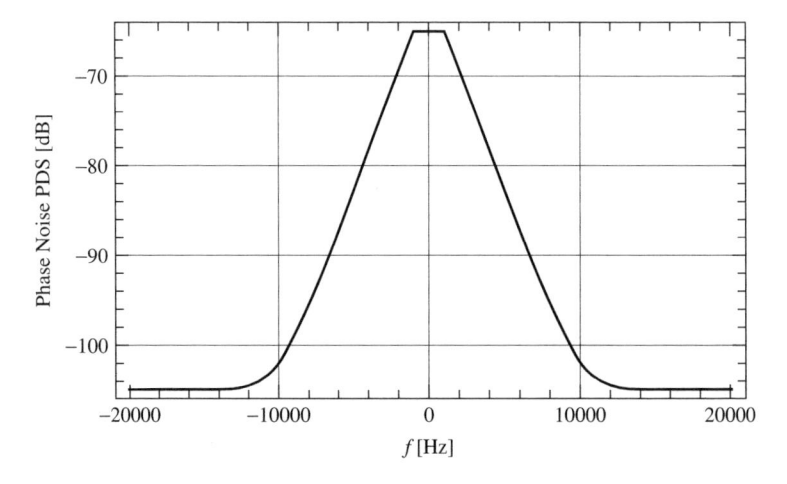

Figure 4-50 Phase noise power density spectrum

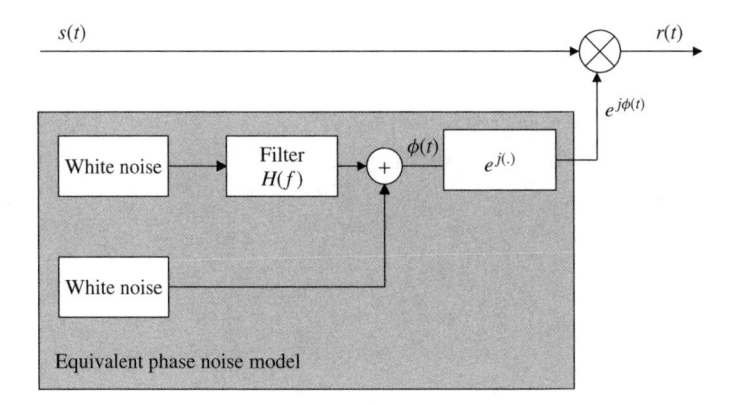

Figure 4-51 Phase noise model

function as shown in Figure 4-50, while the second term gives a phase noise floor which depends on the tuner technology. A digital model of the phase noise process can be obtained by sampling the above analog model at frequency f_{samp}.

Further phase noise models can be found in [2].

4.7.1.2 Effects of Phase Noise in Multi-Carrier Transmission

For reliable demodulation in OFDM systems, orthogonality of the sub-carriers is essential, which is threatened in the receiver by phase noise caused by local oscillator inaccuracies. The local oscillators are applied in receivers for converting the RF signal to a baseband signal. The effects of local oscillator inaccuracies are severe for low-cost mobile receivers.

The complex envelope of an OFDM signal is given by

$$s(t) = \frac{1}{N_c} \sum_{n=0}^{N_c-1} S_n \, e^{j2\pi f_n t}. \tag{4.76}$$

For brevity but without loss of generality, it is assumed that the transmitted signal is only affected by phase noise $\varphi_N(t)$. The oscillator output can be written as

$$r(t) = s(t) \, e^{j\varphi_N(t)}. \tag{4.77}$$

The signal at the FFT output corresponding to sub-carrier n is given by

$$R_n = S_n I_0 + \sum_{\substack{m=0 \\ m \neq n}}^{N_c-1} S_m I_{n-m}, \tag{4.78}$$

where

$$I_n = \frac{1}{T_s} \int_0^{T_s} e^{j2\pi f_n t} \, e^{j\varphi_N(t)} \, \mathrm{d}t. \tag{4.79}$$

According to (4.78), the effects of phase noise can be separated into two parts [69]. The component I_0 in the first term in (4.78) represents a common phase error due to phase noise which is independent of the sub-carrier index and is common to all sub-carriers. The sum of the contributions from the $N_c - 1$ sub-carriers given by the second term in (4.78) represents the ICI caused by phase noise. The ICI depends on the data and channel coefficients of all different $N_c - 1$ sub-channels such that the ICI can be considered as Gaussian noise for large N_c.

The effects of the common phase error and ICI are shown in Figure 4-52, where the mixed time/frequency representation of the total phase error Ψ_E of the sub-carriers per OFDM symbol is shown for an OFDM system with an FFT size of 2048. The mixed time/frequency representation of the total phase error shows in the frequency direction the phase error over all sub-carriers within one OFDM symbol. The time direction is included by illustrating this for 30 subsequent OFDM symbols. It can be shown that each OFDM symbol is affected by a common phase error and noise like ICI. The autocorrelation

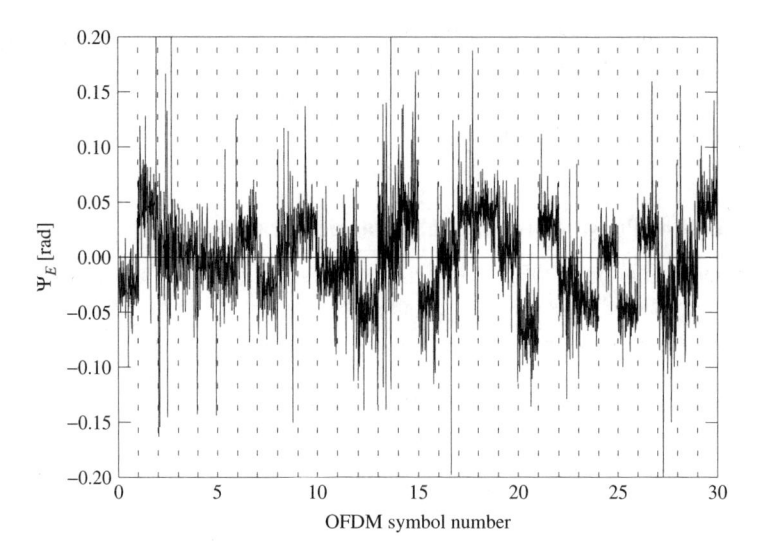

Figure 4-52 Mixed time/frequency representation of the total phase error caused by phase noise

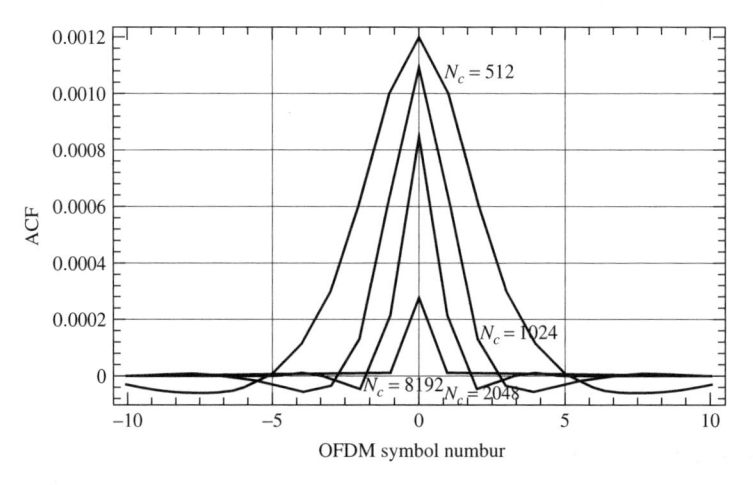

Figure 4-53 Autocorrelation function of the common phase error over OFDM symbols for different numbers of sub-carriers N_c

function (ACF) of the common phase error between adjacent OFDM symbols is shown in Figure 4-53.

The phase noise is modeled as described in Section 4.7.1.1 for the measurement-based power density spectrum. It can be observed that with an increasing number of sub-carriers the correlation of the common phase error between adjacent OFDM symbols decreases.

4.7.1.3 Common Phase Error Correction

The block diagram of a common phase error correction proposed in [69] is shown in Figure 4-54.

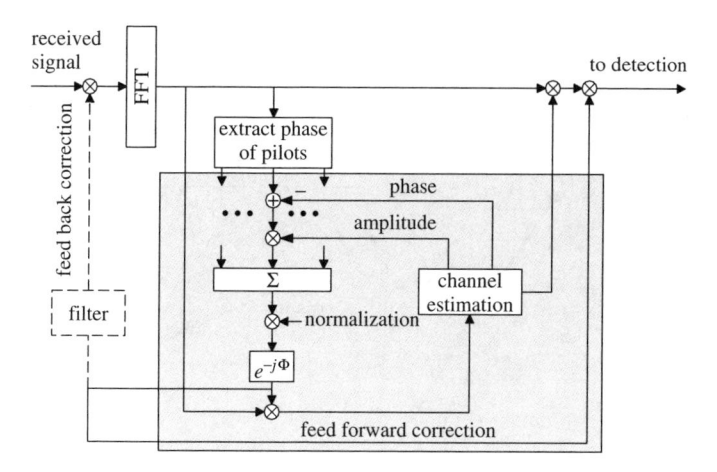

Figure 4-54 Common phase error correction

The common phase error is evaluated and corrected before detection and further processing. The phase of the pilot symbols is extracted after the FFT and the channel phase obtained from the previous OFDM symbol is subtracted on each pilot carrier. After weighting each pilot phase with the amplitude of the previous estimate, all remaining phase errors are added and normalized which results in the common phase error estimate. This estimate is used in the data stream for common phase error correction as well as to correct the pilots used for channel estimation. Since with an increasing number of sub-carriers N_c the common phase error between adjacent OFDM symbols becomes more uncorrelated, it might be necessary to estimate and correct the common phase error within each OFDM symbol, i.e., pilot symbols have to be transmitted in each OFDM symbol.

4.7.1.4 Analysis of the Effects of Phase Noise

OFDM systems are more sensitive to phase noise than single-carrier systems due to the N_c times longer OFDM symbol duration and ICI [62][63]. In OFDM systems, differential modulation schemes are more robust against phase noise than non-differential modulated schemes. Moreover, the OFDM sensitivity to phase noise increases as high-order modulation is applied [9][79].

In Figure 4-55, the degradation due to phase noise is shown for a coded DVB-T transmission since this is a system with high numbers of sub-carriers. The chosen modulation scheme is 64-QAM. OFDM is performed with 2k FFT and the channel fading is Ricean [69]. The performance for a receiver with and without common phase error correction is presented where it can be observed that common phase error correction can significantly improve system performance. Results of the effects of phase noise on MC-CDMA systems can be found in [77] and [81].

4.7.2 Non-Linearities

Multi-carrier modulated systems using OFDM are more sensitive to high power amplifier (HPA) non-linearities than single-carrier modulated systems [75]. The OFDM signal

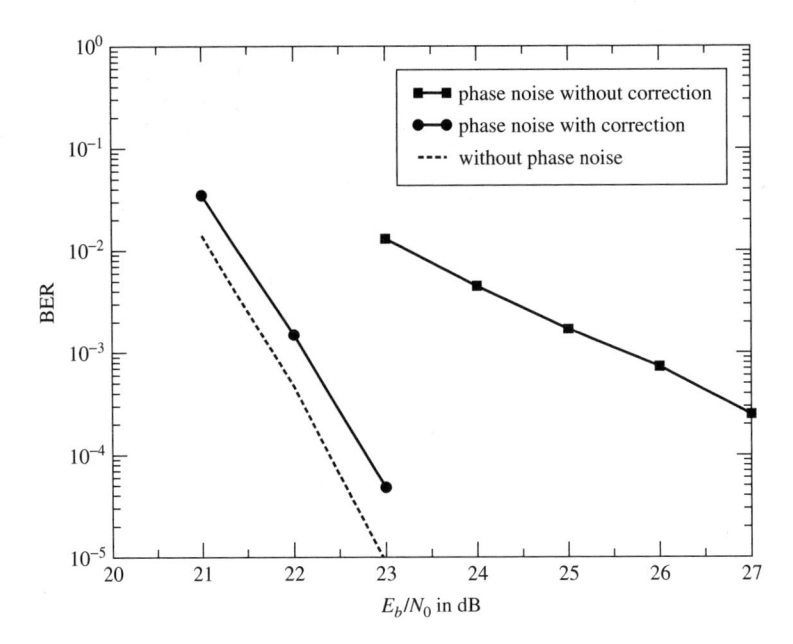

Figure 4-55 Effects of phase noise with and without common phase error correction for a DVB-T transmission with 64-QAM in a Ricean fading channel

requires higher output back-off values to keep an acceptable performance in the presence of non-linear amplifiers. This is due to the presence of a high peak-to-average power ratio (PAPR) in an OFDM signal, leading to severe clipping effects. The PAPR of an OFDM signal is defined as

$$PAPR_{OFDM} = \frac{\max |z_i|^2}{\frac{1}{N_c} \sum_{i=0}^{N_c-1} |z_i|^2}, \quad i = 0, \ldots, N_c - 1, \tag{4.80}$$

where z_i are the transmitted time samples of an OFDM symbol.

The transmitted samples z_i result from the IFFT operation. Following the central limit theorem, the samples z_i have a complex Gaussian distribution for large numbers of sub-carriers N_c, where their amplitudes $|z_i|$ are Rayleigh distributed.

At the output of the HPA, all OFDM signal points with amplitudes higher than the saturation amplitude A_{sat} will be mapped to a point in a circle with radius A_{sat}, which causes high interference, i.e., high degradation of the transmitted signal.

The aim of this section is to analyze the influence of the effects of the non-linearity due to a traveling waves tube amplifier (TWTA) and a solid state power amplifier (SSPA) in an MC-CDMA system. First we present the influence of non-linear distortions on the downlink and uplink transmission, then we examine some techniques to reduce the effects of non-linear distortions. These techniques are based on pre-distortion or the appropriate selection of the spreading codes to reduce the PAPR for an MC-CDMA transmission system.

4.7.2.1 Effects of Non-Linear Distortions in DS-CDMA and MC-CDMA

The fact that multi-carrier transmission is more sensitive to HPA non-linearities than a single-carrier transmission is valid in the single-user case, i.e., in the uplink transmitter. However, for the downlink the situation is different. The downlink transmitted signal is the sum of all active user signals, where the spread signal for both transmission schemes (single-carrier or multi-carrier) will have a high PAPR [21]. Therefore, for the downlink both systems may have quite similar behavior. To justify this, we additionally consider a single-carrier DS-CDMA system, where the transmitter consists of a spreader, transmit filter and the HPA. The receiver is made out of the receive filter and the despreader, i.e., detector. It should be noticed that the transmit and the receive filters are chosen such that the channel is free of inter-chip interference when the HPA is linear. For instance, raised cosine filters with roll-off factor α, equally split between the transmitter and receiver side, can be considered.

For both systems, conventional correlative detection is used. As disturbance, we consider only the effects of HPAs in the presence of additive white Gaussian noise. In order to compensate for the effects of the HPA non-linearities, an automatic gain control (AGC) with phase compensation is used in both systems. This is equivalent to a complex one tape equalizer. Both schemes use BPSK modulation. The processing gain for MC-CDMA is $P_{G,MC} = 64$, and for DS-CDMA $P_{G,DS} = 63$. The total number of users is K.

The non-linear HPA can be modeled as a memory-less device [72]. Let $x(t) = r(t)e^{-j\varphi(t)}$ be the HPA complex input signal with amplitude $r(t)$ and phase $\varphi(t)$. The corresponding output signal can be written as

$$y(t) = R(t)\, e^{-j\phi(t)} \tag{4.81}$$

where $R(t) = f(r(t))$ describes the AM/AM conversion representing the non-linear function between the input and the output amplitudes. The AM/PM distortion $\Phi(t) = g(r(t), \varphi(t))$ produces additional phase modulation. In the following, we consider two types of HPAs, namely a TWTA and a SSPA, which are commonly used in the literature [72].

The non-linear distortions of HPAs depend strongly on the output back-off (OBO)

$$OBO = \frac{P_{sat}}{P_{out}}, \tag{4.82}$$

where $P_{sat} = A_{sat}^2$ represents the saturation power and $P_{out} = E\{|y(t)|^2\}$ is the mean power of the transmitted signal $y(t)$. Small values of the OBO causes the amplifier operation point to be near saturation. In this case a good HPA efficiency is achieved, but as a consequence the HPA output signal is highly distorted.

HPA Models
Traveling Wave Tube Amplifier (TWTA):
For these type of amplifiers the AM/AM and AM/PM conversions are [72]

$$R_n(t) = \frac{2r_n(t)}{1 + r_n^2(t)}$$

$$\phi(t) = \varphi(t) + \pi/3 \frac{r_n^2(t)}{1 + r_n^2(t)} \tag{4.83}$$

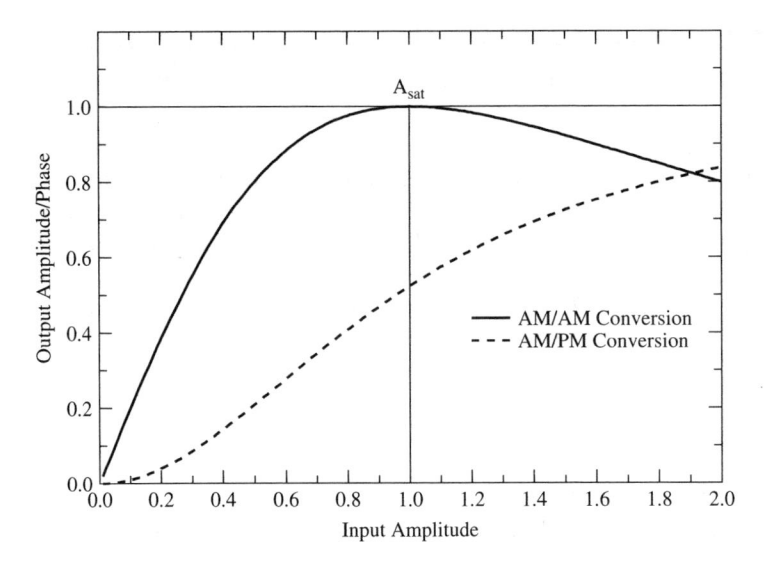

Figure 4-56 TWTA characteristics

where in the above expressions the input and the output amplitudes are normalized by the saturation amplitude A_{sat}. This kind of amplifier has the most critical characteristics due to no one-to-one AM/AM conversion (no bijection) and the AM/PM conversion. In Figure 4-56, the normalized characteristics of a TWTA are illustrated. The AM/PM conversion versus the normalized input amplitude is given in radian.

Solid State Power Amplifier (SSPA):
For these type of amplifiers the AM/AM and AM/PM conversions are given with some specific parameters as follows [67][75]

$$R_n(t) = \frac{r_n(t)}{(1 + r_n^{10}(t))^{1/10}}$$

$$\phi(t) = \varphi(t) \tag{4.84}$$

We can notice that the SSPA adds no phase distortion.

Influence of HPA Non-Linearity
DS-CDMA
After spreading with Gold codes [24] and BPSK modulation, the overall transmitted signal $x(t)$ is obtained.

For low OBO values, the signal $y(t)$ at the output of the amplifier is highly disturbed, where after the receive filter $h(t)$ it leads to a non-linear channel with memory characterized by wrapped output chips containing clusters, which result in inter-chip interference.

Figure 4-57 shows the amplitude distribution (derived by simulations) of the filtered signal before the HPA for the uplink with $L = 63$ and $E_c = -8$ dB. It should be noticed that the distribution is presented in logarithmic scaling. One can see that this signal has low peak amplitudes.

However, for high numbers of users in the downlink, the amplitude distribution of the transmitted signal at the base station in case of BPSK modulation can be approximated

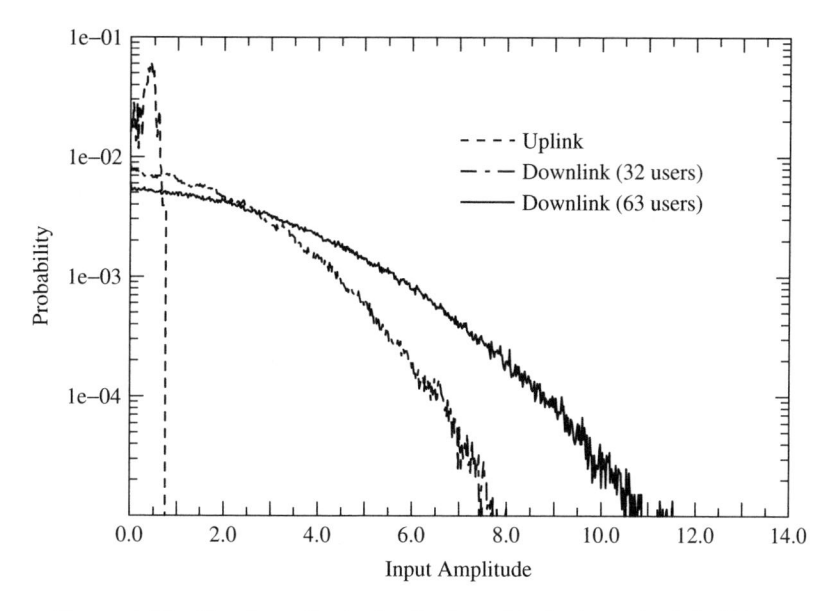

Figure 4-57 Signal amplitude distribution with single-carrier DS-CDMA

by a Gaussian distribution with zero mean and variance Ψ^2. In Figure 4-57 the amplitude distribution (derived by simulations) for the downlink case for $K = 32$ (resp. $K = 63$) active users with $L = 63$ and average power $\Omega = 10\log(K) + E_c = 7$ dB (resp. 10 dB) are presented. One can observe that high numbers of active users result in high peak amplitude values. The presence of high peak amplitudes causes a high degradation in the transmitted signal for low OBO values.

MC-CDMA
The BPSK modulated information bits of each user $k = 0, \ldots, K - 1$, are spread by the corresponding spreading code, where Walsh–Hadamard codes are applied.

For the uplink, the spread data symbol of user k is mapped onto N_c sub-carriers of an OFDM symbol [38]. For the downlink, the spread data of all active users are added synchronously and mapped onto the N_c sub-carriers.

Assuming perfect frequency interleaving, the input signal of the OFDM operation is statistically independent. By using a high number of sub-carriers, the complex-valued OFDM signal can be approximated by a complex-valued Gaussian distribution with zero mean and variance ψ^2. Hence, the amplitude of the OFDM signal is Rayleigh and the phase is uniformly distributed.

Figure 4-58 shows the amplitude distribution of an MC-CDMA signal for the uplink case with a chip energy $E_c = -8$ dB. This curve is derived by simulations with the following parameters: $L = 64$, $N_c = 512$ [25]. The MC-CDMA signal is Rayleigh distributed. Compared to the uplink of a DS-CDMA system, the MC-CDMA signal has a higher peak amplitude. This high signal amplitude leads to a higher degradation for small output back-off values and results in severe clipping effects.

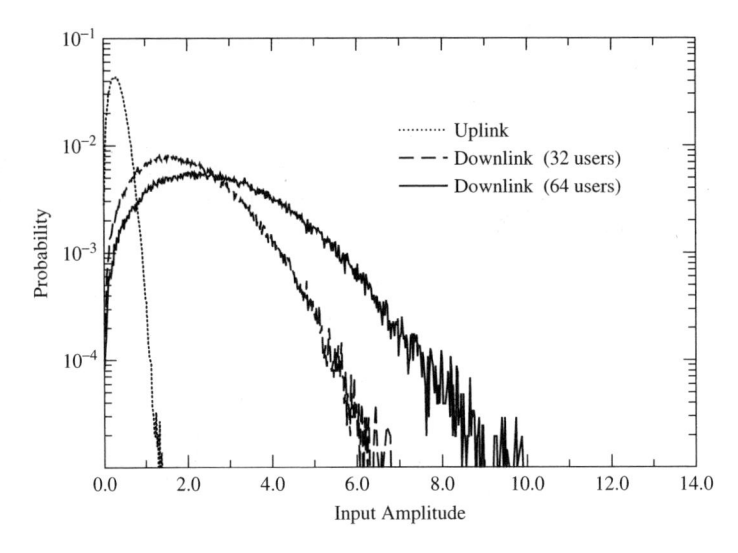

Figure 4-58 Signal amplitude distribution with MC-CDMA

In the same figure, the signal amplitude distribution (derived by simulations) for the downlink case with $L = 64$, $N_c = 512$ and $K = 32$ (resp. $K = 64$) active users is presented. This signal is also Rayleigh distributed with average power $\Omega = 10\log(K) + E_c = 7$ dB (resp. 10 dB). Comparing this distribution to that of the downlink of a DS-CDMA system, one can see that the MC-CDMA system has a lower peak amplitude. Therefore, in this example the MC-CDMA signal is more resistant to non-linearities than the DS-CDMA signal in the downlink.

For low OBO values, the signal $y(t)$ at the output of the amplifier is highly disturbed, where it leads to a non-linear channel with memory characterized by wrapped output chips containing clusters, which results in inter-chip/-carrier interference.

4.7.2.2 Reducing the Influence of HPA Non-Linearities

In this section two techniques to reduce the effects of non-linear amplifiers are discussed. The first one is to adapt the transmitted signal to the HPA characteristics by pre-correction, i.e., pre-distortion methods. The second method concentrates on the choice of spreading code to reduce the transmitted signal PAPR.

Pre-Distortion Techniques

Through the above analysis, we have seen that a high OBO might be needed for the downlink to guarantee a given bit error rate. To make better use of the available HPA power, compensation techniques can be used at the transmitter side. Several methods of data pre-distortion for non-linearity compensation have been previously introduced for single-carrier systems [42][67]. To highlight the gain obtained by pre-distortion techniques for MC-CDMA, we consider a simple method based on the analytical inversion of the HPA characteristics. The presence of extra memory using pre-distortion techniques with memory that may reduce the effects of interference [42] is not considered in this section due to its higher complexity.

RF Issues

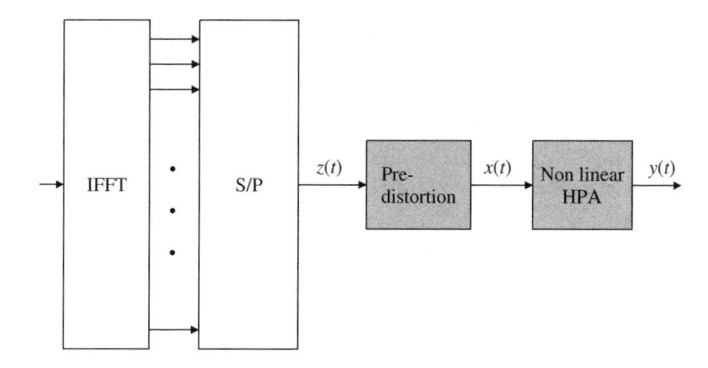

Figure 4-59 Analytical pre-distortion technique for multi-carrier transmission

Let $z(t) = |z(t)| e^{-j\Psi(t)}$ be the signal that has to be amplified and $y(t) = |y(t)| e^{j\phi(t)}$ be the amplified signal (see Figure 4-59). To limit the distortions of a HPA, a device with output signal $x(t) = |x(t)| e^{-j\varphi(t)} = r(t) e^{-j\varphi(t)}$ can be inserted in baseband before the HPA so that the HPA output $y(t)$ becomes as close as possible to the original signal $z(t)$. Hence the pre-distortion function will be chosen such that the global function between $z(t)$ and $y(t)$ will be equivalent to an idealized amplifier

$$y(t) = \begin{cases} z(t) & \text{if } |z(t)| < A_{sat} \\ A_{sat} \dfrac{z(t)}{|z(t)|} & \text{if } |z(t)| \geqslant A_{sat} \end{cases} \tag{4.85}$$

TWTA: The inversion of the TWTA leads to the normalized function

$$r_n(t) = \begin{cases} \dfrac{1}{z_n(t)}\left[1 - \sqrt{1 - z_n^2(t)}\right] & \text{if } |z_n(t)| < 1 \\ \dfrac{z_n(t)}{|z_n(t)|} & \text{if } |z_n(t)| \geqslant 1 \end{cases} \tag{4.86}$$

and

$$\varphi(t) = \begin{cases} \psi(t) - \dfrac{\pi}{3}\dfrac{z_n^2(t)}{(1 + z_n^2(t))} & \text{if } |z_n(t)| < 1 \\ \psi(t) - \dfrac{\pi}{6} & \text{if } |z_n(t)| \geqslant 1 \end{cases} \tag{4.87}$$

SSPA: The inversion of the SSPA equations with the above specific parameters leads to the normalized function

$$r_n(t) = \dfrac{z_n(t)}{(1 - z_n(t)^{10})^{1/10}}$$

$$\varphi(t) = \psi(t). \tag{4.88}$$

Performance of Analytical Pre-Distortion

For performance evaluation, the total degradation (TD) for a given BER is considered as the main criterion. It is given as follows:

$$TD = SNR - SNR_{AWGN} + OBO, \tag{4.89}$$

where SNR is the signal-to-noise ratio in the presence of non-linear distortions, and the SNR_{AWGN} is the signal-to-noise ratio in the case of a linear channel, i.e., only Gaussian noise.

One can see that as the OBO decreases, the HPA is more efficient. But, on the other hand, the non-linear distortion effects increase and a higher SNR is needed to compensate for this effect compared to a linear channel. For high OBO values the HPA works in its linear zone and there are no distortions. However, the loss of HPA efficiency through the high OBO value is taken into account in the total degradation expression. Since the total degradation changes from a decreasing to an increasing function, we can expect an optimal value for the OBO which minimizes the total degradation. Here, in order to be independent from the channel coding, we consider a BER of 10^{-2} and 10^{-3} for our analysis.

The performance of analytical pre-distortion for the MC-CDMA system is presented in Figure 4-60. As Table 4-8 shows, for the uplink one can achieve about 2.1 dB gain with respect to a non-pre-distorted scheme using the TWTA. Similar gains for the downlink case for different BER cases have been obtained.

In Table 4-8 the minimum total degradations TD_{min} at a BER of 10^{-2} for MC-CDMA and DS-CDMA systems are summarized. The lower degradation of the MC-CDMA system for the downlink is due to its lower peak signal amplitude compared to the DS-CDMA signal and to the presence of frequency interleaving. In the case of TWTAs, the gain provided by pre-distortion for MC-CDMA is more than 1.5 dB for both the uplink and the downlink. From this table we conclude that with respect to non-linearities MC-CDMA is adequate for the downlink and MC-DS-CDMA with small number of sub-carriers would be a good choice for the uplink.

Appropriate Selection of Spreading Codes
As we have seen before, MC-CDMA is sensitive to amplifier non-linearity effects. We have used Walsh–Hadamard codes, rather than Gold codes for the downlink in case of

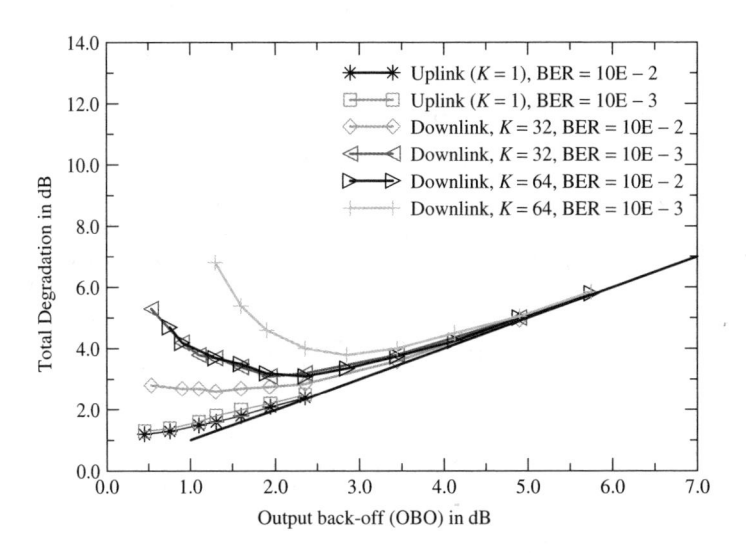

Figure 4-60 Performance of pre-distortion technique for MC-CDMA, TWTA

RF Issues

Table 4-8 Minimum total degradation for different transmission schemes, BER = 10^{-2}

Transmission scheme	Downlink, $K = 32$		Downlink, $K = 64$		Uplink	
	SSPA	TWTA	SSPA	TWTA	SSPA	TWTA
MC-CDMA without pre-distortion	2.6 dB	4.0 dB	3.2 dB	4.7 dB	1.25 dB	3.0 dB
DS-CDMA without pre-distortion	5.0 dB	5.5 dB	5.3 dB	5.8 dB	0.9 dB	1.1 dB
MC-CDMA with pre-distortion	–	2.5 dB	–	3.1 dB	–	0.9 dB

multi-carrier transmission. In this section we highlight the effects of the spreading code selection and detail the appropriate choice for the uplink and the downlink [61][64].

In Section 2.1.4.2, upper bounds for the PAPR of different spreading codes have been presented for MC-CDMA systems. Both uplink and downlink have been analyzed and it is shown that different codes are optimum with respect to the PAPR for up- and downlink.

Exemplary values for these upper bounds are given in Table 4-9 for different spreading code length in the uplink. As it is shown in Table 4-9, Zadoff–Chu codes have the lowest PAPR for the uplink.

For the downlink, the simulation results given in [61] show that for different numbers of active users and a given spreading factor the PAPR in case of Walsh–Hadamard codes decreases as the number of user increases, while for other codes (e.g., Golay codes) it increases. For instance for a spreading factor of $L = 16$, in case of Walsh–Hadamard codes the PAPR is equal to 7.5 and for Golay codes it is about 26. The simulation results given in [61] also confirm that for the downlink the best choice, also for the minimization of multiple access interference, are Walsh–Hadamard codes.

4.7.3 Narrowband Interference Rejection in MC-CDMA

The spread spectrum technique in combination with rake receivers is an interesting approach to remove the effects of interference and multipath propagation [82]. In order to combat strong narrowband interference different techniques of notch-filtering in the time

Table 4-9 PAPR upper bound for different spreading codes in the uplink

Spreading code	$L = 16$	$L = 64$
Walsh–Hadamard	32	128
Gold	~15.5	~31.5
Golay	4	4
Zadoff–Chu	2	2

domain (based on the LMS algorithm) and in the transform domain (based on the FFT) with spread spectrum have been analyzed that provide promising results [52]. However, in a frequency- and time-selective fading channel with coherent detection, its high performance is not longer guaranteed without perfect knowledge about the impulse response of the channel and without performing an optimum detection.

Another interesting approach is based on the MC-CDMA technique [19]. The interference can be considered narrowband multitone *sine* interference. A method for evaluating both interference and the fading process based on frequency domain analysis is studied in the following. The estimated interference and fading process is used for weighting each received chip before despreading.

The narrowband interference can be modeled as consisting of a number of Q continuous sine wave tones [43]

$$Int(t) = \sum_{m=0}^{Q-1} A_m \cos(2\pi f_m t + \phi_m), \tag{4.90}$$

where A_m is the amplitude of the interferer at frequency f_m and ϕ_m is a random phase.

4.7.3.1 Interference Estimation

The interference estimation in multi-carrier systems can be based on the transmission of a null symbol, i.e., one OFDM symbol with a non-modulated signal (see Section 4.2.1). At the receiver side this null symbol contains only interference and noise. The power of interference Int_l in each sub-carrier l can be estimated by performing an FFT operation on this null symbol. The estimated interference power is used for weighting the received sub-carriers.

However, interference can also be detected without the use of a null symbol. It can be done, for instance, by performing an envelope detection of the received signal after the FFT operation. Of course, this method is not accurate, since it suffers from the presence of fading and ICI.

Detection Strategy
In the following, we consider the single-user case. In the presence of fading, interference and noise, the received signal vector $\mathbf{r} = (R_0, R_1, \ldots, R_{L-1})^T$ can be written as

$$\mathbf{r} = \mathbf{Hs} + \mathbf{Int} + \mathbf{n}, \tag{4.91}$$

where $\mathbf{H}$ is a diagonal matrix of dimension $L \times L$ with elements $H_{l,l} = a_l e^{j\varphi_l}$ corresponding to the fading and phase rotation disturbance. $\mathbf{s} = (S_0, S_1, \ldots, S_{L-1})^T$ is the transmitted signal, $\mathbf{Int} = (Int_0, Int_1, \ldots, Int_{L-1})^T$ is the narrowband interference, and $\mathbf{n} = (N_0, N_1, \ldots, N_{L-1})^T$ represents the Gaussian noise. The optimal detection consists of choosing the best transmitted sequence by minimizing the distance between the received sequence and all possible transmitted sequences by using the values a_l, φ_l, Int_l as channel state information (CSI). Let us denote the possible sequences $(v_0^{(i)}, v_1^{(i)}, \ldots, v_{L-1}^{(i)})$, $i = 1, 2$. The information bit is detected, if we maximize the following expression [19]:

$$\Delta_i^2 = \sum_{l=0}^{L-1} \frac{H_{l,l}^*}{\sqrt{1 + Int_l^2}} R_l v_l^{(i)}, \quad i = 1, 2. \tag{4.92}$$

This corresponds to a conventional correlative detection using the CSI where the received chips are weighted by $H_{l,l}^*/\sqrt{1+Int_l^2}$. There are many advantages of such a weighting. The first one is that it is equivalent to *soft-erasure* or *soft switching-off* the sub-carriers containing interference. The second advantage is that no pre-defined threshold for the decision of switching-off the interferers is needed, unlike in the classical transform domain notch-filtering method [52]. Finally, the system does not need extra complexity in terms of hardware, since the FFT operation is already part of the system.

If the number of interference tones is Q, the decision process is equivalent to considering the case that Q chips are erased at the transmitter as we perform erasing of the interferers. Therefore, the above expression can be approximated as

$$\Delta_i^2 \approx \sum_{l=0}^{L-Q-1} \frac{H_{l,l}^*}{\sqrt{1+Int_l^2}} \overline{R_l} v_l^{(i)} \quad i=1,2, \quad (4.93)$$

where $\overline{R_l}$ corresponds to the received chips with low interference.

Performance Evaluation
In Figure 4-61 the performance of an MC-CDMA system in the presence of narrowband interference in case of an uncorrelated Rayleigh fading channel is plotted. OFDM is realized by an inverse FFT with $N_c = 64$ points. The sub-carriers are spaced 20 kHz apart. The useful bandwidth is 1.28 MHz. In all simulations the spreading factor is $L = 64$ where the average received power per chip is normalized to one. One can denote that the interference rejection technique, i.e., soft-erasing, performs well. The analytical curves are also plotted and match with the simulation results.

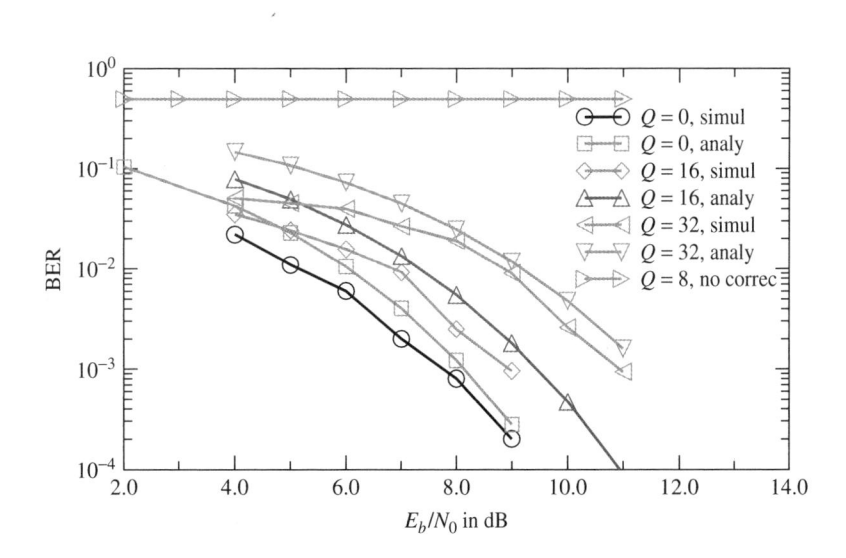

Figure 4-61 Performance of narrowband interference rejection with MC-CDMA

These results show that the combination of spread spectrum with OFDM in the presence of multitone narrowband interference in a frequency- and time-selective fading channel is a promising approach.

4.7.4 Link Budget Evaluation

In order to estimate the transmit power for a given coverage area a link budget analysis is necessary, especially for a cellular mobile communications system. This analysis allows one to estimate the minimum transmit power that a terminal station should transmit to guarantee a given link availability. Especially for mobile terminal stations it is important to save power in order to enable a longer battery life cycle. For any link budget evaluation, especially for the base station, it has to be checked if the transmit power is compliant to the maximum allowed transmission power specified by the regulatory bodies.

The transmit power P_{TX} can be calculated as follows

$$P_{TX} = Path\ loss + P_{Noise} - G_{Antenna} + Offset + Rx_{loss} + C/N, \quad (4.94)$$

where

$$Path\ loss = 10 \log_{10} \left(\frac{4\pi f_c d}{c} \right)^n \quad (4.95)$$

is the propagation path loss. In case of mobile communications n can be estimated to be in the order of 3 to 5 (see Section 1.1.1). In the presence of line of sight, n is equal to 2. d represents the distance between the transmitter and the receiver, f_c is the carrier frequency and c is the speed of light.

$$P_{Noise} = FN_{Thermal} = FKTB \quad (4.96)$$

is the noise power at the receiver input, where F is the receiver noise factor, K is the Blotzmann constant ($K = 1.38 \cdot 10^{-23}$ J/K), T is the temperature in Kelvin and B is the total occupied Nyquist bandwidth. The noise power is expressed in dBm. $G_{Antenna}$ is the sum of the transmit and receive antenna gains expressed in dBi. *Offset* is the margin for all uncertainties such as interference, shadowing, and fading margin reserve. Rx_{loss} is the margin for all implementation uncertainties. C/N is the carrier-to-noise power ratio to guarantee a given BER.

The receiver sensitivity threshold is given by

$$Rx_{th} = P_{Noise} + Rx_{loss} + C/N. \quad (4.97)$$

In Table 4-10, an example of a link budget evaluation for using MC-TDMA in mobile indoor and outdoor communications is given. We consider for the base station transmit antenna a gain of about 16 dBi. However, for the terminal side, we consider an omnidirectional antenna with 0 dBi gain. This table shows that about 25–27 dBm transmit power would be required to achieve a coverage of 10 km with QPSK for outdoor mobile applications at 2 GHz carrier frequency and a coverage of 30 m for WLAN indoor reception at 5 GHz carrier frequency.

Table 4-10 Example of link budget evaluation for mobile and indoor communications

Parameters	Mobile outdoor communications	WLAN indoor communications
Carrier frequency	2 GHz	5 GHz
Coverage	Macro cell (10 km)	Pico cell (30 m)
Path loss power, n	2.2	3
Path loss	130 dB	112 dB
Indoor penetration	N.A.	10 dB
Noise factor	6 dB	6 dB
Bandwidth	2 MHz	20 MHz
Base station antenna gain (90°)	~16 dBi	~16 dBi
Terminal station antenna gain	0 dBi	0 dBi
Offset	~5 dB	~5 dB
C/N for coded QPSK (rate 1/2)	6 dB	6 dB
Receiver sensitivity	−94 dBm	−84 dBm
Needed TX power	25 dBm	27 dBm

4.8 References

[1] Alard M. and Lassalle R., "Principles of modulation and channel coding for digital broadcasting for mobile receivers," *European Broadcast Union Review*, no. 224, pp. 168–190, Aug. 1987.

[2] Armada A.G., "Understanding the effects of phase noise in orthogonal frequency division multiplexing (OFDM)," *IEEE Transactions on Broadcasting*, vol. 47, pp. 153–159, June 2001.

[3] Berens F., Jung P., Plechinger J. and Baier P.W., "Multi-carrier joint detection CDMA mobile communications," in *Proc. IEEE Vehicular Technology Conference (VTC'97)*, Phoenix, USA, pp. 1897–1901, May 1997.

[4] Berrou C., Glavieux A. and Thitimajshima P., "Near Shannon limit error-correcting coding and decoding: Turbo-codes (1)," in *Proc. IEEE International Conference on Communications (ICC'93)*, Geneva, Switzerland, pp. 1064–1070, May 1993.

[5] Bölcskei H., Heath R.W. and Paulraj A.J., "Blind channel identification and equalization in OFDM-based multi-antenna systems," *IEEE Transactions on Signal Processing*, vol. 50, pp. 96–109, Jan. 2002.

[6] Chase D., "A class of algorithms for decoding block codes with channel measurement information," *IEEE Transactions on Information Theory*, vol. 18, pp. 170–182, Jan. 1972.

[7] Classen F. and Meyer H., "Frequency synchronization for OFDM systems suitable for communication over frequency selective fading channels," in *Proc. IEEE Vehicular Technology Conference (VTC'94)*, Stockholm, Sweden, pp. 1655–1659, June 1994.

[8] Choi Y.-S., Voltz P.J. and Cassara F.A., "On channel estimation and detection for multicarrier signals in fast and selective Rayleigh fading channels," *IEEE Transactions on Communications*, vol. 49, pp. 1375–1387, Aug. 2001.

[9] Costa E. and Pupolin S., "M-QAM-OFDM system performance in the presence of a nonlinear amplifier and phase noise," *IEEE Transactions on Communications*, vol. 50, pp. 462–472, March 2002.

[10] Daffara F. and Chouly A., "Maximum likelihood frequency detectors for orthogonal multi-carrier systems," in *Proc. IEEE International Conference on Communications (ICC'93)*, Geneva, Switzerland, pp. 766–771, May 1993.
[11] Daffara F. and Adami O., "A new frequency detector for orthogonal multi-carrier transmission techniques," in *Proc. IEEE Vehicular Technology Conference*, Chicago, USA, pp. 804–809, July 1995.
[12] De Bot, P., Le Floch B., Mignone V. and Schütte H.-D., "An overview of the modulation and channel coding schemes developed for digital terrestrial TV broadcasting within the dTTb project," in *Proc. International Broadcasting Convention (IBC)*, Amsterdam, The Netherlands, pp. 569–576, Sept. 1994.
[13] Edfors O., Sandell M., van de Beek J.-J., Wilson S.K. and Börjesson P.O., "OFDM channel estimation by singular value decomposition," *IEEE Transactions on Communications*, vol. 46, pp. 931–939, July 1998.
[14] ETSI DAB (EN 300 401), "Radio broadcasting systems; digital audio broadcasting (DAB) to mobile, portable and fixed receivers," Sophia Antipolis, France, April 2000.
[15] ETSI DVB-RCT (EN 301 958), "Interaction channel for digital terrestrial television (RCT) incorporating multiple access OFDM," Sophia Antipolis, France, March 2001.
[16] ETSI DVB-T (EN 300 744), "Digital video broadcasting (DVB); framing structure, channel coding and modulation for digital terrestrial television," Sophia Antipolis, France, July 1999.
[17] ETSI HIPERLAN (TS 101 475), "Broadband radio access networks HIPERLAN Type 2 functional specification – Part 1: Physical layer," Sophia Antipolis, France, Sept. 1999.
[18] ETSI HIPERMAN (Draft TS 102 177), "High performance metropolitan area network, Part A1: Physical layer," Sophia Antipolis, France, Feb. 2003.
[19] Fazel K., "Narrow-band interference rejection in orthogonal multi-carrier spread-spectrum communications," in *Proc. IEEE International Conference on Universal Personal Communications (ICUPC'94)*, San Diego, USA, pp. 46–50, Oct. 1994.
[20] Fazel K., Decanis C., Klein J., Licitra G., Lindh L. and Lebert Y.Y., "An overview of the ETSI-BRAN HA physical layer air interface specification," in *Proc. IEEE International Symposium on Personal, Indoor and Mobile Radio Communications (PIMRC 2002)*, Lisbon, Portugal, pp. 102–106, Sept. 2002.
[21] Fazel K. and Kaiser S., "Analysis of non-linear distortions on MC-CDMA", in *Proc. IEEE International Conference on Communications (ICC'98)*, Atlanta, USA, pp. 1028–1034, June 1998.
[22] Fazel K., Kaiser S. and Robertson P., "OFDM: A key component for terrestrial broadcasting and cellular mobile radio," in *Proc. International Conference on Telecommunication (ICT'96)*, Istanbul, Turkey, pp. 576–583, April 1996.
[23] Fazel K., Kaiser S., Robertson P. and Ruf M.J., "A concept of digital terrestrial television broadcasting," *Wireless Personal Communications*, vol. 2, nos. 1 & 2, pp. 9–27, 1995.
[24] Fazel K., Kaiser S. and Schnell M., "A flexible and high-performance cellular mobile communications system based on multi-carrier SSMA," *Wireless Personal Communications*, vol. 2, nos. 1 & 2, pp. 121–144, 1995
[25] Fazel K. and Papke L., "On the performance of convolutionally-coded CDMA/OFDM for mobile communications system," in *Proc. IEEE International Symposium on Personal, Indoor and Mobile Radio Communications (PIMRC'93)*, Yokohama, Japan, pp. 468–472, Sept. 93.
[26] Frenger P.K. and Svensson N.A.B., "Decision-directed coherent detection in multi-carrier systems on Rayleigh fading channels," *IEEE Transactions on Vehicular Technology*, vol. 48, pp. 490–498, March 1999.
[27] Haas E. and Kaiser S., "Two-dimensional differential demodulation for OFDM," *IEEE Transactions on Communications*, vol. 51, pp. 580–586, April 2003.
[28] Hagenauer J. and Höher P., "A Viterbi algorithm with soft-decision outputs and its applications," in *Proc. IEEE Global Telecommunications Conference (GLOBECOM'89)*, Dallas, USA, pp. 1680–1686, Nov. 1989.
[29] Hagenauer J., Offer E. and Papke L., "Iterative decoding of binary block and convolutional codes," *IEEE Transactions on Information Theory*, vol. 42, pp. 429–445, Mar. 1996.
[30] Hagenauer J., Robertson R. and Papke L., "Iterative ('turbo') decoding of systematic convolutional codes with the MAP and SOVA algorithms," in *Proc. ITG Conference on Source and Channel Coding*, Munich, Germany, pp. 21–29, Oct. 1994.
[31] Heath R.W. and Giannakis G.B., "Exploiting input cyclostationarity for blind channel identification in OFDM systems," *IEEE Transactions on Signal Processing*, vol. 47, pp. 848–856, Mar. 1999.
[32] Höher P., "TCM on frequency-selective land-mobile fading channels," in *Proc. 5th Tirrenia International Workshop on Digital Communications*, Tirrenia, Italy, pp. 317–328, Sept. 1991.

References

[33] Höher P., Kaiser S. and Robertson P., "Two-dimensional pilot-symbol-aided channel estimation by Wiener filtering," in *Proc. IEEE International Conference on Acoustics, Speech and Signal Processing (ICASSP'97)*, Munich, Germany, pp. 1845–1848, April 1997.

[34] Höher P., Kaiser S. and Robertson P., "Pilot-symbol-aided channel estimation in time and frequency," in *Proc. IEEE Global Telecommunications Conference (GLOBECOM'97), Communication Theory Mini Conference*, Phoenix, USA, pp. 90–96, Nov. 1997.

[35] Hsieh M.H. and Wei C.H., "A low complexity frame synchronization and frequency offset compensation scheme for OFDM systems over fading channels," *IEEE Transactions on Communications*, vol. 48, pp. 1596–1609, Sept. 1999.

[36] Kaiser S., "Trade-off between channel coding and spreading in multi-carrier CDMA systems," in *Proc. IEEE International Symposium on Spread Spectrum Techniques and Applications (ISSSTA'96)*, Mainz, Germany, pp. 1366–1370, Sept. 1996.

[37] Kaiser S., "OFDM code division multiplexing in fading channels," *IEEE Transactions on Communications*, vol. 50, pp. 1266–1273, Aug. 2002.

[38] Kaiser S. and Fazel K., "A flexible spread-spectrum multi-carrier multiple-access system for multimedia applications," in *Proc. IEEE International Symposium on Personal, Indoor and Mobile Radio Communications (PIMRC'97)*, Helsinki, Finland, pp. 100–104, Sept. 1997.

[39] Kaiser S. and Höher P., "Performance of multi-carrier CDMA systems with channel estimation in two dimensions," in *Proc. IEEE International Symposium on Personal, Indoor and Mobile Radio Communications (PIMRC'97)*, Helsinki, Finland, pp. 115–119, Sept. 1997.

[40] Kamal S.S. and Lyons R.G., "Unique-word detection in TDMA: Acquisition and retention," *IEEE Transactions on Communications*, vol. 32, pp. 804–817, 1984.

[41] Kammeyer K.-D., Petermann T. and Vogeler S., "Iterative blind channel estimation for OFDM receivers," in *Proc. International Workshop on Multi-Carrier Spread-Spectrum & Related Topics (MC-SS 2001)*, Oberpfaffenhofen, Germany, pp. 283–292, Sept. 2001.

[42] Karam G., *Analysis and Compensation of Non-linear Distortions in Digital Microwaves Systems*, ENST-Paris, 1989, PhD thesis.

[43] Ketchum J.W. and Proakis J.G., "Adaptive algorithms for estimating and suppressing narrow band interference in PN spread-spectrum systems," *IEEE, Transactions on Communications*, vol. 30, pp. 913–924, May 1982.

[44] Li Y., "Simplified channel estimation for OFDM systems with multiple transmit antennas," *IEEE Transactions on Wireless Communications*, vol. 1, pp. 67–75, Jan. 2002.

[45] Li Y., Cimini L.J. and Sollenberger N.R., "Robust channel estimation for OFDM systems with rapid dispersive fading channels," *IEEE Transactions on Communications*, vol. 46, pp. 902–915, July 1998.

[46] Li Y., Seshadri N. and Ariyavisitakul S., "Channel estimation for OFDM systems with transmitter diversity in mobile wireless channels," *IEEE Journal on Selected Areas in Communications*, vol. 17, pp. 461–471, March 1999.

[47] Li Y., Winters J.H. and Sollenberger N.R., "MIMO-OFDM for wireless communications: Signal detection with enhanced channel estimation, "*IEEE Transactions on Communications*, vol. 50, pp. 1471–1477, Sept. 2002.

[48] Lin S. and Costello D, *Error Control Coding: Fundamentals and Applications*. Englewood Cliffs, NJ: Prentice Hall, 1983.

[49] Luise M., Reggiannini R. and Vitetta G.M., "Blind equalization/detection for OFDM signals over frequency-selective channels," *IEEE Journal on Selected Areas in Communications*, vol. 16, pp. 1568–1578, Oct. 1998.

[50] Massey J., "Optimum frame synchronization," *IEEE Transactions on Communications*, vol. 20, pp. 115–119, April 1997.

[51] Mignone V. and Morello A., "CD3-OFDM: A novel demodulation scheme for fixed and mobile receivers," *IEEE Transactions on Communications*, vol. 44, pp. 1144–1151, Sept. 1996.

[52] Milstein L.B., "Interference rejection techniques in spread spectrum communications," *Proceedings of the IEEE*, vol. 76, pp. 657–671, June 1988.

[53] Minn H. and Bhargava V.J., "An investigation into time-domain approach for OFDM channel estimation," *IEEE Transactions on Broadcasting*, vol. 46, pp. 240–248, Dec. 2000.

[54] Minn H., Kim D.I. and Bhargava V.K., "A reduced complexity channel estimation for OFDM systems with transmit diversity in mobile wireless channels," *IEEE Transactions on Communications*, vol. 50, pp. 799–807, May 2002.

[55] Moose P.H., "A technique for orthogonal frequency division multiplexing frequency offset correction", *IEEE Transactions on Communications*, vol. 42, pp. 2908–2914, Oct. 1994.
[56] Morelli M. and Mengali U., "A comparison of pilot-aided channel estimation methods for OFDM systems," *IEEE Transactions on Signal Processing*, vol. 49, pp. 3065–3073, Dec. 2001.
[57] Müller A., "Schätzung der Frequenzabweichung von OFDM-Signalen," in *Proc. ITG-Fachtagung Mobile Kommunikation, ITG-Fachberichte 124*, Neu-Ulm, Germany, pp. 89–101, Sept. 1993.
[58] Müller A., European Patent 0529421A2-1993, Priority date 29/08/1991.
[59] Muquet B., de Courville M. and Duhamel P., "Subspace-based blind and semi-blind channel estimation for OFDM systems," *IEEE Transactions on Signal Processing*, vol. 50, pp. 1699–1712, July 2002.
[60] Necker M, Sanzi F. and Speidel J., "An adaptive Wiener filter for improved channel estimation in mobile OFDM-systems," in *Proc. IEEE International Symposium on Signal Processing and Information Technology*, Cairo, Egypt, pp. 213–216, Dec. 2001.
[61] Nobilet S., Helard J.-F. and Mottier D., "Spreading sequences for uplink and downlink MC-CDMA systems: PAPR and MAI minimization", *European Transactions on Telecommunications (ETT)*, vol. 13, pp. 465–474, Sept. 2002.
[62] Pollet T., Moeneclaey M., Jeanclaude I. and Sari H., "Effect of carrier phase jitter on single-carrier and multi-carrier QAM systems," in *Proc. IEEE International Conference on Communications (ICC'95)*, Seattle, USA, pp. 1046–1050, June 1995.
[63] Pollet T., van Bladel M. and Moeneclaey M., "BER sensitivity of OFDM systems to carrier frequency offset and Wiener phase noise," *IEEE Transactions on Communications*, vol. 43, pp. 191–193, Feb./Mar./Apr. 1995.
[64] Popovic B.M., "Spreading sequences for multi-carrier CDMA systems," *IEEE Transactions on Communications*, vol. 47, pp. 918–926, June 1999.
[65] Proakis J.G. *Digital Communications*. New York: McGraw-Hill, 1995.
[66] Pyndiah R., "Near-optimum decoding of product codes: Block Turbo codes," *IEEE Transactions on Communications*, vol. 46, pp. 1003–1010, Aug. 1999.
[67] Rapp C., *Analyse der nichtlinearen Verzerrungen modulierter Digitalsignale–Vergleich codierter und uncodierterter Modulationsverfahren und Methoden der Kompensation durch Vorverzerrung*. Düsseldorf: VDI Verlag, Fortschritt-Berichte VDI, series 10, no. 195, 1991, PhD thesis.
[68] Robertson P., "Effects of synchronization errors on multi-carrier digital transmission systems," *DLR Internal Report*, April 1994.
[69] Robertson P. and Kaiser S., "Analysis of the effects of phase-noise in orthogonal frequency division multiplex (OFDM) systems, "in *Proc. IEEE International Conference on Communications (ICC'95)*, Seattle, USA, pp. 1652–1657, June 1995.
[70] Robertson P. and Kaiser S., "Analysis of the loss of orthogonality through Doppler spread in OFDM systems," in *Proc. IEEE Global Telecommunications Conference (GLOBECOM'99)*, Rio de Janeiro, Brazil, pp. 701–706, Dec. 1999.
[71] Robertson P. and Kaiser S., "Analysis of Doppler spread perturbations in OFDM(A) systems," *European Transactions on Telecommunications (ETT)*, vol. 11, pp. 585–592, Nov./Dec. 2000.
[72] Saleh A.M., "Frequency-independent and frequency-dependent non-linear models of TWTA," *IEEE Transactions on Communications*, vol. 29, pp. 1715–1720, Nov. 1981.
[73] Sandall M., *Design and Analysis of Estimators for Multi-Carrier Modulations and Ultrasonic Imaging*, Lulea University, Sweden, Sept. 1996, PhD thesis.
[74] Sanzi F. and ten Brink S., "Iterative channel estimation and detection with product codes in multi-carrier systems," in *Proc. IEEE Vehicular Technology Conference (VTC 2000-Fall)*, Boston, USA, Sept. 2000.
[75] Schilpp M., Sauer-Greff W., Rupprecht W. and Bogenfeld E., "Influence of oscillator phase noise and clipping on OFDM for terrestrial broadcasting of digital HDTV", in *Proc. IEEE International Conference on Communications (ICC'95)*, Seattle, USA, pp. 1678–1682, June 1995.
[76] Schmidl T.M. and Cox D.C., "Robust frequency and timing synchronization for OFDM," *IEEE Transactions on Communications*, vol. 45, pp. 1613–1621, Dec. 1997.
[77] Steendam H. and Moeneclaey M., "The effect of carrier phase jitter on MC-CDMA performance," *IEEE Transactions on Communications*, vol. 47, pp. 195–198, Feb. 1999.
[78] Steiner B., "Time domain uplink channel estimation in multi-carrier-CDMA mobile radio system concepts," in *Proc. International Workshop on Multi-Carrier Spread-Spectrum (MC-SS'97)*, Oberpfaffenhofen, Germany, pp. 153–160, April 1997.

[79] Tomba L., "On the effect of Wiener phase noise in OFDM systems," *IEEE Transactions on Communications*, vol. 46, pp. 580–583, May 1998.
[80] Tomba L. and Krzymien W.A., "On the use of chip-level differential encoding for the uplink of MC-CDMA systems," in *Proc. IEEE Vehicular Technology Conference (VTC'98)*, Ottawa, Canada, pp. 958–962, May 1998.
[81] Tomba L. and Krzymien W.A., "Sensitivity of the MC-CDMA access scheme to carrier phase noise and frequency offset," *IEEE Transactions on Vehicular Technology*, vol. 48, pp. 1657–1665, Sept. 1999.
[82] Turin G.L., "Introduction to spread-spectrum anti-multi-path techniques and their application to urban digital radio," *Proceedings of the IEEE*, vol. 68, pp. 328–353, March 1980.
[83] Wang X. and Liu K.J.R., "Adaptive channel estimation using cyclic prefix in multi-carrier modulation system," *IEEE Communications Letters*, vol. 3, pp. 291–293, Oct. 1999.
[84] Yang B., Cao Z. and Letaief K.B., "Analysis of low-complexity windowed DFT-based MMSE channel estimator for OFDM systems," *IEEE Transactions on Communications*, vol. 49, pp. 1977–1987, Nov. 2001.
[85] Yang B., Letaief K.B., Cheng R.S. and Cao Z., "Channel estimation for OFDM transmission in multipath fading channels based on parametric channel modeling," *IEEE Transactions on Communications*, vol. 49, pp. 467–479, March 2001.
[86] Yeh C.-S. and Lin Y., "Channel estimation using pilot tones in OFDM systems," *IEEE Transactions on Broadcasting*, vol. 45, pp. 400–409, Dec. 1999.
[87] Yeh C.-S., Lin Y. and Wu Y., "OFDM system channel estimation using time-domain training sequence for mobile reception of digital terrestrial broadcasting," *IEEE Transactions on Broadcasting*, vol. 46, pp. 215–220, Sept. 2000.
[88] Zhou S. and Giannakis G.B., "Finite-alphabet based channel estimation for OFDM and related multicarrier systems," *IEEE Transactions on Communications*, vol. 49, pp. 1402–1414, Aug. 2001.

5

Applications

5.1 Introduction

The deregulation of the telecommunications industry, creating pressure on new operators to innovate in service provision in order to compete with existing traditional telephone service providers, is and will be an important factor for an efficient use of the spectrum. It is certain that most of the information communicated over future digital networks will be *data* rather than purely *voice*. Hence, the demand for high-rate packet-oriented services such as mixed data, voice, and video services, which exceed the bandwidth of conventional systems, will increase.

Multimedia applications and computer communications are often bursty in nature. A typical user will expect to have an instantaneous high bandwidth available delivered by his access provides when needed. It means that the average bandwidth required to deliver a given service will be low, even though the instantaneous bandwidth required is high.

Properly designed broadband systems instantly allocate capacity to specific users and, given a sufficiently large number of users, take advantage of statistical multiplexing to serve each user with a fraction of the bandwidth needed to handle the peak data rate. The emergence of internet protocol (IP) and asynchronous transfer mode (ATM) networks exemplifies this trend.

As the examples given in Table 5-1 show, the average user rate varies for different multimedia services. Generally, the peak data rate for a single user is required only for short periods (high peak-to-mean ratio). Therefore, the data rate that will be supported by future systems will be variable on demand up to a peak of at least 25 Mbit/s in uplink and downlink directions delivered at the user network interface. It may be useful in some systems to allow only lower data rates to be supported, thereby decreasing the overall traffic requirement, which could reduce costs and lead to longer ranges.

The user's demand for high bandwidth packet-oriented services with current delivery over low-bandwidth wireline copper loops (e.g., PSTN, ISDN, xDSL) might be adequate today but certainly will not be in the future.

Wireless technologies are currently limited to some restricted services, but by offering *high mobility*, wireless technologies will offer new alternatives. In Figure 5-1 the data rate versus mobility for current and future standards (4G) is plotted. The current 2G GSM system provides high mobility but a low data rate. 3G systems provide similar mobility as

Table 5-1 Examples of average and peak data rates for different services

Service	Average rate	Peak rate
Video telephony and video conferencing	384 kbit/s to 2 Mbit/s	384 kbit/s to 2 Mbit/s
Video on demand (downlink only)	3 Mbit/s (typical)	6 Mbit/s
Computer gaming	10 kbit/s	25 Mbit/s
POTS	64 kbit/s	64 kbit/s
ISDN	144 kbit/s	144 kbit/s
Internet	10 kbit/s	25 Mbit/s
Remote LAN	10 kbit/s	25 Mbit/s
Compressed Voice	10 kbit/s	100 kbit/s

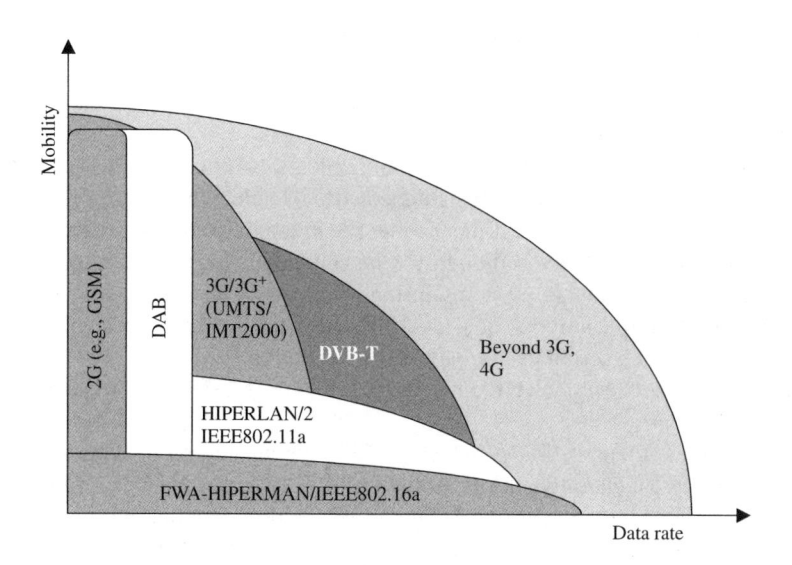

Figure 5-1 Data rate versus mobility in wireless standards

GSM but can deliver higher data rates as mobility decreases, i.e., up to 2 Mbps for pico cells. The HIPERLAN/2 and IEEE 802.11a standards have been designed for high-rate data services with low mobility and low coverage (indoor environments). On the other hand, the HIPERMAN and IEEE 802.16a standards provide high data rates for fixed positioned wireless terminals with high coverage. HIPERLAN, IEEE 802.11a, HIPERMAN and IEEE 802.16a can provide high peak data rates of up to 50 Mbit/s.

On the broadcast side, DAB offers similar mobility as GSM, however, with a much higher broadcast data rate. Although the DVB-T standard was originally designed for fixed or portable receivers, the results of several recent field trials have demonstrated its robustness at high speeds as well [4].

Introduction

The common feature of the current wireless standards that offer a high data rate is the use of *multi-carrier transmission*, i.e., OFDM [5][6][7][8][9][11][12]. In addition, these standards employ adaptive technologies by using several transmission modes, i.e., allowing different combinations of channel coding and modulation together with power control.

A simple adaptive strategy was introduced in DAB using multi-carrier differential QPSK modulation (and also in GSM, using single-carrier GMSK modulation) with several punctured convolutional code rates. By applying a simple combination of source and channel coding, the primary goal was to protect the most important audio/speech message part with the most robust FEC scheme and to transmit the less important source-coded data even without FEC. This technique allows one to receive the highest quality sound/speech in most reception conditions and an acceptable quality in the worst reception areas, where it should be noted that in analog transmission no signal would be received.

DVB-T employs different concatenated FEC coding rates with high-order modulation up to 64-QAM and different numbers of sub-carriers and guard times. Here the objective is to provide different video quality versus distance and different cell-planning flexibility, i.e., country-wide single frequency network or regional network, for instance, using so-called taboo channels (free channels that cannot be used for analog transmission due to the high level of co-channel interference).

In UMTS, besides using different FEC coding rates, a variable spreading factor (VSF) with adaptive power control is introduced. As in GSM, the combination of FEC with source coding is exploited. The variable spreading code allows a good trade-off between coverage, single-cell/multi-cell environments, and mobility. For high coverage areas with high delay spread, large spreading factors can be applied and for low coverage areas with low delay spread, the smallest spreading factor can be used.

In HIPERLAN/2, IEEE 802.11a, and draft HIPERMAN and IEEE 802.16a standards, a solution is adopted based on the combination of multi-carrier transmission with high order modulation (up to 64-QAM), adaptive FEC (variable rate convolutional coding or concatenated coding) and adaptive power control. For each user, according to its required data rate and channel conditions the best combination of FEC, modulation scheme, and the number of time slot is allocated. The main objective is to offer the best trade-off between data rate and coverage, where the mobility is not of great importance. These standards also allow different guard times adapted to different cell coverages.

Offering a trade-off between coverage, data rate, and mobility with a generic air interface architecture is the primary goal of the next generation of wireless systems. Users having no mobility and the lowest coverage distance (pico cells) with an ideal channel condition will be able to receive the highest data rate, where on the other hand subscribers with the highest mobility conditions and highest coverage area (macro-cells) will be able to receive the necessary data rate to establish the required communication link. A combination of MC-CDMA with variable spreading codes or OFDM with adaptive technologies (adaptive FEC, modulation, and power control) can be considered as potential candidates for 4G.

The aim of this chapter is to examine in detail the different application fields of multi-carrier transmission for multiuser environments. This chapter gives an overview of the important technical parameters, and highlights the strategy behind their choices. First, a concrete example of the application of MC-CDMA for a future 4G cellular mobile radio system is given. Then, the OFDM-based HIPERLAN/2 and IEEE 802.11a standards are studied. The application of OFDM and OFDMA in fixed wireless access is then examined. Finally, the DVB-T return channel (DVB-RCT) specification is presented.

5.2 Cellular Mobile Communications Beyond 3G

5.2.1 Objectives

Besides the introduction of new technologies to cover the need for higher data rates and new services, the *integration* of existing technologies in a common platform, as illustrated in Figure 5-2, is an important objective of the next generation of wireless systems.

Hence, the design of a *generic* multiple access scheme for new wireless systems is challenging. This new multiple access scheme should enable i) the integration of existing technologies, ii) higher data rates in a given spectrum, i.e., maximizing the spectral efficiency, iii) different cell configurations to be supported and automatic adaptation to the channel conditions, iv) simple protocol and air interface layers, and finally, v) a seamless adaptation of new standards and technologies in the future.

Especially for the downlink of a cellular mobile communications system, the need for data rates exceeding 2 Mbit/s is commonly recognized. The study on high speed downlink packet access (HSDPA) physical layer is currently under investigation within the 3rd Generation Partnership Project (3GPP) [1]. To gain spectral efficiency, i.e., data rate, the objective of HSDPA is to combine new techniques such as adaptive coding and modulation, hybrid automatic repeat request (H-ARQ), and fast scheduling with the W-CDMA air interface. However, even by adopting such techniques, a significant increase in data rate cannot be expected, since the spectral efficiency of W-CDMA is limited by multi-access interference (see Chapter 1).

Therefore, new physical layer and multiple access technologies are needed to provide high-speed data rates with flexible bandwidth allocation. A low cost *generic radio interface*, operational in mixed-cell and in different environments with scalable bandwidth and data rate, is expected to have a better acceptance.

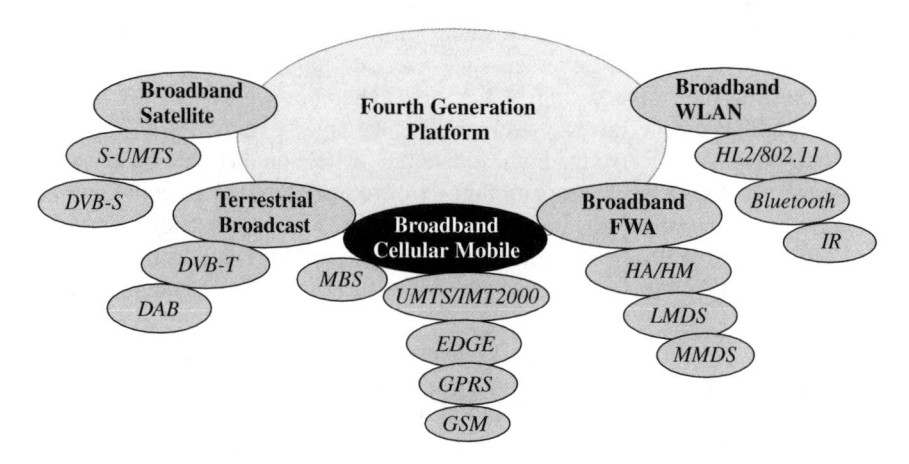

Figure 5-2 Beyond 3G: Integrated perspective

5.2.2 Network Topology and Basic Concept

An advanced 4G system with a point to multi-point topology for a cellular system based on multi-carrier transmission has been proposed by NTT DoCoMo (see Figure 5-3) and successful demonstrations have been carried out in the NTT DoCoMo testbed [2]. High-rate multimedia applications with an asymmetrical data rate are the main objective. The generic architecture allows a capacity optimization with seamless transition from a single cell to a multi-cell environment. This broadband packet-based air interface applies variable spreading factor orthogonal frequency and code division multiplexing (VSF-OFCDM) with two-dimensional spreading in the downlink and MC-DS-CDMA for the uplink [2][3]. The target maximum throughput is over 100 Mbit/s in the downlink and 20 Mbit/s in the uplink. The proposal mainly focuses on asymmetric FDD in order to avoid the necessity of inter-cell synchronization in multi-cell environments and to accommodate independent traffic assignment in the up- and downlink according to traffic.

An application of TDD for special environments is also foreseen. In both cases (FDD and TDD) the same air interface is used.

Figure 5-4 illustrates the generic architecture proposed by NTT DoCoMo. The use of a two-dimensional variable spreading code together with adaptive channel coding and M-QAM modulation in an MC-CDMA system allows an automatic adaptation of the radio link parameters to different traffic, channel, and cellular environment conditions. Furthermore, by appropriate selection of the transmission parameters (FEC, constellation, frame length, FFT size, RF duplex, i.e., TDD/FDD, etc.), this concept can support different multi-carrier or spread spectrum-based transmission schemes. For instance, by choosing a spreading factor of one in both the time and frequency direction, one may obtain a pure OFDM transmission system. However, if the spreading factor in the frequency direction and the number of sub-carriers are set to one, we can configure the system to a classical DS-CDMA scheme. Hence, such a flexible architecture could be seen as a *basic platform* for the integration of the existing technologies as well.

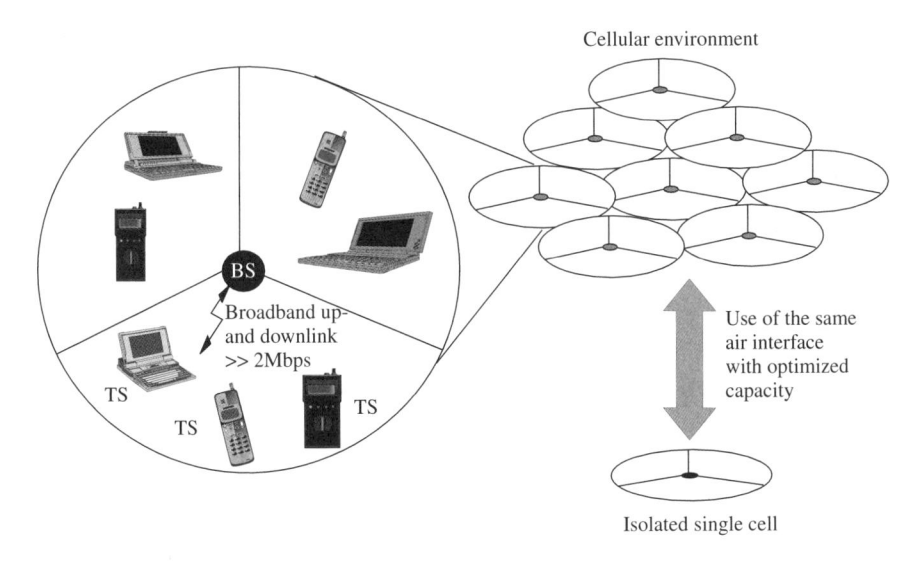

Figure 5-3 Basic concept of NTT DoCoMo for 4G

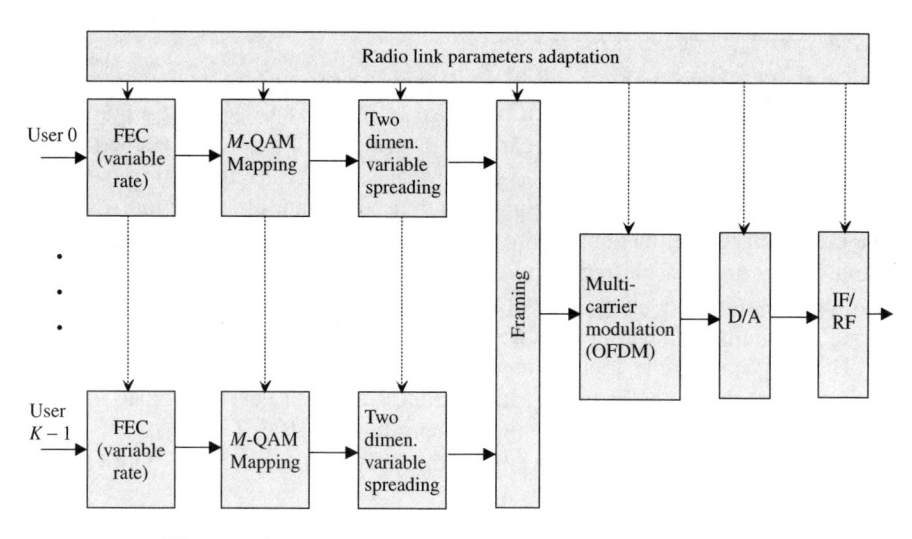

Figure 5-4 Generic architecture concept of NTT DoCoMo

5.2.3 System Parameters

5.2.3.1 Downlink

As depicted in Figure 5-5, by using VSF-OFCDM for the downlink one can apply variable spreading code lengths L and different spreading types. In multi-cell environments, spreading codes of length $L > 1$ are chosen in order to achieve a high link capacity by using a frequency reuse factor of one. Two-dimensional spreading has a total spreading

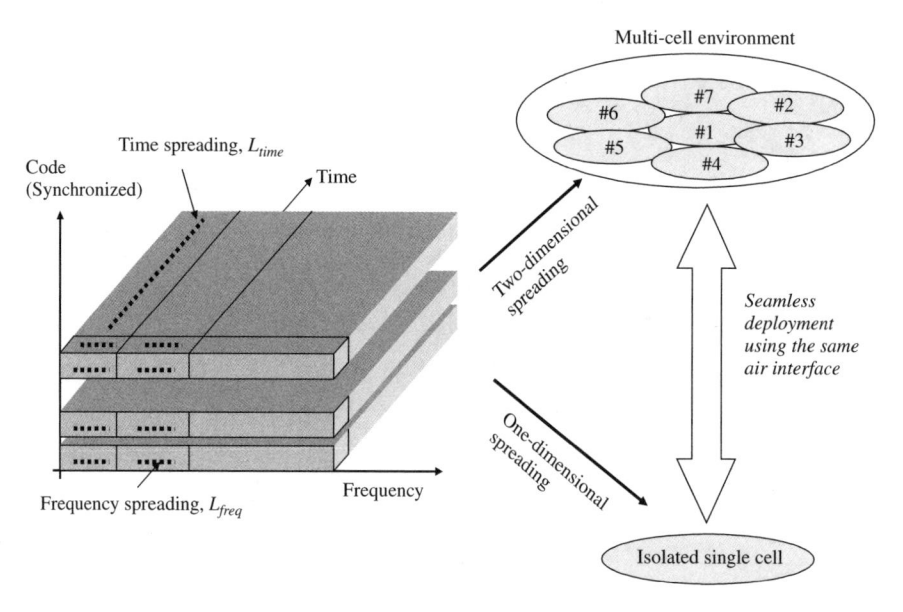

Figure 5-5 Downlink transmission based on VSF-OFCDM

code length of
$$L = L_{time}L_{freq}. \tag{5.1}$$

Two-dimensional spreading with priority for time domain spreading rather than frequency domain spreading is used. The motivation is that in frequency-selective fading channels it is easier to maintain orthogonality among the spread user signals by spreading in the time direction than in the frequency direction. The concept of two-dimensional spreading is described in detail in Section 2.1.4.3. Additional frequency domain spreading in combination with interleaving together with time domain spreading is used for channels which have low SNR such that additional frequency diversity can enhance the transmission quality. The spreading code lengths L_{time} and L_{freq} are adapted to the radio link conditions such as delay spread, Doppler spread, and inter-cell interference, and to the link parameters such as symbol mapping. In isolated areas (hot-spots or indoor offices) only one-dimensional spreading in the time direction is used in order to maintain orthogonality between the spread user signals. Finally, spreading can be completely switched off with $L = 1$ if a single user operates in a isolated cell with a high data rate.

For channel estimation, two different frame formats have been defined. The first format is based on a time multiplexed pilot structure where two subsequent OFDM symbols with reference data are transmitted periodically over predefined distances. The second format applies a code multiplexed pilot structure where the reference data is spread by a reserved spreading code and multiplexed with the spread data symbols so that no explicit pilot symbols or carriers are required. The assumption for this channel estimation method is that the whole spreading code is faded flat and the different spreading codes remain orthogonal.

Table 5-2 summarizes the downlink system parameters. Note that for signal detection at the terminal station side, single-user detection with MMSE equalization is proposed before despreading, which is a good compromise between receiver complexity and performance achievement.

Furthermore, high-order modulation such as 16-QAM or 64-QAM is used with no frequency or even time spreading. In a dense cellular system with high interference and frequency selectivity the lowest order modulation QPSK with highest spreading factor in both directions is employed.

The throughput of a VSF-OFCDM system in the downlink is shown in Figure 5-6 [2]. The throughput in Mbit/s versus the SNR per symbol in a Rayleigh fading channel is plotted. The system applies a spreading code length of $L = 16$, where 12 codes are used. The symbol timing is synchronized using a guard interval correlation and the channel estimation is realized with a time-multiplexed pilot channel within a frame. It can be observed from Figure 5-6 that an average throughput over 100 Mbit/s can be achieved at an SNR of about 13 dB when using QPSK with rate 1/2 Turbo coding.

5.2.3.2 Uplink

In contrast to the downlink, a very low number of sub-carriers in an asynchronous MC-DS-CDMA has been chosen by NTT DoCoMo for the uplink. MC-DS-CDMA guarantees a low-power mobile terminal since it has a lower PAPR reducing the back-off of the amplifier compared to MC-CDMA or OFDM. A code-multiplexed pilot structure is applied for channel estimation based on the principle described in the previous section. To combat the

Table 5-2 NTT DoCoMo system parameters for the downlink

Parameters	Characteristics/Values
Multiple access	VSF-OFCDM
Bandwidth B	101.5 MHz
Data rate objective	>100 Mbits/s
Spreading code	Walsh–Hadamard codes
Spreading code length L	1–256
Number of sub-carriers N_c	768
Sub-carrier spacing F_s	131.8 kHz
OFDM symbol duration T_s	7.585 µs
Guard interval duration T_g	1.674 µs
Total OFDM symbol duration T'_s	9.259 µs
Number of OFDM symbols per frame N_s	54
OFDM frame length T_{fr}	500 µs
Symbol mapping	QPSK, 16-QAM, 64-QAM
Channel code	Convolutional Turbo code, memory 4
Channel code rate R	1/3–3/4

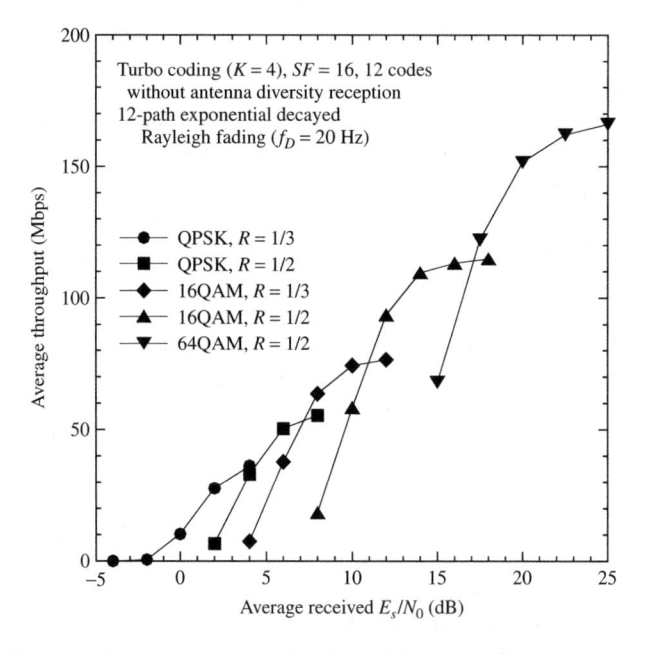

Figure 5-6 Throughput with VSF-OFCDM in the downlink [2]

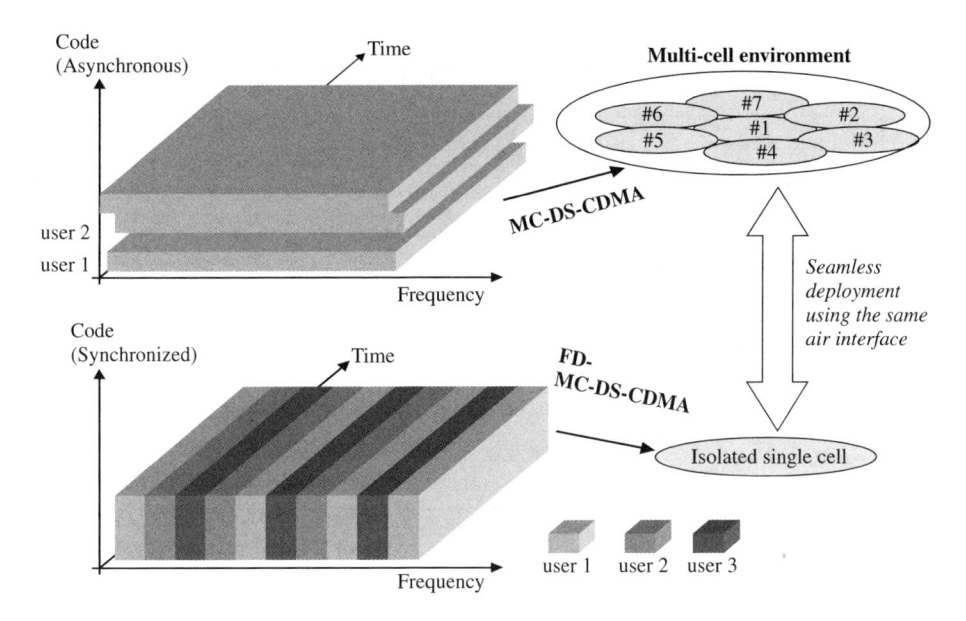

Figure 5-7 Uplink transmission based on MC-DS-CDMA and with an FD-MC-DS-CDMA option

multiple access interference, a rake receiver with interference cancellation in conjunction with adaptive array antenna at the base station is proposed. As shown in Figure 5-7, the capacity can be optimized for each cell configuration.

In a multi-cell environment, MC-DS-CDMA with complex interference cancellation at the base station is used, where in a single-cell environment an orthogonal function in the frequency (FD-MC-DS-CDMA) or time direction (TD-MC-DS-CDMA) is introduced into DS-CDMA. In addition, this approach allows a seamless deployment from a multi-cell to a single cell with the same air interface. The basic system parameters for the uplink are summarized in Table 5-3.

Note that high-order modulation such as 16-QAM or 64-QAM is used even in a single cell with no spreading and good reception conditions. However, in a dense cellular system with high frequency selectivity and high interference, the lowest-order modulation QPSK with the highest spreading factor is deployed.

In Figure 5-8, the throughput of an MC-DS-CDMA system in the uplink is shown [2]. The throughput in Mbit/s versus the SNR per symbol in a Rayleigh fading channel is plotted. The system applies a spreading code length of $L = 4$, where 3 codes are used. Receive antenna diversity with 2 antennas is exploited. The channel estimation is realized with a code-multiplexed pilot channel within a frame. It can be observed from Figure 5-8 that an average throughput of over 20 Mbit/s can be achieved at an SNR of about 9 dB when using QPSK with rate 1/2 Turbo coding.

5.3 Wireless Local Area Networks

Local area networks typically cover a story or building and their wireless realization should avoid complex installation of a wired infrastructure. WLANs are used in public

Table 5-3 NTT DoCoMo system parameters for the uplink

Parameters	Characteristics/Values
Multiple access	MC-DS-CDMA
Bandwidth B	40 MHz
Data rate objective	>20 Mbit/s
Spreading code length L	1 – 256
Number of sub-carriers N_c	2
Sub-carrier spacing F_s	20 MHz
Chip rate per sub-carrier	16.384 Mcps
Roll-off factor	0.22
Total OFDM symbol duration T'_s	9.259 µs
Number of chips per frame	8192
Frame length T_{fr}	500 µs
Symbol mapping	QPSK, 16-QAM, 64-QAM
Channel code	Convolutional Turbo code, memory 4
Channel code rate R	1/16–3/4

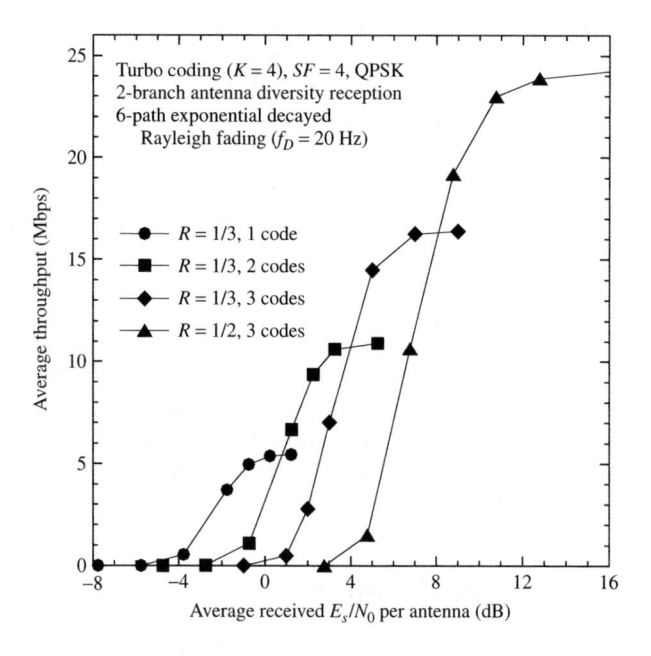

Figure 5-8 Throughput with MC-DS-CDMA in the uplink [2]

and private environments and support high data rates. They are less expensive than wired networks for the same data rate, are simple and fast to install, offer flexibility and mobility, and are cost-efficient due to the possibility of license exempt operation.

5.3.1 Network Topology

WLANs can be designed for infrastructure networks, ad hoc networks or combinations of both. The mobile terminals in infrastructure networks communicate via the base stations (BSs) which control the multiple access. The base stations are linked to each other by a wireless (e.g., FWA) or wired backbone network. Infrastructure networks have access to other networks, including the internet. The principle of an infrastructure network is illustrated in Figure 5-9. Soft handover between different base stations can be supported by WLANs such as HIPERLAN/2.

In ad hoc networks, the mobile terminals communicate directly with each other. These networks are more flexible than infrastructure networks, but require a higher complexity in the mobile terminals since they have to control the complete multiple access as base station does. Communication within ad hoc networks is illustrated in Figure 5-10.

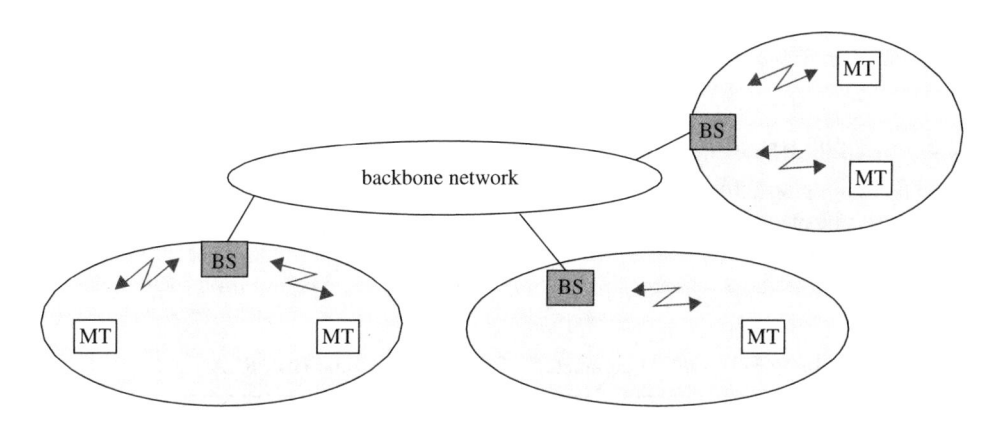

Figure 5-9 WLAN as an infrastructure network

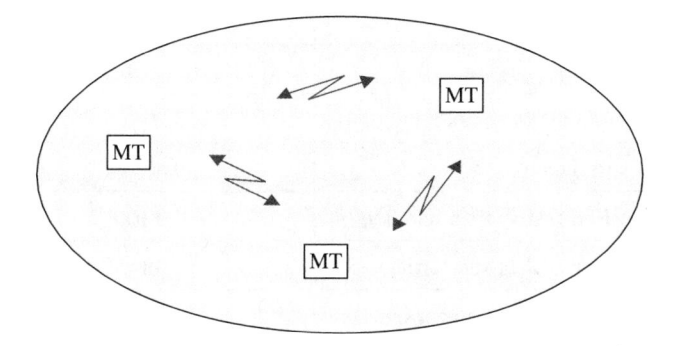

Figure 5-10 WLAN as an ad hoc network

5.3.2 Channel Characteristics

WLAN systems often use the license-exempt 2.4 GHz and 5 GHz frequency bands which have strict limitations on the maximum transmit power since these frequency bands are also used by many other communications systems. This versatile use of the frequency band results in different types of narrowband and wideband interference, such as a microwave oven, which the WLAN system has to cope with.

WLAN cell size is up to several 100 m and multipath propagation typically results in maximum delays of less than 1 μs. Mobility in WLAN cells is low and corresponds to a walking speed of about 1 m/s. The low Doppler spread in the order of 10–20 Hz makes OFDM very interesting for high-rate WLAN systems.

5.3.3 IEEE 802.11a, HIPERLAN/2, and MMAC

The physical layer of the OFDM-based WLAN standards IEEE 802.11a, HIPERLAN/2, and MMAC are harmonized, which enables the use of the same chip set for products of different standards. These WLAN systems operate in the 5 GHz frequency band. All standards apply MC-TDMA for user separation within one channel and FDMA for cell separation. Moreover, TDD is used as a duplex scheme for the separation of uplink and downlink. The basic OFDM parameters of IEEE 802.11a and HIPERLAN/2 are summarized in Table 5-4 [8][11].

5.3.3.1 Frame structure

The TDD frame structure of HIPERLAN/2 is shown in Figure 5-11. One MAC frame includes the header followed by the downlink (DL) phase, an optional direct link (DiL) phase and the uplink (UL) phase. The MAC frame ends with a random access slot (RCH), where users can request resources for the next MAC frame. The duration of the DL, DiL,

Table 5-4 OFDM parameters of IEEE 802.11a and HIPERLAN/2

Parameter	Value
IFFT/FFT length	64
Sampling rate	20 MHz
Sub-carrier spacing	312.5 kHz (= 20 MHz/64)
Useful OFDM symbol duration	3.2 μs
Guard duration	0.8 μs
Total OFDM symbol duration	4.0 μs
Number of data sub-carriers	48
Number of pilot sub-carriers	4
Total number of sub-carriers	52

Wireless Local Area Networks 207

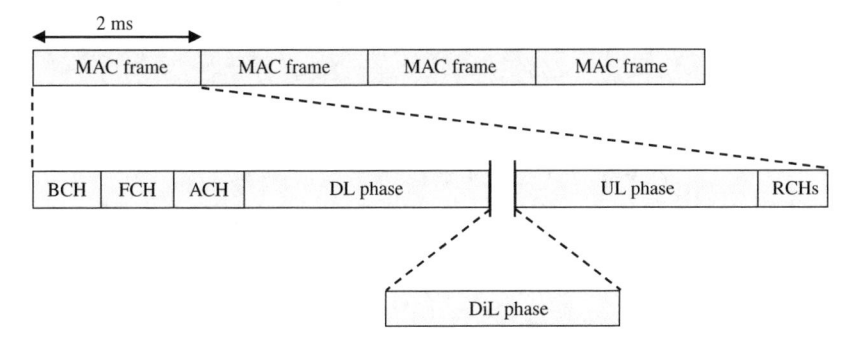

Figure 5-11 TDD frame structure of HIPERLAN/2

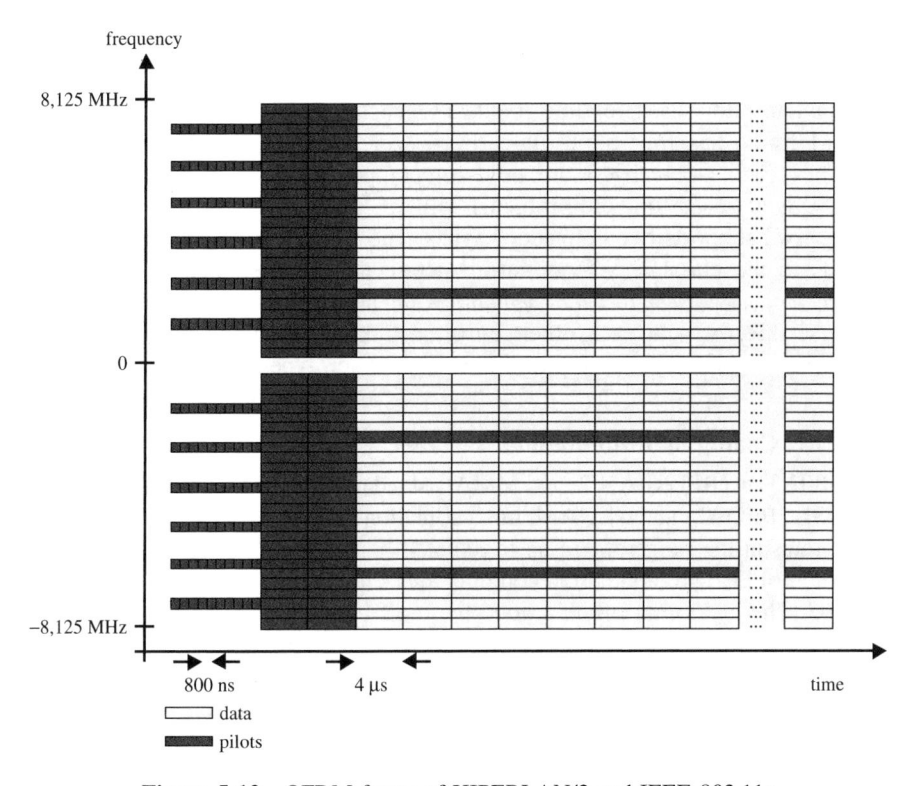

Figure 5-12 OFDM frame of HIPERLAN/2 and IEEE 802.11a

and UL phases depends on the resources requested by the users and can vary from frame to frame. A MAC frame has a duration of 2 ms and consists of 500 OFDM symbols.

MC-TDMA is applied as a multiple access scheme within IEEE 802.11a and HIPERLAN/2, where within the DL and UL phase different time slots are allocated to different users. Each time slot consists of several OFDM symbols.

The OFDM frame structure specified by HIPERLAN/2 and IEEE 802.11a is shown in Figure 5-12. The frame of 2 ms duration starts with up to 10 short pilot symbols,

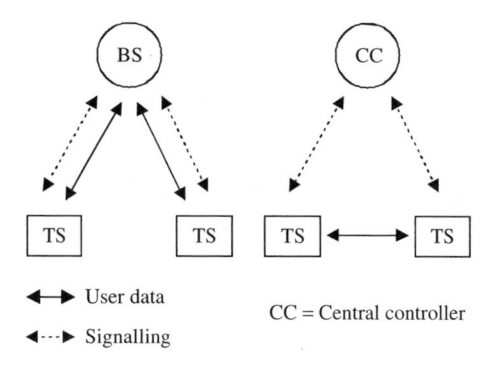

Figure 5-13 Connection types supported by HIPERLAN/2

depending on the frame type. These pilot symbols are used for coarse frequency synchronization, frame detection and automatic gain control (AGC). The following two OFDM symbols contain pilots used for fine frequency synchronization and channel estimation. The OFDM frame has four pilot sub-carriers, which are the sub-carriers -21, -7, 7 and 21. These pilot sub-carriers are used for compensation of frequency offsets. The sub-carrier 0 is not used to avoid problems with DC offsets.

HIPERLAN/2 supports two connection types. The first is called *centralized mode* and corresponds to the classical WLAN infrastructure network connection. The second is called *direct mode, i.e., peer-to-peer communication*, and enables that two mobile terminals communicate directly with each other; only the link control is handled by a so-called central controller (CC). The principle of both connection types is shown in Figure 5-13.

5.3.3.2 FEC Coding and Modulation

The IEEE 802.11a, HIPERLAN/2, and MMAC standards support the modulation schemes BPSK, QPSK, 16-QAM and 64-QAM, in combination with punctured convolutional codes (CC) with rates in the range of 1/2 up to 3/4.

The different FEC and modulation combinations supported by IEEE 802.11a are shown in Table 5-5. This flexibility offers a good trade-off between coverage and data rate.

5.3.4 Transmission Performance

5.3.4.1 Transmission Capacity

As shown in Table 5-6, the use of flexible channel coding and modulation in the IEEE 802.11a standard provides up to 8 physical modes (PHY modes), i.e., combinations of FEC and modulation. The data rates that can be supported are in the range of 6 Mbit/s up to 54 Mbit/s and depend on the coverage and channel conditions.

Note that the data rates supported by HIPERLAN/2 differ only slightly from those of Table 5-6. The data rate 24 Mbit/s is replaced by 27 Mbit/s and the data rate of 48 Mbit/s is not defined.

It should be emphasized that the overall data rate in a cellular system is limited by the coverage distance and the amount of interference due to a dense frequency reuse. Indeed, a global capacity optimization per cell (or per sector) can be achieved if the PHY mode

Table 5-5 FEC and modulation parameters of IEEE 802.11a

Modulation	Code rate R	Coded bits per sub-channel	Coded bits per OFDM symbol	Data bits per OFDM symbol
BPSK	1/2	1	48	24
BPSK	3/4	1	48	36
QPSK	1/2	2	96	48
QPSK	3/4	2	96	72
16-QAM	1/2	4	192	96
16-QAM	3/4	4	192	144
64-QAM	2/3	6	288	192
64-QAM	3/4	6	288	216

Table 5-6 Data rates of IEEE 802.11a

	PHY Mode	Data rate (Mbit/s)
1	BPSK, CC1/2	6
2	BPSK, CC3/4	9
3	QPSK, CC1/2	12
4	QPSK, CC3/4	18
5	16-QAM, CC1/2	24
6	16-QAM, CC3/4	36
7	64-QAM, CC2/3	48
8	64-QAM, CC3/4	54

is adapted to each terminal station link condition individually. Results in [10] show that compared to a single PHY mode, the areal spectral efficiency can be at least doubled if adaptive PHY modes are employed.

5.3.4.2 Link Budget

The transmit power, depending on the coverage distance, is given by

$$P_{T_x} = Path\ loss + P_{Noise} - G_{Antenna} + Fade\ Margin + Rx_{loss} + \frac{C}{N}, \quad (5.2)$$

where

$$Path\ loss = 10\log_{10}\left(\frac{4\pi f_c d}{c}\right)^n \quad (5.3)$$

Table 5-7 Minimum receiver sensitivity thresholds for HIPERLAN/2

Nominal bit rate [Mbit/s]	Minimum sensitivity
6	−85 dBm
9	−83 dBm
12	−81 dBm
18	−79 dBm
27	−75 dBm
36	−73 dBm
54	−68 dBm

is the propagation path loss, d represents the distance between the transmitter and the receiver, f_c is the carrier frequency, and c is the speed of light. In case of WLANs, n can be estimated to be in the order of 3 to 4.

$$P_{Noise} = FN_{Thermal} = FKTB \qquad (5.4)$$

is the noise power at the receiver input, where F is the receiver noise factor (about 6 dB), K is the Blotzman constant ($K = 1.38 \cdot 10^{-23}$ J/K), T is the temperature in Kelvin, and B is the total occupied Nyquist bandwidth. The noise power is expressed in dBm. $G_{Antenna}$ is the sum of the transmit and receive antenna gains, expressed in dBi. In WLAN, the terminal station antenna can be omni-directional with 0 dBi gain, but the base station antenna may have a gain of about 14 dBi. *FadeMargin* is the margin needed to counteract the fading and is about 5 to 10 dB. Rx_{loss} is the margin for all implementation losses and all additional uncertainties such as interference. This margin can be about 5 dB. C/N is the carrier-to-noise power ratio (equivalent to E_s/N_0) for BER $= 10^{-6}$. By considering a transmission power of about 23 dBm and following the above parameters for an omni-directional antenna, the maximum coverage for the robust PHY mode at 2.4 GHz carrier frequency can be estimated to be about 300 m.

The minimum receiver sensitivity thresholds for HIPERLAN/2, depending on the PHY mode, i.e., data rate for a BER of 10^{-6}, are given in Table 5-7. The receiver sensitivity threshold Rx_{th} is defined by

$$Rx_{th} = P_{Noise} + \frac{C}{N} + Rx_{loss}. \qquad (5.5)$$

5.4 Fixed Wireless Access below 10 GHz

The aim of the fixed broadband wireless access (FWA) systems HIPERMAN and IEEE 802.16a is to provide wireless high speed services, e.g., IP to fixed positioned residential customer premises and to small offices/home offices (SOHO) with a coverage area up to 20 km. To maintain reasonably low RF costs for the residential market as well as good penetration of the radio signals, the FWA systems should typically use below 10 GHz carrier frequencies, e.g., the MMDS band (2.5–2.7 GHz) in the USA or around the 5 GHz band in Europe and other countries.

Advantages of FWA include rapid deployment, high scalability, lower maintenance and upgrade costs compared to cable. Nevertheless, the main goal of a future-proof FWA system for the residential market has to be an increase in spectral efficiency, in coverage, in flexibility for the system/network deployment, in simplification of the installation and, above all, reliable communication even in non-line of sight (NLOS) conditions has to be guaranteed. In a typical urban or suburban deployment scenario, at least 30% of the subscribers have an NLOS connection to the base station. In addition, for most users LOS is obtained through rooftop positioning of the antenna that requires very accurate pointing, thereby making the installation both time- and skill-consuming. Therefore, a system operating in NLOS conditions enabling self-installation will play an important role in the success of FWA for the residential market.

In response to these trends under the ETSI-Broadband Radio Access Networks (BRAN) project the HIPERMAN (HIgh PErfoRmance Metropolitan Area Networks, HM) and under the IEEE 802.16 project the WirelessMan (Wireless Metropolitan Area Networks, WMAN) specification are currently under standardization. Both standards will offer a wide range of data services (especially IP) for residential (i.e., single- or multi-dwelling household) customers and for small to medium-sized enterprises by adopting multi-carrier transmission for radio frequencies (RF) below 10 GHz.

5.4.1 Network Topology

As shown in Figure 5-14, the FWA system will be deployed to connect user network interfaces (UNIs) physically fixed in customer premises to a service node interface (SNI) of a broadband core network (e.g., IP), i.e., for last mile connections. The base station

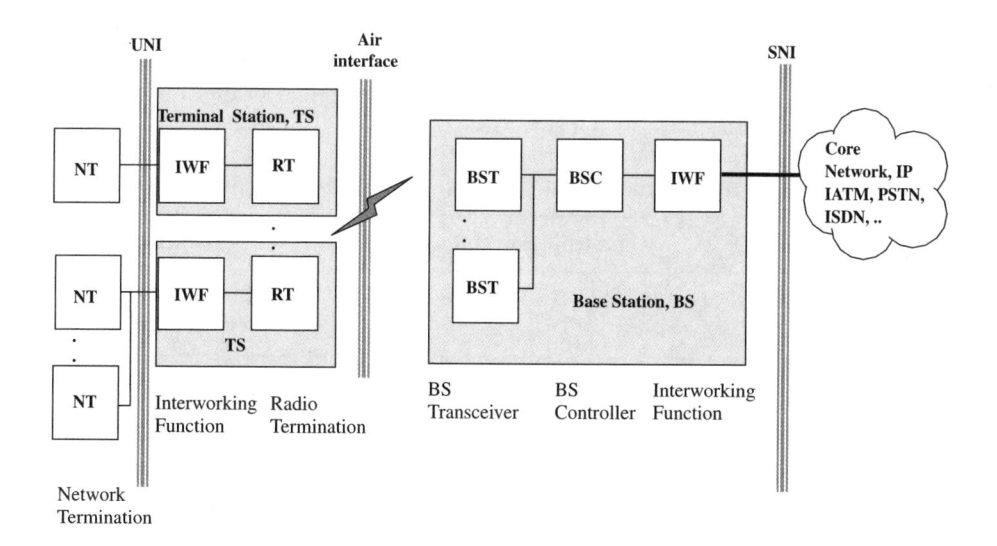

Figure 5-14 Simplified FWA reference model

typically manages communications of more than one carrier or sector. For each base station sector one antenna or more is positioned to cover the deployment region. The terminal station antenna can be directional or omni-directional. At the terminal station side the network termination (NT) interface connects the terminal station with the local user network (i.e., LAN).

The FWA network deployments will potentially cover large areas (i.e., cities, rural areas) [9][12]. Due to the large capacity requirements of the network, a high amount of spectrum with high transmission ranges (up to 20 km) is needed. For instance a typical network may therefore consist of several cells each covering a part of the designated deployment area. Each cell will operate in a point- to multi-point (PMP) or mesh manner.

Two duplex schemes can be used: i) frequency division duplex (FDD) and ii) time division duplex (TDD). The channel size is between 1.5 to 28 MHz wide in both the FDD and the TDD case. The downlink data stream transmitted to different terminal stations is multiplexed in the time domain by MC-TDM (Time Division Multiplexing) using OFDM or OFDMA transmission. In the uplink case, MC-TDMA (Time Division Multiple Access) will be used with OFDM or OFDMA.

5.4.2 Channel Characteristics

Table 5-8 lists some target frequency bands below 10 GHz carrier frequency. The channel bandwidths depend on the used carrier frequency as well. The use of these radio bands provides a physical environment where, due to its wavelength characteristics, line of sight (LOS) is not necessary but multipath may be significant (delay spread is similar to DVB-T up to 0.2 ms). Doppler effects are negligible due to the fixed positioned terminals. Therefore, multi-carrier transmission to combat the channel frequency selectivity (NLOS conditions) is an excellent choice for FWA below 10 GHz, i.e., HIPERMAN and WirelessMan.

In order to maximize the capacity, i.e., the spectral efficiency, and coverage per cell/sector, several advanced technologies will be adopted [9][12]: i) adaptive coding, ii) adaptive modulation, and iii) adaptive power control mechanisms.

5.4.3 Multi-Carrier Transmission Schemes

The draft physical layer of the these standards supports multi-carrier transmission modes. The basic transmission mode is OFDM. Depending on the selected time/frequency

Table 5-8 Example of some target frequency bands for HIPERMAN and WirelessMan

Frequency bands (GHz)	Allocated Channel Spacing	Remarks
2.150–2.162 2.500–2.690	125 kHz to (n × 6) MHz	USA CFR 47 part 21.901, part 74.902 (MMDS)
3.400–4.200	1.75 to 30 MHz paired with 1.75 to 30 MHz (FDD)	CEPT/ERC Rec.12-08 E/ITU-R F.1488, Annex II
3.400–3.700	n × 25 MHz (single or paired) (FDD or TDD)	ITU-R F.1488, Annex I, Canada SRSP-303.4
5.470–5.725	n × 20 MHz	CEPT/ERC Rec.70-03

parameters, the system can support TDMA as well as OFDMA. This flexibility ensures that the system can be optimized for short bursty applications as well as more streaming applications. The main advantage of using OFDMA with high numbers of sub-carriers with the same data rate as the OFDM mode is to provide higher coverage, i.e., a longer guard time.

In the pure OFDM mode, a total of 256 sub-carriers will be transmitted at once. The downlink applies time division multiplexing (TDM) and the uplink uses time division multiple access (TDMA). In the OFDMA mode, the channel bandwidth is divided into 2048 sub-carriers, where each user is assigned to a given group of sub-carriers. Therefore, the number of sub-carriers varies from 256 to 2048.

As shown in Figure 5-15, there are several sub-carrier types:

— data sub-carriers,
— pilot sub-carriers (boosted and used for channel estimation purposes),
— null sub-carriers (used for guard bands and DC sub-carrier).

5.4.3.1 OFDM Mode

In Figure 5-16, the OFDM frame structure for the downlink (DL) and the uplink (UL) in case of FDD is illustrated. The frame has a nominal duration between 2–5 ms. The total frame length is an integer multiple of OFDM symbols, such that the actual frame duration is nearest to the nominal frame duration.

Important parameters of the OFDM mode are summarized in Tables 5-9 and 5-10.

The downlink is a TDM transmission. Every downlink frame starts with a preamble. The preamble is used for synchronization purposes. It is followed by a control channel zone and downlink data bursts. Each burst uses different physical modes and each downlink burst consists of an integer number of OFDM symbols.

The uplink is a TDMA transmission. Every uplink burst emanating from each terminal is preceded by a preamble. Each uplink burst, independent of channel coding and modulation, transmits an integer number of OFDM symbols as well.

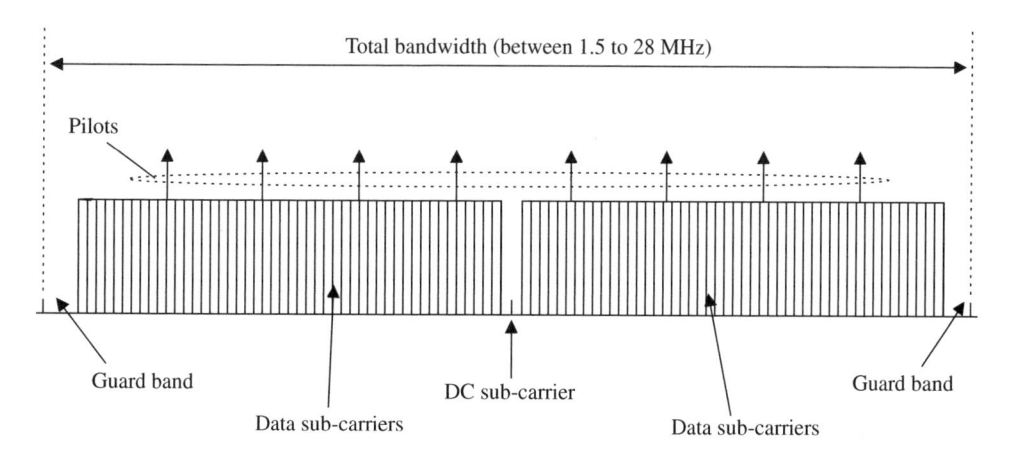

Figure 5-15 Example of sub-carrier allocation

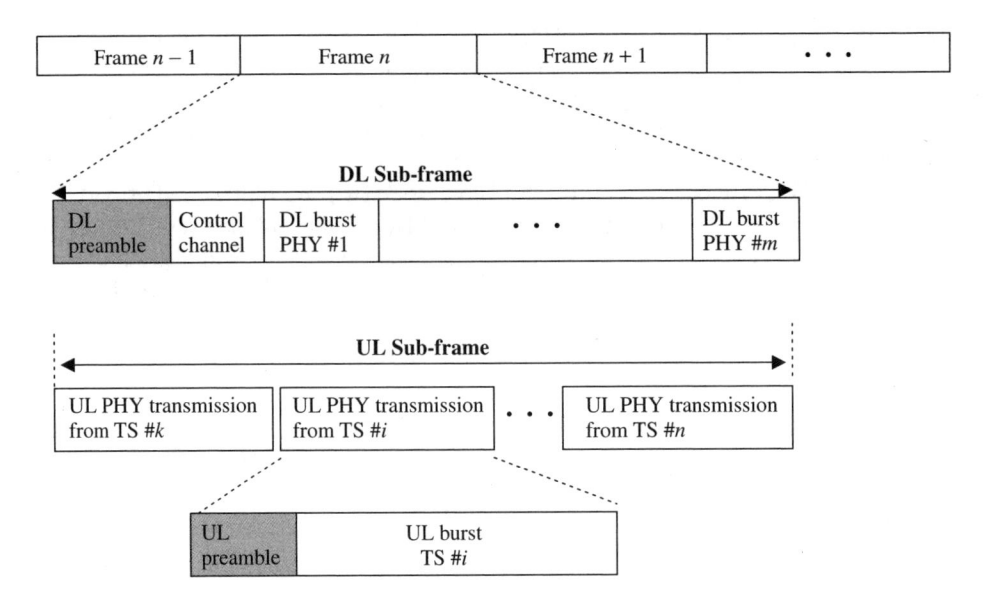

Figure 5-16 Downlink and uplink frame structure for FDD mode

Table 5-9 OFDM mode parameters

Parameter	Value
Number of DC sub-carriers	1
Number of guard sub-carriers, left/right	28/27
Number of used sub-carriers	200
Total number of sub-carriers	256
Number of fixed located pilot sub-carriers	8

Table 5-10 OFDM parameters for ETSI channelization with 256 sub-carriers

Bandwidth (MHz)	T_s (µs)	T_g (µs)			
1.75	128	4	8	16	32
3.5	64	2	4	8	16
7	32	1	2	4	8
14	16	1/2	1	2	4
28	8	1/4	1/2	1	2

Fixed Wireless Access below 10 GHz

The uplink preamble consists of 2 × 128 samples with guard time (= one OFDM symbol). The downlink preamble is made up of two OFDM symbols: the first one carries 4 × 64 samples and the second one transmits 2 × 128 samples. These reference samples have good correlation properties which eases the synchronization tasks. The power of the uplink and downlink preambles is boosted by 3 dB compared to the data part.

5.4.3.2 OFDMA Mode

As described in Chapter 3, in OFDMA only a part of the sub-carriers may be used for data transmission. A set of sub-carriers, called a sub-channel, will be assigned to each user (see Figure 5-17). For both uplink and downlink the used sub-carriers are allocated to pilot and data sub-carriers. However, there is a small difference between the uplink and the downlink sub-carrier allocation. In the downlink, there is one set of common pilot carriers spread over all the bandwidth, whereas in the uplink each sub-channel contains its own pilot sub-carriers. This is since the downlink is broadcast to all terminal stations, but in the uplink each sub-channel is transmitted from a different terminal station. The goal of these pilot sub-carriers is to estimate the channel characteristics.

For OFDMA with FDD, the frame duration is an integer number of three OFDM symbols, where the actual frame duration is nearest to the nominal frame duration between 2–5 ms.

In addition to the sub-channel dimension (set of sub-carriers), OFDMA uses the time dimension for data transmission. An uplink or downlink burst in OFDMA has a two-dimensional allocation: a transmit burst is mapped to a group of contiguous sub-channels and to contiguous OFDM symbols. Each data packet is first segmented into blocks sized to fit into one FEC block. Then, each FEC block spans one OFDMA sub-channel in the sub-channel axis and three OFDM symbols in time axis. The FEC blocks are mapped such that the lowest numbered FEC block occupies the lowest numbered sub-channel in the lowest numbered OFDM symbol. The mapping is continued such that the OFDMA sub-channel index is increased for each FEC block mapped. When the edge of the data region is reached, the mapping will continue from the lowest numbered OFDMA sub-channel in the next OFDM symbol (see Figure 5-18).

For the uplink transmission a number of sub-channels over a number of OFDM symbols is assigned per terminal station. The number of OFDM symbols shall be equal to $1 + 3N$,

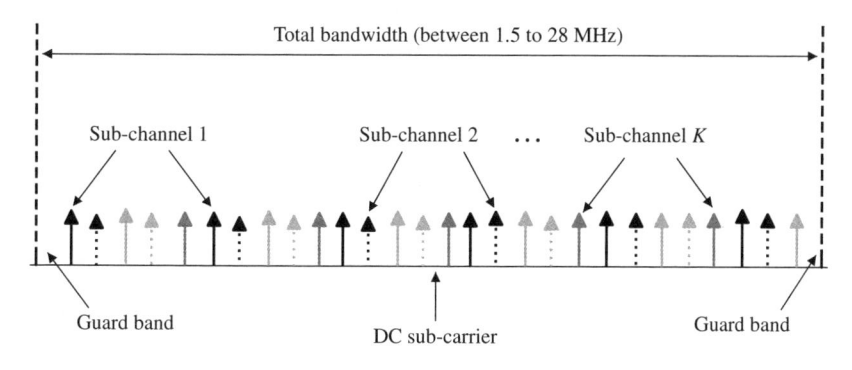

Figure 5-17 Example of OFDMA frequency allocation for K users

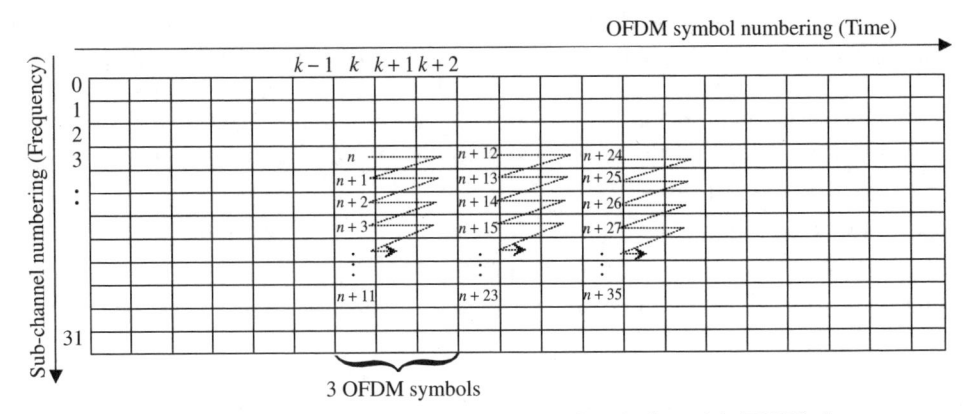

Figure 5-18 Example of mapping of FEC blocks to OFDMA sub-channels and symbols

where N is a positive integer. In other words the smallest number of allocated sub-channels per terminal station is one sub-channel for a duration of four OFDM symbols, where the first OFDM symbol is a *preamble*.

The transmission of the downlink is performed on the sub-channels of the OFDM symbol. The number of sub-channels needed for different coding and modulation is transmitted in the downlink control channel.

OFDMA Downlink Sub-Carrier Allocation
As shown in Figure 5-19, for the downlink the pilots will have both fixed and variable positions [12]. The variable pilot location structure is repeated every four symbols. The allocated data sub-carriers are partitioned into groups of contiguous sub-carriers. The number of groups is therefore equal to the number of sub-carriers per sub-channel.

In Table 5-11, the basic OFDMA downlink parameters are given.

OFDMA Uplink Sub-Carrier Allocation
The total number of sub-carriers used are first partitioned into sub-channels (see Figure 5-20). Within each sub-channel, there are 48 data sub-carriers, one fixed located pilot sub-carrier and four variable located pilot sub-carriers. The fixed located pilot is always at sub-carrier 26 within each sub-channel. The variable located pilot sub-carriers are repeated every 13 symbols, whereas the fixed and the variable positioned pilots will never coincide.

In Tables 5-12 and 5-13 the OFDMA uplink parameters and guard times are given, respectively. Note that for OFDMA with 2048 sub-carriers the symbol duration and guard times will be four times longer than with 256 sub-carriers.

5.4.3.3 FEC Coding and Modulation

The FEC consists of the concatenation of a Reed–Solomon (RS) outer code and a punctured convolutional inner code. Block Turbo codes and convolutional Turbo codes can also be used. Different modulation schemes with Gray mapping (QPSK, 16-QAM, and 64-QAM) are employed.

Fixed Wireless Access below 10 GHz

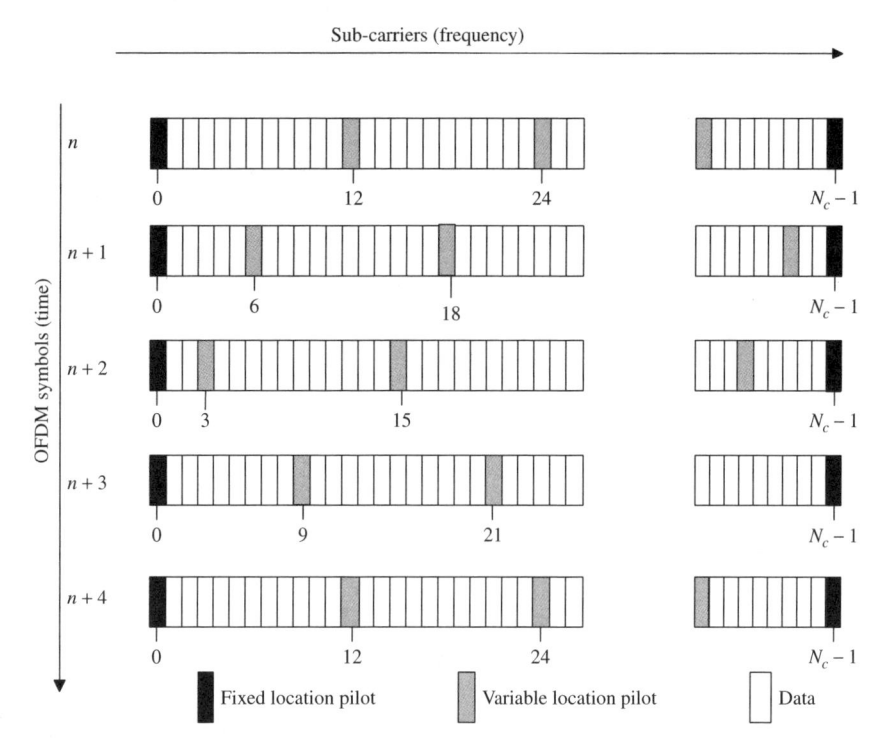

Figure 5-19 Example of sub-carrier allocation in the downlink

Table 5-11 Example of OFDMA downlink sub-carrier allocation

Parameter	Value
Number of DC sub-carriers	1
Number of guard sub-carriers, left/right	173/172
Number of used sub-carriers	1702
Total number of sub-carriers	2048
Number of variable located pilots	142
Number of fixed located pilots	32
Total number of pilots	166 (where 8 fixed and variable pilots coincide)
Number of data sub-carriers	1536
Number of sub-channels	32
Number of sub-carriers per sub-channel	48

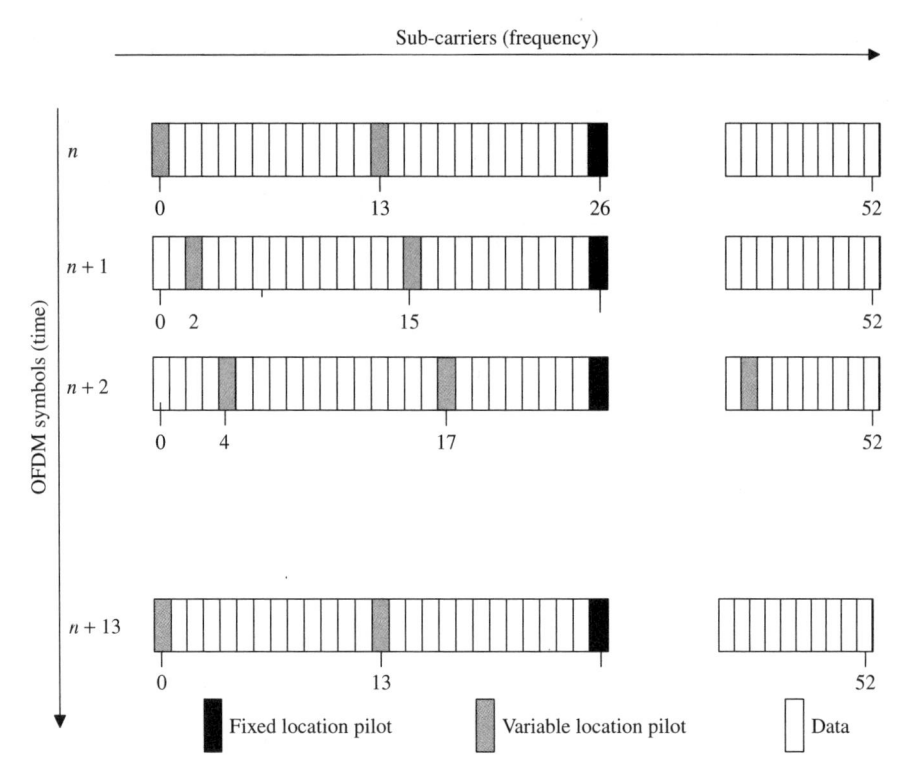

Figure 5-20 Example of sub-carrier allocation in the uplink

Table 5-12 Example of OFDMA uplink sub-carriers allocation

Parameter	Value
Number of DC sub-carriers	1
Number of guard sub-carriers, left/right	176/175
Number of used sub-carriers	1696
Total number of sub-carriers	2048
Number of sub-channels	32
Number of data sub-carriers per sub-channel	53
Number of pilot sub-carriers per sub-channel	5

The outer RS code can be shortened and punctured to enable variable block sizes and variable error correction capability. The RS mother code is a RS(255, 239, $t = 8$) code. The inner convolutional code can be punctured as well to provide several inner code rates. The mother convolutional code is based on memory 6, rate 1/2. Each RS block is extended with zero tail bits for trellis termination within the convolutional code.

Table 5-13 OFDMA parameters for ETSI channelization with 2048 sub-carriers

Bandwidth (MHz)	T_s (µs)	T_g (µs)			
1.75	1024	32	64	128	256
3.5	512	16	32	64	128
7	256	8	16	32	64
14	128	4	8	16	32
28	64	2	4	8	16

Table 5-14 Channel coding and modulation parameters for uplink and downlink (OFDM)

PHY mode #n	Modulation	Inner coding	Outer coding	Overall coding rate	Efficiency (bit/s/Hz)
1	QPSK	CC 2/3	RS(32,24,4)	1/2	1.0
2	QPSK	CC 5/6	RS(40,36,2)	3/4	1.5
3	16-QAM	CC 2/3	RS(60,48,8)	1/2	2.0
4	16-QAM	CC 5/6	RS(80,72,4)	3/4	3.0
5 (optional)	64-QAM	CC 3/4	RS(108,96,6)	2/3	4.0
6 (optional)	64-QAM	CC 5/6	RS(120,108,6)	3/4	4.5

Table 5-15 Channel coding and modulation parameters for uplink and downlink (OFDMA)

PHY mode #n	Modulation	Inner coding	Outer coding	Overall coding rate	Efficiency (bit/s/Hz)
1	QPSK	CC 2/3	RS(24,18,3)	1/2	1.0
2	QPSK	CC 5/6	RS(30,26,2)	~3/4	~1.5
3	16-QAM	CC 2/3	RS(48,36,6)	1/2	2.0
4	16-QAM	CC 5/6	RS(60,54,3)	3/4	3.0
5 (optional)	64-QAM	CC 3/4	RS(80,72,4)	2/3	4.0
6 (optional)	64-QAM	CC 5/6	RS(90,82,4)	~3/4	~4.5

Tables 5-14 and 5-15 show the detailed FEC parameters for the OFDM and OFDMA modes. The resulting number of bytes per FEC block matches an integer number of OFDM symbols. As 64-QAM is optional, the codes for this modulation shall only be implemented if the modulation is implemented. As these tables show, different coding and modulation schemes are supported. The lowest concatenated coding scheme with code rate 1/2 is used for control information.

5.4.4 Transmission Performance
5.4.4.1 Transmission Capacity

The maximum data rate achievable for both OFDM and OFDMA modes depend on the used channel bandwidth, PHY mode, and guard time. In Tables 5-16 and 5-17 the data

Table 5-16 Data rate in Mbit/s for OFDM mode with 7 MHz channelization (256 sub-carriers)

T_G	PHY#1	PHY#2	PHY#3	PHY#4	PHY#5	PHY#6
1/32 T_s	5.94	8.91	11.88	17.82	23.76	26.73
1/4 T_s	4.9	7.35	9.80	14.70	19.60	22.05

Table 5-17 Data rate in Mbit/s for OFDMA mode with 7 MHz channelization (2048 sub-carrier)

T_G	PHY#1	PHY#2	PHY#3	PHY#4	PHY#5	PHY#6
1/32 T_s	5.82	8.73	11.64	17.45	23.27	26.18
1/4 T_s	4.8	7.2	9.6	14.40	19.20	21.6

Table 5-18 Example of minimum receiver threshold sensitivity (dBm)

Bandwidth (MHz)	QPSK		16-QAM		64-QAM	
	1/2	3/4	1/2	3/4	2/3	3/4
1.5	−91	−89	−84	−82	−78	−76
1.75	−90	−87	−83	−81	−77	−75
3	−88	−86	−81	−79	−75	−73
3.5	−87	−85	−80	−78	−74	−72
5	−86	−84	−79	−77	−72	−71
6	−85	−83	−78	−76	−72	−70
7	−84	−82	−77	−75	−71	−69
10	−83	−81	−76	−74	−69	−68
12	−82	−80	−75	−73	−69	−67
14	−81	−79	−74	−72	−68	−66
20	−80	−78	−73	−71	−66	−65

Table 5-19 Estimated SNR in AWGN

Modulation	Coding rate	Receiver E_s/N_0 (dB)
QPSK	1/2	9.4
	3/4	11.2
16-QAM	1/2	16.4
	3/4	18.2
64-QAM	2/3	22.7
	3/4	24.4

rate for the OFDM and OFDMA modes for different PHY modes in a 7 MHz ETSI channelization raster is given. Both systems provide the same data rate. However, at the expense of a slightly lower data rate, the OFDMA coverage could be higher.

5.4.4.2 Link Budget

For the link budget evaluation a bit error rate (BER) of 10^{-6} is considered. The receiver power threshold is shown in Table 5-18. These values are derived assuming a 5 dB implementation loss, a noise factor of 7 dB and receiver E_s/N_0 values which are listed in Table 5-19. The remaining parameters are assumed as follows:

— The propagation path loss power can be estimated to be $n = 2$ to 3;
— The terminal station antenna is omni-directional with 0 dBi gain and the base station antenna with 90° sectorization has a 14 dBi gain.
— The fade margin to counteract the fading is 5 to 10 dB.

By considering a transmission power of 20 dBm and following the above parameters, for an omni-directional antenna, the maximum coverage can be about 10 km. However, with a narrowband directional antenna, the coverage can be increased to over 20 km.

5.5 Interaction Channel for DVB-T: DVB-RCT

The VHF and UHF frequency bands (typically from 120–860 MHz) are reserved for broadcast TV services and to convey uplink information from the subscriber premises, i.e., terminal stations, to the base station for interactive services. In addition to the uplink channel, interactive services also require a downlink channel for transmitting messages, uplink channel access control commands, and other data. The downlink interactive information can be either embedded in the broadcast channels or defined as a standalone channel specifically devoted to interaction with subscribers.

5.5.1 *Network Topology*

A DVB-T interactive network architecture is illustrated in Figure 5-21 [6]. This could be seen as a *point to multi-point (PMP)* network topology consisting of a base station and several subscribers (terminal stations). Like other digital access networks, DVB-RCT networks are also intended to offer a variety of services requiring different data rates. Therefore, the multiple access scheme needs to be flexible in terms of data rate assignment to each subscriber. Furthermore, another major constraint for the choice of

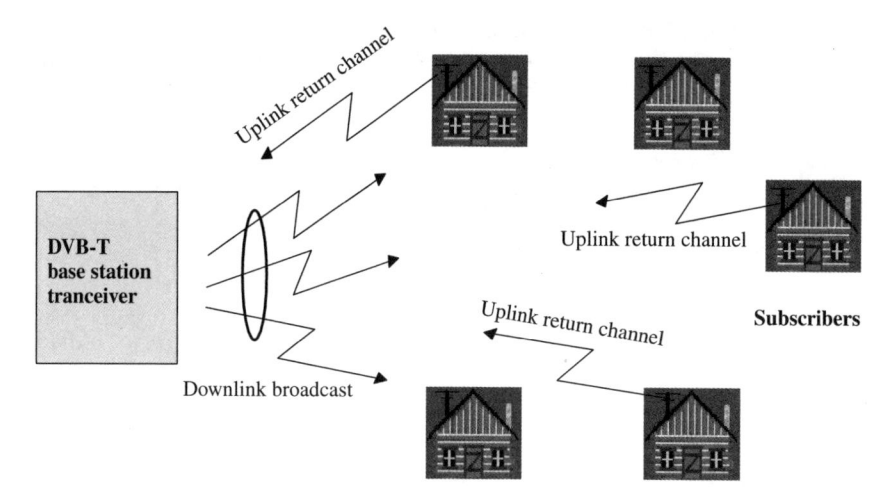

Figure 5-21 DVB-RCT network architecture

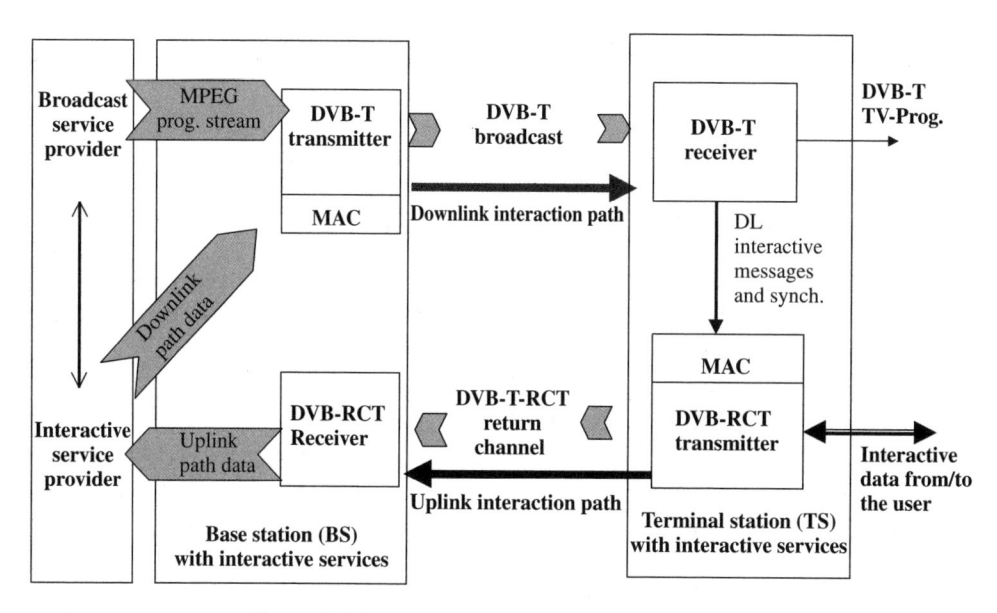

Figure 5-22 Overview of the DVB-RCT standard

parameters of the DVB-RCT specification is to employ the existing infrastructure used for broadcast DVB-T services.

As shown in Figure 5-22, the interactive downlink path is embedded in the broadcast channel, exploiting the existing DVB-T infrastructure [7]. The access for the uplink interactive channels carrying the return interaction path data is based on a combination of OFDMA and TDMA type of multiple access scheme [6].

The downlink interactive information data is made up of MPEG-2 transport stream packets with a specific header that carries the medium access control (MAC) management data. The MAC messages control the access of the subscribers, i.e., terminal stations, to the shared medium. These embedded MPEG-2 transport stream packets are carried in the DVB-T broadcast channel (see Figure 5-22).

The uplink interactive information is mainly made up of ATM cells mapped onto physical bursts. ATM cells include application data messages and MAC management data.

To allow access by multiple users, the VHF/UHF radio frequency return channel is partitioned both in the frequency and time domain, using frequency and time division. Each subscriber can transmit his data for a given period of time on a given sub-carrier, resulting in a combination of OFDMA and TDMA multiple access.

A global synchronization signal, required for the correct operation of the uplink demodulator at the base station, is transmitted to all users via global DVB-T timing signals. Time synchronization signals are conveyed to all users through the broadcast channel, either within the MPEG2 transport stream or via global DVB-T timing signals. In other words, the DVB-RCT frequency synchronization is derived from the broadcast DVB-T signal whilst the time synchronization results from the use of MAC management packets conveyed through the broadcast channel. Furthermore, the so-called periodic *ranging signals* are transmitted from the base station to individual terminal stations for timing misalignment adjustment and power control purposes.

The DVB-RCT OFDMA based system employs either 1024 (1k) or 2048 (2k) sub-carriers and operates as follows:

— Each terminal station transmits one or several low bit rate modulated sub-carriers towards the base station;
— The sub-carriers are frequency-locked and power-ranged and the timing of the modulation is synchronized by the base station. In other words, the terminal stations derive their system clock from the DVB-T downstream. Accordingly, the transmission mode parameters are fixed in a strict relationship with the DVB-T downstream;
— On the reception side, the uplink signal is demodulated, using an FFT process, like the one performed in a DVB-T receiver.

5.5.2 Channel Characteristics

As in the downlink terrestrial channel, the return channels suffer from high multipath propagation delays [7].

In the DVB-RCT system, the downlink interaction data and the uplink interactive data are transmitted in the same radio frequency bands, i.e., VHF/UHF bands III, IV, and V. Hence, the DVB-T and DVB-RCT systems may form a bi-directional FDD communication system which shares the same frequency bands with sufficient duplex spacing. Thus, it is possible to benefit from common features in regard to RF devices and parameters (e.g., antenna, combiner, propagation conditions). The return channel (RCT) can be also located in any free segment of an RF channel, taking into account existing national and regional analog television assignments, interference risks, and future allocations for DVB-T.

5.5.3 Multi-Carrier Uplink Transmission

The method used to organize the DVB-RCT channel is inspired by the DVB-T standard. The DVB-RCT RF channel provides a grid of time-frequency slots, each slot usable by any terminal station. Hence, the concept of DVB uplink channel allocation is based on a combination of OFDMA with TDMA. Thus, the uplink is divided into a number of time slots. Each time slot is divided in the frequency domain into groups of sub-carriers referred to as sub-channels. The MAC layer controls the assignment of sub-channels and time slots by resource requests and grant messages.

The DVB-RCT standard provides two types of sub-carrier shaping, where out of these only one is used at any time. The shaping functions are:

— *Nyquist shaping* in the time domain on each sub-carrier to provide immunity against both ICI and ISI. A square root raised cosine pulse with a roll-off factor $\alpha = 0.25$ is employed. The total symbol duration is 1.25 times the inverse of the sub-carrier spacing.
— *Rectangular shaping* with guard interval T_g that has a possible value of $T_s/4$, $T_s/8$, $T_s/16$, $T_s/32$, where T_s is the useful symbol duration (without guard time).

5.5.3.1 Transmission Modes

The DVB-RCT standard provides six transmission modes characterized by a dedicated combination of the maximum number of sub-carriers used and their sub-carrier spacings [6]. Only one transmission mode is implemented in a given RCT radio frequency channel, i.e., transmission modes are not mixed.

The sub-carrier spacing governs the robustness of the system in regard to the possible synchronization misalignment of any terminal station. Each value implies a given maximum transmission cell size and a given resistance to the Doppler shift experienced when the terminal station is in motion, i.e., in case of portable receivers. The three targeted DVB-RCT sub-carrier spacing values are defined in Table 5-20.

Table 5-21 gives the basic DVB-RCT transmission mode parameters applicable for the 8 MHz and 6 MHz radio frequency channels with 1024 or 2048 sub-carriers. Due to the combination of the above parameters, the DVB-RCT final bandwidth is a function of sub-carrier spacing and FFT size. Each combination has a specific trade-off between frequency diversity and time diversity, and between coverage range and portability/mobility capability.

5.5.3.2 Time and Frequency Frames

Depending on the transmission mode in operation, the total number of allocated sub-carriers for uplink data transmission is 1024 carriers (1k mode) or 2048 carriers (2k mode) (see Figure 5-23). Table 5-22 shows the main parameters.

Table 5-20 DVB-RCT targeted sub-carrier spacing for 8 MHz channel

Sub-carrier spacing	Targeted sub-carrier spacing
Sub-carrier spacing 1	≈ 1 kHz (symbol duration ≈ 1000 μs)
Sub-carrier spacing 2	≈ 2 kHz (symbol duration ≈ 500 μs)
Sub-carrier spacing 3	≈ 4 kHz (symbol duration ≈ 250 μs)

Interaction Channel for DVB-T: DVB-RCT

Table 5-21 DVB-RCT transmission mode parameters for the 8 and 6 MHz DVB-T systems

Parameters	8 MHz DVB-T system		6 MHz DVB-T system	
Total number of sub-carriers	2048 (2k)	1024 (1k)	2048 (2k)	1024 (1k)
Used sub-carriers	1712	842	1712	842
Useful symbol duration	896 µs	896 µs	1195 µs	1195 µs
Sub-carrier spacing	1.116 kHz	1.116 kHz	0.837 kHz	0.837 kHz
RCT channel bandwidth	1.911 MHz	0.940 MHz	1.433 MHz	0.705 MHz
Useful symbol duration	448 µs	448 µs	597 µs	597 µs
Sub-carrier spacing	2.232 kHz	2.232 kHz	1.674 kHz	1.674 kHz
RCT channel bandwidth	3.821 MHz	1.879 MHz	2.866 MHz	1.410 MHz
Useful symbol duration	224 µs	224 µs	299 µs	299 µs
Sub-carrier spacing	4.464 kHz	4.464 kHz	3.348 kHz	3.348 kHz
RCT channel bandwidth	7.643 MHz	3.759 MHz	5.732 MHz	2.819 MHz

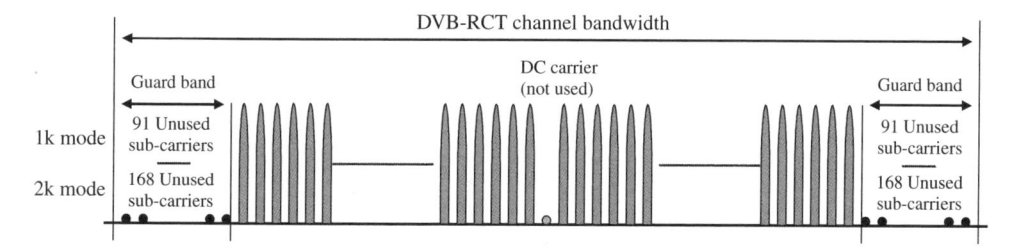

Figure 5-23 DVB-RCT channel organization for the 1k and 2k mode

Table 5-22 Sub-carrier organization for the 1k and 2k mode

Parameters	1k Mode structure	2k Mode structure
Number of FFT points	1024	2048
Overall usable sub-carriers	842	1712
Overall used sub-carriers – With burst structure 1 and 2 – With burst structure 3	840 841	1708 1711
Lower and upper channel guard band	91 sub-carriers	168 sub-carriers

Two types of transmission frames (TFs) are defined:

— *TF1*: The first frame type consists of a set of OFDM symbols which contain several data sub-channels, a null symbol and a series of synchronization/ranging symbols;
— *TF2*: The second frame type is made up of a set of general purpose OFDM symbols which contain either data or synchronization/ranging sub-channels.

Furthermore, three different burst structures are specified as follows:

— *Burst structure 1* uses one unique sub-carrier to carry the total data burst over time, with an optional frequency hopping law applied within the duration of the burst;
— *Burst Structure 2* uses four sub-carriers simultaneously, each carrying a quarter of the total data burst over time;
— *Burst structure 3* uses 29 sub-carriers simultaneously, each carrying one twenty-ninth of the total data burst over time.

These three burst structures provide a pilot-aided modulation scheme to allow coherent detection in the base station. The defined pilot insertion ratio is approximately 1/6, which means one pilot carrier is inserted for approximately every five data sub-carriers. Furthermore, they give various combinations of time and frequency diversity, thereby providing various degrees of robustness, burst duration and a wide range of bit rates to the system.

Each burst structure makes use of a set of sub-carriers called a sub-channel. One or several sub-channels can be used simultaneously by a given terminal station depending on the allocation performed by the MAC process.

Figure 5-24 depicts the organization of a TF1 frame in the time domain. It should be noted that the burst structures are symbolized regarding their duration and not regarding their occupancy in the frequency domain. The corresponding sub-carrier(s) of burst structure 1 and burst structure 2 are spread over the whole RCT channel.

Null symbol and ranging symbols always use rectangular shaping. The user symbols of TF1 use either rectangular shaping or Nyquist shaping. If the user part employs rectangular shaping, the guard interval value is identical for any OFDM symbol embedded in the whole TF1 frame. If the user part performs Nyquist shaping, the guard interval value to apply onto the Null symbol and ranging symbols is $T_s/4$. The user part of the TF1 frame is suitable to carry one burst structure 1 or four burst structure 2. The burst structures are not mixed in a given DVB-RCT channel.

The time duration of a transmission frame depends on the number of consecutive OFDM symbols and on the time duration of the OFDM symbol. The time duration of an OFDM symbol depends on

— the reference downlink DVB-T system clock,
— the sub-carrier spacing, and
— the rectangular filtering of the guard interval (1/4, 1/8, 1/16, 1/32 times T_s).

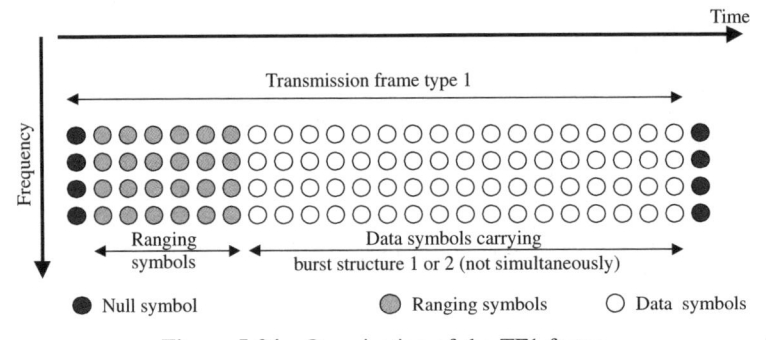

Figure 5-24 Organization of the TF1 frame

Table 5-23 Transmission frame duration in seconds with burst structure 1 and with rectangular filtering with $T_g = T_s/4$ or Nyquist filtering and for reference clock 64/7 MHz

Shaping scheme	Number of consecutive OFDM symbols	Sub-carrier spacing 1	Sub-carrier spacing 2	Sub-carrier spacing 3
Rectangular	187	0.20944 s	0.10472 s	0.05236 s
Nyquist w/o FH	195	0.2184 s	0.1092 s	0.0546 s
Nyquist with FH	219	0.24528 s	0.12264 s	0.06132 s

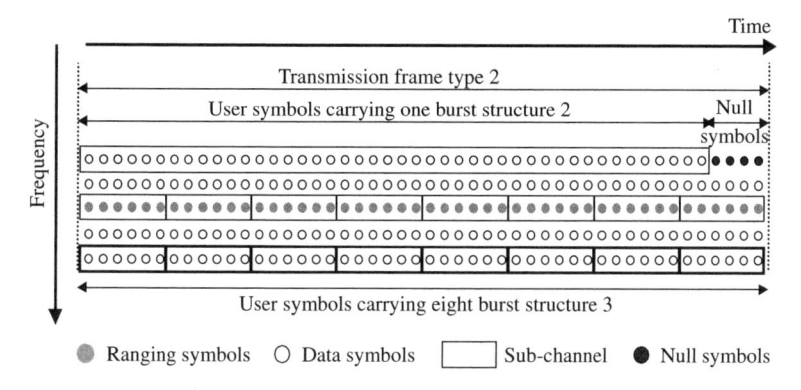

Figure 5-25 Organization of the TF2 frame

In Table 5-23, the values of the frame durations in seconds for TF1 using burst structure 1 is given.

Figure 5-25 depicts the organization of the TF2 in the time domain. The corresponding sub-carrier(s) of burst structures 2 and 3 are spread on the whole RCT channel. TF2 will be used only in the rectangular pulse shaping case. The guard interval applied on any OFDM symbol embedded in the whole TF2 is the same (i.e., either 1/4, 1/8, 1/16 or 1/32 of the useful symbol duration). The user part of the TF2 allows the usage of burst structure 3 or, optionally, burst structure 2. When one burst structure 2 is transmitted, it shall be completed by a set of four null modulated symbols to have a duration equal to the duration of eight burst structure 3.

5.5.3.3 FEC Coding and Modulation

Channel coding is based on a concatenation of a Reed–Solomon outer code and a rate-compatible convolutional inner code. Convolutional Turbo codes can also be used. Different modulation schemes (QPSK, 16-QAM, and 64-QAM) with Gray mapping are employed.

Whatever FEC is used, the data bursts produced after the encoding and mapping processes have a fixed length of 144 modulated symbols. Table 5-24 defines the original sizes of the useful data payloads to be encoded in relation to the selected physical modulation and encoding rate.

Table 5-24 Number of useful data bytes per burst

Parameters	QPSK		16-QAM		64-QAM	
FEC encoding rate	$R = 1/2$	$R = 3/4$	$R = 1/2$	$R = 3/4$	$R = 1/2$	$R = 3/4$
Number of data bytes in 144 symbols	18	27	36	54	54	81

Under the control of the base station, a given terminal station can use different successive bursts with different combinations of encoding rates. Here, the use of adaptive coding and modulation is aimed to provide flexible bit rates to each terminal station in relation to the individual reception conditions encountered in the base station.

The outer Reed–Solomon encoding process uses a shortened systematic RS(63, 55, $t = 4$) encoder over a Galois field GF(64), i.e., each RS symbol consists of 6 bits. Data bits issued from the Reed–Solomon encoder are fed to the convolutional encoder of constraint length 9. To produce the two overall coding rates expected (1/2 and 3/4), the RS and convolutional encoder have implemented the coding rates defined in Table 5-25.

The terminal station uses the modulation scheme determined by the base station through MAC messages. The encoding parameters defined in Table 5-26 are used to produce the desired coding rate in relation with the modulation schemes. It should be noted that the number of channel symbols per burst in all combinations remains constant, i.e., 144 modulated symbols per burst.

Table 5-25 Overall encoding rates

Outer RS encoding rate R_{outer}	Inner CC encoding rate R_{inner}	Overall code rate $R_{total} = R_{outer} \cdot R_{inner}$
3/4	2/3	1/2
9/10	5/6	3/4

Table 5-26 Coding parameters for combination of coding rate and modulation

Modulation code rate	RS input	CC input	Number of CC output bits
QPSK1/2	144 bits = 24 RS Symb.	32 RS Symb. = 192 bits	288
QPSK3/4	216 bits = 36 RS Symb.	40 RS Symb. = 240 bits	288
16-QAM1/2	288 bits = 48 RS Symb.	2 × 32 RS Symb. = 384 bits	576
16-QAM3/4	432 bits = 72 RS Symb.	2 × 40 RS Symb. = 480 bits	576
64-QAM1/2	432 bits = 72 RS Symb.	3 × 32 RS Symb. = 576 bits	864
64-QAM3/4	648 bits = 108 RS Symb.	3 × 40 RS Symb. = 720 bits	864

Pilot sub-carriers are inserted into each data burst in order to constitute the burst structure and are modulated according to their sub-carrier location. Two power levels are used for these pilots, corresponding to +2.5 dB or 0 dB relative to the mean useful symbol power. The selected power depends on the position of the pilot inside the burst structure.

5.5.4 Transmission Performance

5.5.4.1 Transmission Capacity

The transmission capacity depends on the used M-QAM modulation density, error control coding and the used mode with Nyquist or rectangular pulse shaping.

The net bit rate per sub-carrier for burst structure 1 is given in Table 5-27 with and without frequency hopping (FH).

5.5.4.2 Link Budget

The service range given for the different transmission modes and configurations can be calculated using the RF figures derived from the DVB-T implementation and propagation models for rural and urban areas. In order to limit the terminal station RF power to reasonable limits, it is recommended to put the complexity on the base station side by using high-gain sectorized antenna schemes and optimized reception configurations.

To define mean service ranges, Table 5-28 details the RF configurations for sub-carrier spacing 1 and QPSK 1/2 modulation levels for 800 MHz in transmission modes with burst structure 1 and 2. The operational C/N is derived from [7] and considers +2 dB implementation margin, +1 dB gain due to block Turbo code/concatenated RS and convolutional codes, and +1 dB gain when using time interleaving in Rayleigh channels.

Table 5-27 Net bit rate in kbit/s per sub-carrier for burst structure 1 using rectangular shaping

Channel spacing, modulation and coding parameters			Rectangular shaping with/without FH		Nyquist shaping without FH
			$T_G = 1/4\ T_s$	$T_G = 1/32\ T_s$	$\alpha = 0.25$
Channel spacing 1	QPSK	1/2	0.66	0.69	0.83
		3/4	0.99	1.03	1.25
	16-QAM	1/2	1.32	1.37	1.67
		3/4	1.98	2.06	2.50
	64-QAM	1/2	1.98	2.06	2.50
		3/4	2.97	3.09	3.75
Channel spacing 1	QPSK	1/2	2.63	2.75	3.33
		3/4	3.95	4.12	5.00
	16-QAM	1/2	5.27	5.50	6.67
		3/4	7.91	8.25	10.00
	64-QAM	1/2	7.91	8.25	10.00
		3/4	11.87	12.38	15.00

Table 5-28 Parameters for service range simulations

Transmission modes	Outdoor	Indoor
Antenna location	Rural/fixed	Indoor urban/portable
Frequency	800 MHz	800 MHz
Sub-carrier spacing	1 kHz	1 kHz
Modulation scheme C/N [7] Operational C/N	QPSK1/2 3.6 dB 5 dB	QPSK1/2 3.6 dB 5 dB
BS receiver antenna gain	16 dBi (60 degree)	16 dBi (60 degree)
Antenna height (user side)	Outdoor 10 m	Indoor 10 m (2nd floor)
TS Antenna gain	13 dBi (directive)	3 dBi (~omnidir.)
Cable loss	4 dB	1 dB
Duplexer loss	4 dB	1 dB (separate antennas/switch)
Indoor penetration loss	/	10 dB (mean 2nd floor)
Propagation models	ITU-R 370	OKUMURA-HATA suburban
Standard deviation for location variation	−10 dB for BS1 −5 dB for BS-2 and BS-3 (spread multi-carrier)	−10 dB for BS1 −5 dB for BS-2 and BS-3 (spread multi-carrier)

Reasonable dimensioning of the output amplifier in terms of bandwidth and intermodulation products (linearity) indicates that a transmit power of the order of 25 dBm could be achievable at low cost. It is shown in [6] that with 24 dBm transmit power, indoor reception would be possible up to a distance of 15 km, while outdoor reception would be offered up to 40 km or more.

5.6 References

[1] 3GPP (TR25.858), "High speed downlink packet access: Physical layer aspects," *Technical Report*, 2001.
[2] Atarashi H., Maeda N., Abeta S. and Sawahashi M., "Broadband packet wireless access based on VSF-OFCDM and MC/DS-CDMA," in *Proc. IEEE International Symposium on Personal, Indoor and Mobile Radio Communications (PIMRC 2002)*, Lisbon, Portugal, pp. 992–997, Sept. 2002.
[3] Atarashi H. and Sawahashi M., "Variable spreading factor orthogonal frequency and code division multiplexing (VSF-OFCDM)," in *Proc. International Workshop on Multi-Carrier Spread-Spectrum & Related Topics (MC-SS 2001)*, Oberpfaffenhofen, Germany, pp. 113–122, Sept. 2001.
[4] Burow R., Fazel K., Höher P., Kussmann H., Progrzeba P., Robertson P. and Ruf M., "On the Performance of the DVB-T system in mobile environments," in *Proc. IEEE Global Telecommunications Conference (GLOBECOM'98), Communication Theory Mini Conference*, Sydney, Australia, Nov. 1998.

References

[5] ETSI DAB (EN 300 401), "Radio broadcasting systems; digital audio broadcasting (DAB) to mobile, portable and fixed receivers," Sophia Antipolis, France, April 2000.

[6] ETSI DVB RCT (EN 301 958), "Interaction channel for digital terrestrial television (RCT) incorporating multiple access OFDM," Sophia Antipolis, France, March 2001.

[7] ETSI DVB-T (EN 300 744), "Digital video broadcasting (DVB); framing structure, channel coding and modulation for digital terrestrial television," Sophia Antipolis, France, July 1999.

[8] ETSI HIPERLAN (TS 101 475), "Broadband radio access networks HIPERLAN Type 2 functional specification – Part 1: Physical layer," Sophia Antipolis, France, Sept. 1999.

[9] ETSI HIPERMAN (Draft TS 102 177), "High performance metropolitan area network, Part A1: Physical Layer," Sophia Antipolis, France, Feb. 2003.

[10] Fazel K., Decanis C., Klein J., Licitra G., Lindh L. and Lebret Y.Y., "An overview of the ETSI-BRAN HA physical layer air interface specification," *in Proc. IEEE International Symposium on Personal, Indoor and Mobile Radio Communications (PIMRC 2002)*, Lisbon, Portugal, pp. 102–106, Sept. 2002.

[11] IEEE 802.11 (P802.11a/D6.0), "LAN/MAN specific requirements – Part 2: Wireless MAC and PHY specifications – high speed physical layer in the 5 GHz band," IEEE 802.11, May 1999.

[12] IEEE 802.16ab-01/01, "Air interface for fixed broadband wireless access systems – Part A: Systems between 2 and 11 GHz," IEEE 802.16, June 2000.

6

Additional Techniques for Capacity and Flexibility Enhancement

6.1 Introduction

As shown in Chapter 1, wireless channels suffer from attenuation due to the destructive addition of multipath propagation paths and interference. Severe attenuation makes it difficult for the receiver to detect the transmitted signal unless some additional, less-attenuated replica of the transmitted signal are provided. This principle is called diversity and it is the most important factor in achieving reliable communications. Examples of diversity techniques are:

— *Time diversity*: Time interleaving in combination with channel coding provides replicas of the transmitted signal in the form of redundancy in the temporal domain to the receiver.
— *Frequency diversity*: The signal transmitted on different frequencies induces different structures in the multipath environment. Replicas of the transmitted signal are provided to the receiver in the form of redundancy in the frequency domain. Best examples of how to exploit the frequency diversity are the technique of multi-carrier spread spectrum and coding in the frequency direction.
— *Spatial diversity*: Spatially separated antennas provide replicas of the transmitted signal to the receiver in the form of redundancy in the spatial domain. This can be provided with no penalty in spectral efficiency.

Exploiting all forms of diversity in future systems (e.g., 4G) will ensure the highest performance in terms of capacity and spectral efficiency.

Furthermore, the future generation of broadband mobile/fixed wireless systems will aim to support a wide range of services and bit rates. The transmission rate may vary from voice to very high rate multimedia services requiring data rates up to 100 Mbit/s. Communication channels may change in terms of their grade of mobility, cellular infrastructure, required symmetrical or asymmetrical transmission capacity, and whether they are indoor or outdoor. Hence, air interfaces with the highest flexibility are demanded in order to maximize the area spectrum efficiency in a variety of communication environments. The adaptation and integration of existing and new systems to emerging new standards would be feasible if both the receiver and the transmitter are reconfigurable using software-defined radio (SDR).

The aim of this last chapter is to look at new antenna diversity techniques (e.g., *space time coding* (STC), *space frequency coding* (SFC) and at the concept of *software-defined radio* (SDR) which will all play a major role in the realization of 4G.

6.2 General Principle of Multiple Antenna Diversity

In conventional wireless communications, spectral and power efficiency is achieved by exploiting time and frequency diversity techniques. However, the spatial dimension so far only exploited for cell sectorization will play a much more important role in future wireless communication systems. In the past most of the work has concentrated on the design of intelligent antennas, applied for *space division multiple access (SDMA)*. In the meantime, more general techniques have been introduced where arbitrary antenna configurations at the transmit and receive sides are considered.

If we consider M transmit antennas and L receive antennas, the overall system channel defines the so-called *multiple input/multiple output (MIMO)* channel (see Figure 6-1). If the MIMO channel is assumed to be linear and time-invariant during one symbol duration, the channel impulse response $h(t)$ can be written as

$$h(t) = \begin{bmatrix} h_{0,0}(t) & \cdots & h_{0,L-1}(t) \\ \vdots & \ddots & \vdots \\ h_{M-1,0}(t) & \cdots & h_{M-1,L-1}(t) \end{bmatrix} \quad (6.1)$$

where $h_{m,l}(t)$ represents the impulse response of the channel between the transmit (Tx) antenna m and the receive (Rx) antenna l.

From the above general model, two possibilities exist: i) case $M = 1$, resulting in a single input/multiple output (SIMO) channel and ii) case $L = 1$, resulting in a multiple input/single output (MISO) channel. In the case of SIMO, conventional receiver diversity

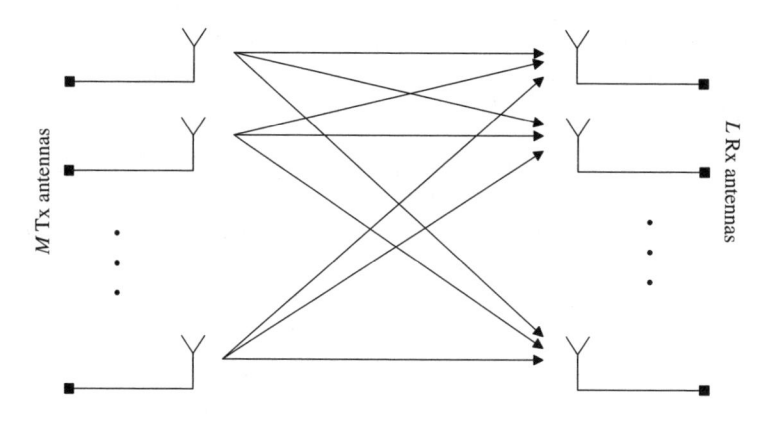

Figure 6-1 MIMO channel

techniques such as MRC can be realized, which can improve power efficiency, especially if the channels between the Tx and the Rx antennas are independently faded paths (e.g., Rayleigh distributed), where the multipath diversity order is identical to the number of receiver antennas [15].

With diversity techniques, a frequency- or time-selective channel tends to become an AWGN channel. This improves the power efficiency. However, there are two ways to increase the spectral efficiency. The first one, which is the trivial way, is to increase the symbol alphabet size and the second one is to transmit different symbols in parallel in space by using the MIMO properties.

The capacity of MIMO channels for an uncoded system in flat fading channels with perfect channel knowledge at the receiver is calculated by Foschini [11] as

$$C = \log_2 \left[\det \left(I_L + \frac{E_s/N_o}{M} h(t) h^{*T}(t) \right) \right], \qquad (6.2)$$

where "det" means determinant, I_L is an $L \times L$ identity matrix, and $(\cdot)^{*T}$ means the conjugate complex of the transpose matrix. Note that this formula is based on the Shannon capacity calculation for a simple AWGN channel.

Two approaches exist to exploit the capacity in MIMO channels. The information theory shows that with M transmit antennas and $L = M$ receive antennas, M independent data streams can be simultaneously transmitted, hence, reaching the channel capacity. As an example, the BLAST (Bell-Labs Layered Space Time) architecture can be referred to [11][20]. Another approach is to use a MISO scheme to obtain diversity, where in this case sophisticated techniques such as space–time coding (STC) can be realized. All transmit signals occupy the same bandwidth, but they are constructed such that the receiver can exploit spatial diversity, as in the Alamouti scheme [1]. The main advantage of STCs especially for mobile communications is that they do not require multiple receive antennas.

6.2.1 BLAST Architecture

The basic concept of the BLAST architecture is to exploit channel capacity by increasing the data rate through simultaneous transmission of independent data streams over M transmit antennas. In this architecture, the number of receive antennas should at least be equal to the number of transmit antennas $L \geqslant M$ (see Figure 6-1).

For m-array modulation, the receiver has to choose the most likely out of m^M possible signals in each symbol time interval. Therefore, the receiver complexity grows exponentially with the number of modulation constellation points and the number of transmit antennas. Consequently, suboptimum detection techniques such as those proposed in BLAST can be applied. Here, in each step only the signal transmitted from a single antenna is detected, whereas the transmitted signals from the other antennas are canceled using the previously detected signals or suppressed by means of zero-forcing or MMSE equalization.

Two basic variants of BLAST are proposed [11][20]: D-BLAST (diagonal BLAST) and V-BLAST (vertical BLAST). The only difference is that in V-BLAST transmit antenna m corresponds all the time to the transmitted data stream m, where in D-BLAST the assignment of the antenna to the transmitted data stream is hopped periodically. If the

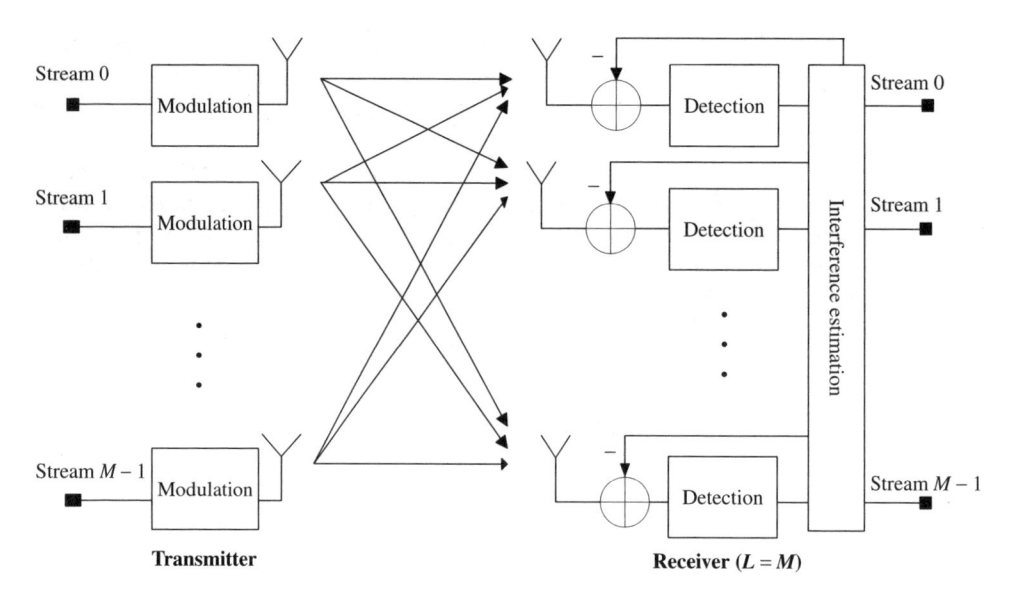

Figure 6-2 V-BLAST transceiver

channel does not vary during transmission, in V-BLAST, the different data streams may suffer from asymmetrical performance. Furthermore, in general the BLAST performance is limited due to the error propagation issued by the multistage decoding process.

As it is illustrated in Figure 6-2, for detection of data stream 0, the signals transmitted from all other antennas are estimated and suppressed from the received signal of the data stream 0. In [2][3] an iterative decoding process for the BLAST architecture is proposed, which outperforms the classical approach.

However, the main disadvantages of the BLAST architecture for mobile communications is the need of high numbers of receive antennas, which is not practical in a small mobile terminal. Furthermore, high system complexity may prohibit the large-scale implementation of such a scheme.

6.2.2 Space–Time Coding

An alternative approach is to obtain transmit diversity with M transmit antennas, where the number of received antennas is not necessarily equal to the number of transmit antennas. Even with one receive antenna the system should work. This approach is more suitable for mobile communications.

The basic philosophy with STC is different from the BLAST architecture. Instead of transmitting independent data streams, the same data stream is transmitted in an appropriate manner over all antennas. This could be, for instance, a downlink mobile communication, where in the base station M transmit antennas are used while in the terminal station only one or few antennas might be applied.

The principle of STC is illustrated in Figure 6-3. The basic idea is to provide through coding *constructive superposition* of the signals transmitted from different antennas. Constructive combining can be achieved for instance by modulation diversity, where

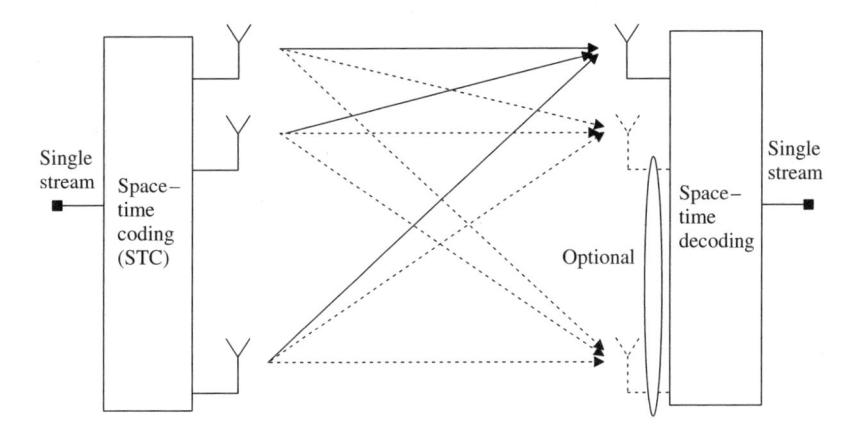

Figure 6-3 General principle of space–time coding (STC)

orthogonal pulses are used in different transmit antennas. The receiver uses the respective matched filters, where the contributions of all transmit antennas can be separated and combined with MRC.

The simplest form of modulation diversity is delay diversity, a special form of space–time trellis codes. The other alternative of STC is space–time block codes. Both spatial coding schemes are described in the following.

6.2.2.1 Space–Time Trellis Codes (STTC)

The simplest form of STTCs is the delay diversity technique (see Figure 6-4). The idea is to transmit the same symbol with a delay of iT_s from transmit antenna $i = 0, \ldots, M - 1$.

The delay diversity can be viewed as a rate $1/M$ repetition code. The detector could be a standard equalizer. Replacing the repetition code by a more powerful code, additional coding gain on top of the diversity advantage can be obtained [16]. However, there is no general rule how to obtain good space–time trellis codes for arbitrary numbers of transmit antennas and modulation methods. Powerful STTCs are given in [18] and obtained from an exhaustive search. However, the problem of STTCs is that the detection complexity measured in the number of states grows exponentially with m^M.

In Figure 6-5, an example of a STTC for two transmit antennas $M = 2$ in case of QPSK $m = 2$ is given. This code has four states with spectral efficiency of 2 bit/s/Hz.

Assuming ideal channel estimation, the decoding of this code at the receive antenna j can be performed by minimizing the following metric:

$$D = \sum_{j=0}^{L-1} \left| r_j - \sum_{i=0}^{M-1} h_{i,j} x_i \right|^2, \qquad (6.3)$$

where r_j is the received signal at receive antenna j and x_i is the branch metric in the transition of the encoder trellis. Here, the Viterbi algorithm can be used to choose the best path with the lowest accumulated metric. The results in [18] show the coding advantages obtained by increasing the number of states as the number of received antennas is increased.

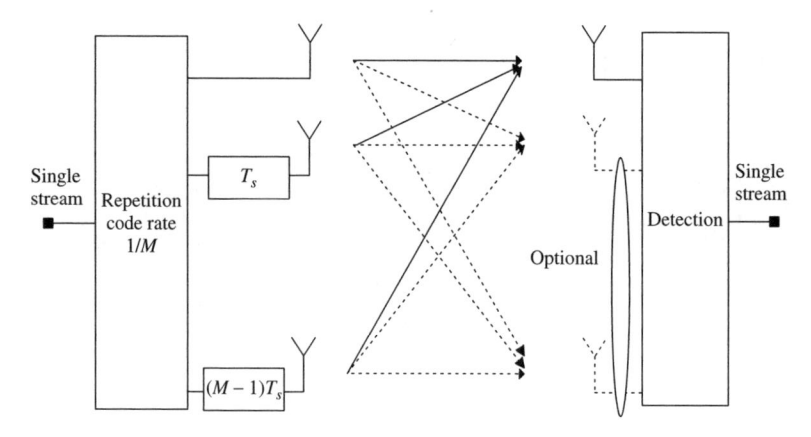

Figure 6-4 Space–time trellis code with delay diversity technique

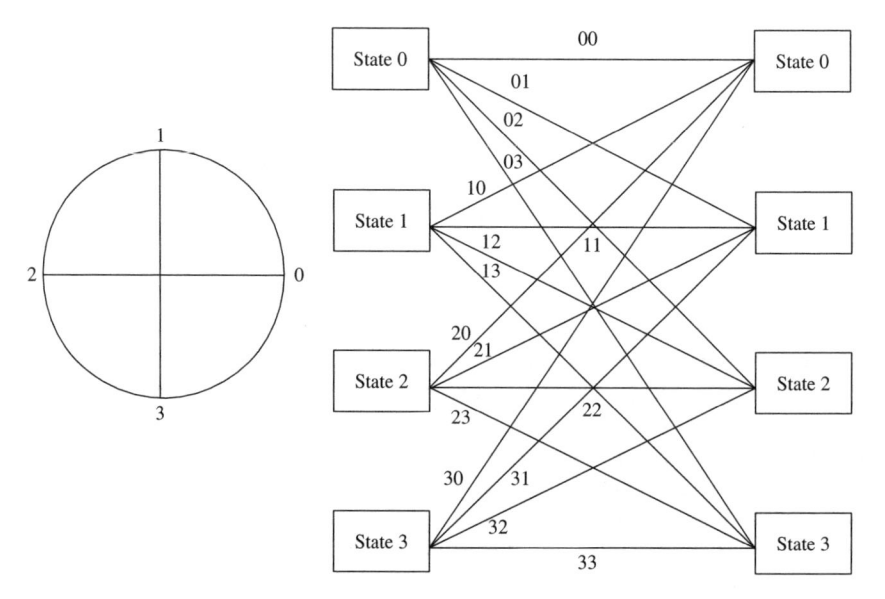

Figure 6-5 Space–time trellis code with four states

6.2.2.2 Space–Time Block Codes (STBC)

A simple transmit diversity scheme for two transmit antennas using STBCs was introduced by Alamouti in [1] and generalized to an arbitrary number of antennas by Tarokh et al. [17]. Basically, STBCs are designed as pure diversity schemes and provide no additional coding gain as with STTCs. In the simplest Alamouti scheme with $M = 2$ antennas, the transmitted symbols x_i are mapped to the transmit antenna with the mapping

$$B = \begin{bmatrix} x_0 & x_1 \\ -x_1^* & x_0^* \end{bmatrix}, \tag{6.4}$$

where the row corresponds to the time index and the column to the transmit antenna index. In the first symbol time interval x_0 is transmitted from antenna 0 and x_1 is transmitted from antenna 1 simultaneously, where in the second symbol time interval antenna 0 transmits $-x_1^*$ and simultaneously antenna 1 transmits x_0^*.

The coding rate of this STBCs is one, meaning that no bandwidth expansion will take place (see Figure 6-6). Due to the orthogonality of the space–time block codes, the symbols can be separated at the receiver by a simple linear combining (see Figure 6-7). The spatial diversity combining with block codes applied for multi-carrier transmission is described in more detail in Section 6.3.4.1.

6.2.3 Achievable Capacity

For STBCs of rate R the channel capacity is given by [2]

$$C = R \log_2 \left[\left(1 + \frac{E_s/N_o}{M} \sum_{i=0}^{M-1} \sum_{j=0}^{L-1} |h_{i,j}|^2 \right) \right]. \quad (6.5)$$

For $R = 1$ and $L = 1$, this is equivalent to the channel capacity of a MISO scheme. However, for $L > 1$, the capacity curve is only shifted, but the asymptotic slope is not

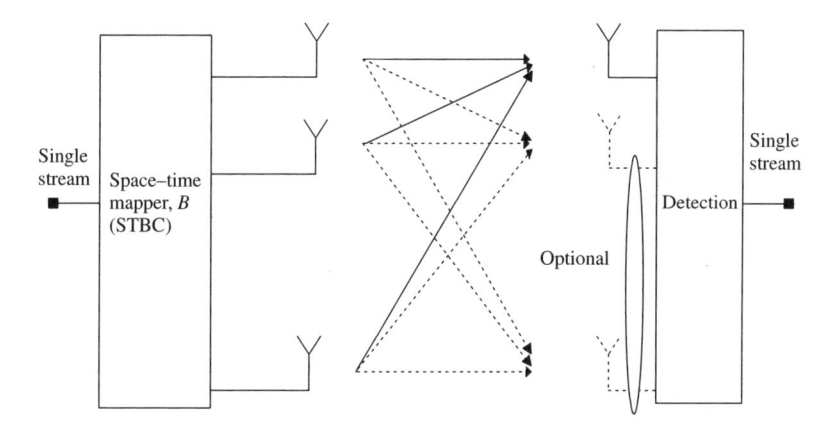

Figure 6-6 Space–time block code transceiver

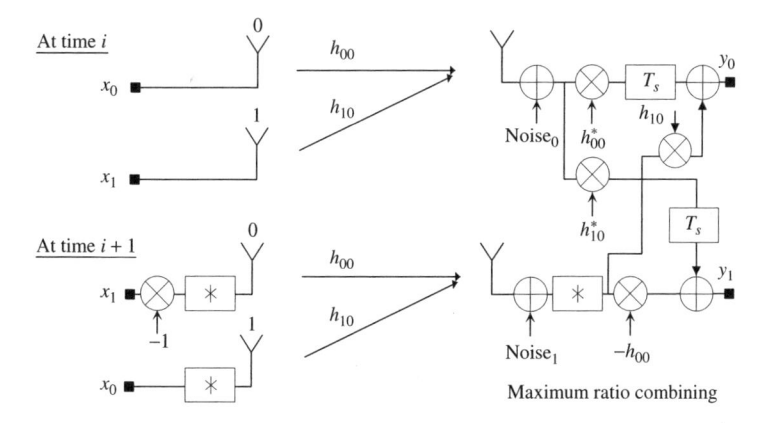

Figure 6-7 Principle of space–time block coding

increased, therefore, the MIMO capacity will not be achieved [3]. This also corresponds to results for STTCs.

From an information theoretical point of view it can be concluded that STCs should be used in systems with $L = 1$ receive antennas. If multiple receive antennas are available, the data rate can be increased by transmitting independent data from different antennas as in the BLAST architecture.

6.3 Diversity Techniques for Multi-Carrier Transmission

6.3.1 Transmit Diversity

Several techniques to achieve spatial transmit diversity in OFDM systems are discussed in this section. The number of used transmit antennas is M. OFDM is realized by an IFFT and the OFDM blocks shown in the following figures also include a frequency interleaver and a guard interval insertion/removal. It is important to note that the total transmit power Ω is the sum of the transmit power Ω_m of each antenna, i.e.,

$$\Omega = \sum_{m=0}^{M-1} \Omega_m. \qquad (6.6)$$

In the case of equal transmit power per antenna, the power per antenna is

$$\Omega_m = \frac{\Omega}{M}. \qquad (6.7)$$

6.3.1.1 Delay Diversity

As discussed before, the principle of delay diversity (DD) is to artificially increase the frequency selectivity of the mobile radio channel by introducing additional constructive delayed signals. Delay diversity can be considered a simple form of STTC. Increased frequency selectivity can enable a better exploitation of diversity which results in an improved system performance. With delay diversity, the multi-carrier modulated signal itself is identical on all M transmit antennas and differs only in an antenna-specific delay $\delta_m, m = 1, \ldots, M - 1$ [14]. The block diagram of an OFDM system with spatial transmit diversity applying delay diversity is shown in Figure 6-8.

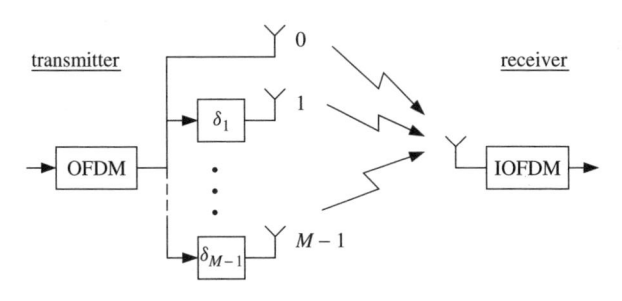

Figure 6-8 Delay diversity

In order to achieve frequency selective fading within the transmission bandwidth B, the delay has to fulfill the condition

$$\delta_m \geqslant \frac{1}{B}. \qquad (6.8)$$

To increase the frequency diversity by multiple transmit antennas, the delay of the different antennas should be chosen as

$$\delta_m \geqslant \frac{km}{B}, \quad k \geqslant 1, \qquad (6.9)$$

where k is a constant factor introduced for the system design which has to be chosen large enough ($k \geqslant 1$) in order to guarantee a diversity gain. A factor of $k = 2$ seems to be sufficient to achieve promising performance improvements in most scenarios. This result is verified by the simulation results presented in Section 6.3.1.2.

The disadvantage of delay diversity is that the additional delays $\delta_m, m = 1, \ldots, M - 1$, increase the total delay spread at the receiver antenna and require an extension of the guard interval duration by the maximum $\delta_m, m = 1, \ldots, M - 1$, which reduces the spectral efficiency of the system. This disadvantage can be overcome by phase diversity presented in the next section.

6.3.1.2 Phase Diversity

Phase diversity (PD) transmits signals on M antennas with different phase shifts, where $\Phi_{m,n}, m = 1, \ldots, M - 1, n = 0, \ldots, N_c - 1$, is an antenna- and sub-carrier specific phase offset [12][13]. The phase shift is efficiently realized by a phase rotation before OFDM, i.e., before the IFFT. The block diagram of an OFDM system with spatial transmit diversity applying phase diversity is shown in Figure 6-9.

In order to achieve frequency selective fading within the transmission bandwidth of the N_c sub-channels, the phase $\Phi_{m,n}$ has to fulfill the condition

$$\begin{aligned}\Phi_{m,n} &\geqslant \frac{2\pi f_n}{B} \\ &\geqslant \frac{2\pi n}{N_c}\end{aligned} \qquad (6.10)$$

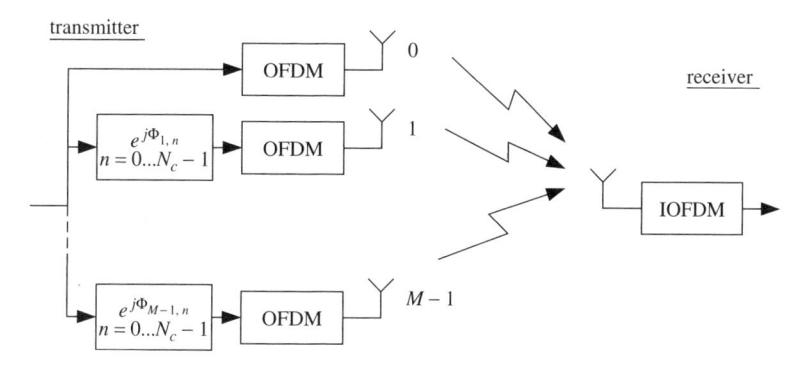

Figure 6-9 Phase diversity

where $f_n = n/T_s$ is the nth sub-carrier frequency, T_s is the OFDM symbol duration without guard interval and $B = N_c/T_s$. To increase the frequency diversity by multiple transmit antennas, the phase offset of the nth sub-carrier at the mth antenna should be chosen as

$$\Phi_{m,n} = \frac{2\pi kmn}{N_c}, \quad k \geqslant 1, \tag{6.11}$$

where k is a constant factor introduced for the system design which has to be chosen large enough ($k \geqslant 1$) to guarantee a diversity gain. The constant k corresponds to k introduced in Section 6.3.1.1. Since no delay of the signals at the transmit antennas occurs with phase diversity, no extension of the guard interval is necessary compared to delay diversity.

In Figure 6-10, the SNR gain to reach a BER of $3 \cdot 10^{-4}$ with 2 transmit antennas applying delay diversity and phase diversity compared to a 1 transmit antenna scheme over the parameter k introduced in (6.9) and (6.11) is shown for OFDM and OFDM-CDM. The results are presented for an indoor and outdoor scenario. The performance of delay diversity and phase diversity is the same for the chosen system parameters, since the guard interval duration exceeds the maximum delay of the channel and the additional delay due to delay diversity. The curves show that gains of more than 5 dB in the indoor scenario and of about 2 dB in the outdoor scenario can be achieved for $k \geqslant 2$ and justify the selection of $k = 2$ as a reasonable value. It is interesting to observe that even in an outdoor environment, which already has frequency selective fading, significant performance improvements are achievable.

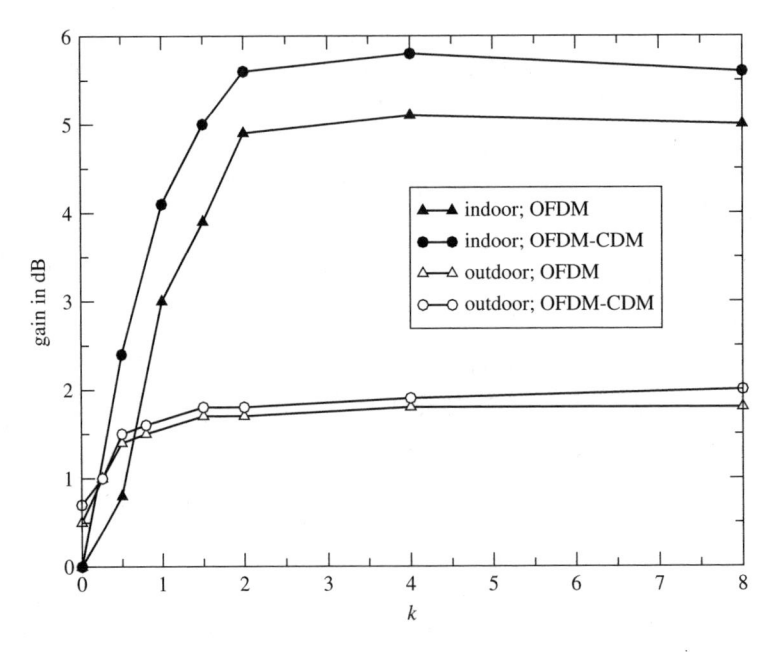

Figure 6-10 Performance gains with delay diversity and phase diversity over k; $M = 2$; BER $= 3 \cdot 10^{-4}$

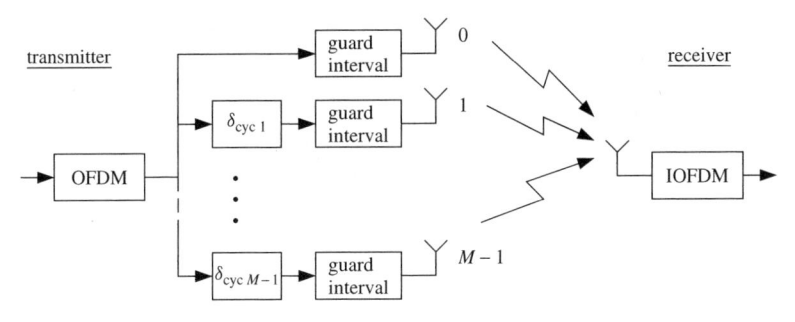

Figure 6-11 Cyclic delay diversity

An efficient implementation of phase diversity is cyclic delay diversity (CDD) [6], which instead of M OFDM operations requires only one OFDM operation in the transmitter. The signals constructed by phase diversity and by cyclic delay diversity are equal. Signal generation with cyclic delay diversity is illustrated in Figure 6-11. With cyclic delay diversity, $\delta_{cycl\ m}$ denotes cyclic shifts [7]. Both phase diversity and cyclic delay diversity are performed before guard interval insertion.

6.3.1.3 Time-Variant Phase Diversity

The spatial transmit diversity concepts presented in the previous sections introduce only frequency diversity. Time-variant phase diversity (TPD) can additionally exploit time diversity. It can be used to introduce time diversity or to introduce both time and frequency diversity. The block diagram shown in Figure 6-9 is still valid, only the phase offsets $\Phi_{m,n}$ have to be replaced by the time-variant phase offsets $\Theta_{m,n}(t), m = 1, \ldots, M-1$, $n = 0, \ldots, N_c - 1$, which are given by [13]

$$\Theta_{m,n}(t) = \Phi_{m,n} + 2\pi t F_m. \tag{6.12}$$

The frequency shift F_m at transmit antenna m has to be chosen such that the channel can be considered as time-invariant during one OFDM symbol duration, but appears time-variant over several OFDM symbols. It has to be taken into account in the system design that the frequency shift F_m introduces ICI which increases with increasing F_m.

The gain in SNR to reach the BER of $3 \cdot 10^{-4}$ with 2 transmit antennas applying time-variant phase diversity compared to time-invariant phase diversity with 2 transmit antennas over the frequency shift F_1 is shown in Figure 6-12. The frequency shifts F_m should be less than a few percent of the sub-carrier spacing to avoid non-negligible degradations due to ICI.

6.3.1.4 Sub-Carrier Diversity

With sub-carrier diversity (SCD), the sub-carriers used for OFDM are clustered in M smaller blocks and each block is transmitted over a separate antenna [5]. The principle of sub-carrier diversity is shown in Figure 6-13.

After serial-to-parallel (S/P) conversion, each OFDM block processes N_c/M complex-valued data symbols out of a sequence of N_c. Each of the M OFDM blocks maps its N_c/M data symbols on its exclusively assigned set of sub-carriers. The sub-carriers of

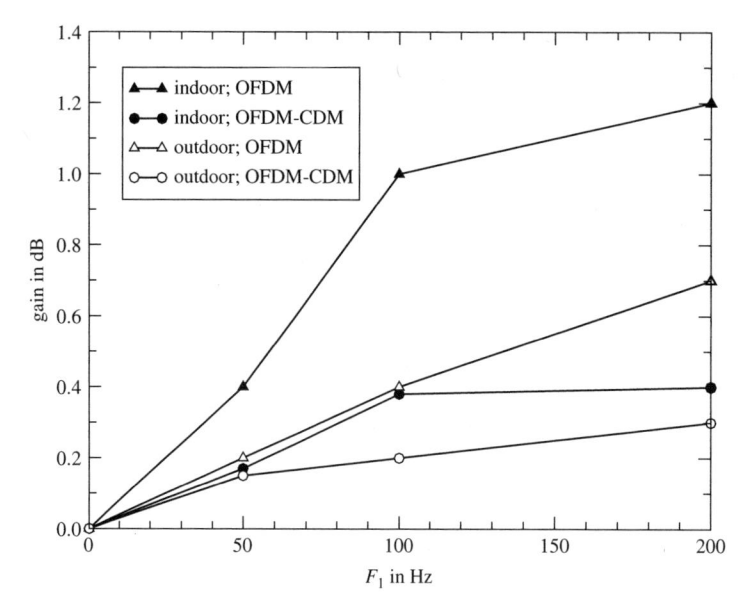

Figure 6-12 Performance improvements due to time-variant phase diversity; $M = 2$; $k = 2$; BER $= 3 \cdot 10^{-4}$

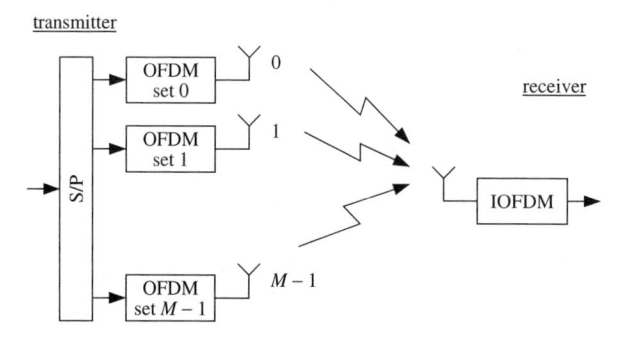

Figure 6-13 Sub-carrier diversity

one block should be spread over the entire transmission bandwidth in order to increase the frequency diversity per block, i.e., the sub-carriers of the individual blocks should be interleaved.

The advantage of sub-carrier diversity is that the peak-to-average power ratio per transmit antenna is reduced compared to a single antenna implementation since there are fewer sub-channels per transmit antenna.

6.3.2 Receive Diversity

6.3.2.1 Maximum Ratio Combining (MRC)

The signals at the output of the L receive antennas are combined linearly so that the SNR is maximized. The optimum weighting coefficient is the conjugate complex of the assigned channel coefficient illustrated in Figure 6-14.

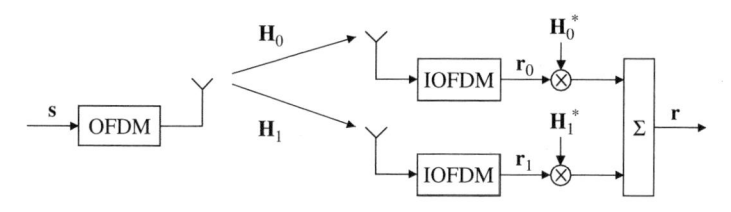

Figure 6-14 OFDM with MRC receiver; $L = 2$

With the received signals

$$r_0 = H_0 s + n_0$$
$$r_1 = H_1 s + n_1 \quad , \quad (6.13)$$

the diversity gain achievable with MRC can be observed as follows:

$$r = H_0^* r_0 + H_1^* r_1$$
$$= (|H_0|^2 + |H_1|^2)s + H_0^* n_0 + H_1^* n_1. \quad (6.14)$$

6.3.2.2 Delay and Phase Diversity

The transmit diversity techniques delay, phase, and time-variant phase diversity presented in Section 6.3.1 can also be applied in the receiver, achieving the same diversity gains plus an additional gain due to the collection of the signal power from multiple receive antennas. A receiver with phase diversity is shown in Figure 6-15.

6.3.3 Performance Analysis

The gain in SNR due to different transmit diversity techniques to reach the BER of $3 \cdot 10^{-4}$ with M transmit antennas compared to 1 transmit antenna over the number of antennas M is shown in Figure 6-16. The results are presented for a rate 1/2 coded OFDM system in an indoor environment. Except for sub-carrier diversity without interleaving, promising performance improvements are already obtained with 2 transmit antennas. The optimum choice of the number of antennas M is a trade-off between cost and performance.

The BER performance of the presented spatial transmit diversity concepts is shown in Figure 6-17 for an indoor environment with 2 transmit antennas. Simulation results are shown for coded OFDM and OFDM-CDM systems. The performance of the OFDM

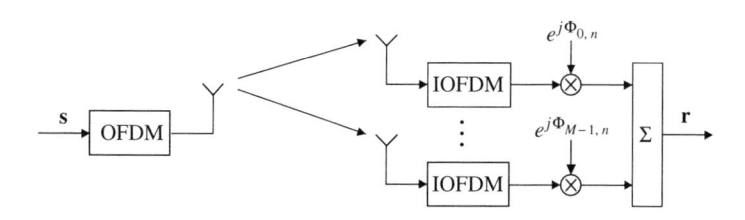

Figure 6-15 Phase diversity at the receiver

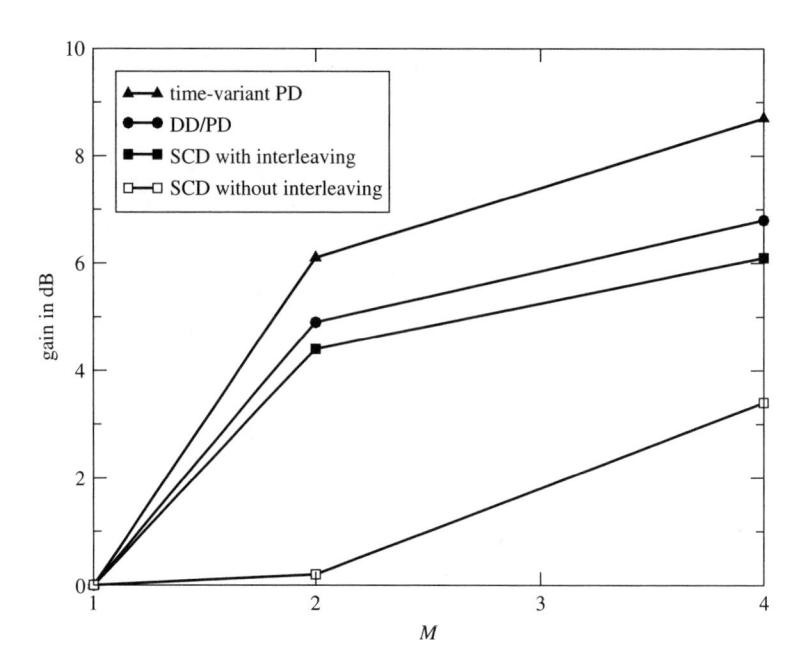

Figure 6-16 Spatial transmit diversity gain over the number of antennas M; $k = 2$; $F_1 = 100$ Hz for time-variant phase diversity; indoor; BER $= 3 \cdot 10^{-4}$

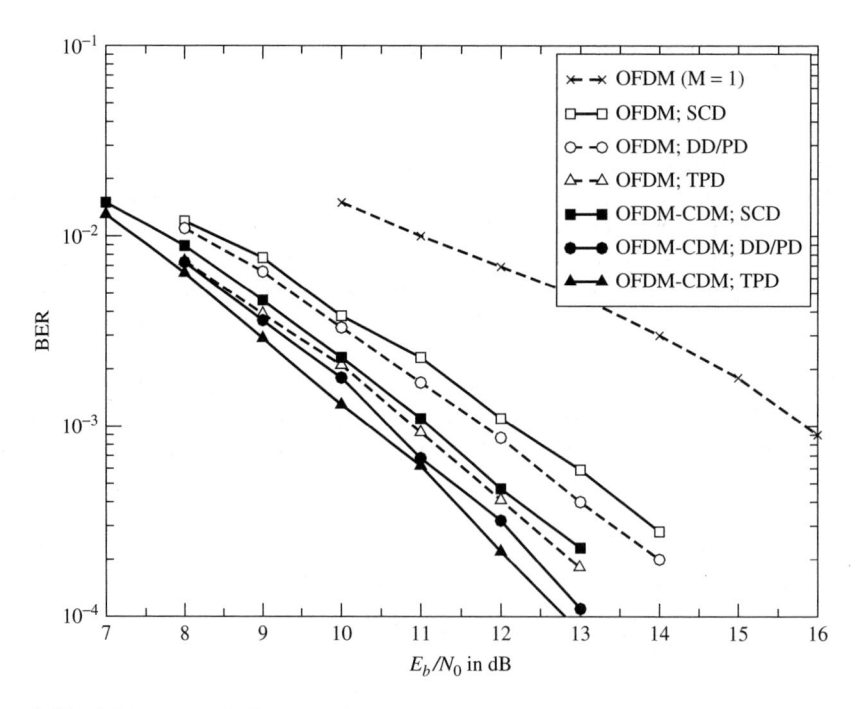

Figure 6-17 BER versus SNR; $M = 2$; $k = 2$; $F_1 = 100$ Hz for time-variant phase diversity; indoor

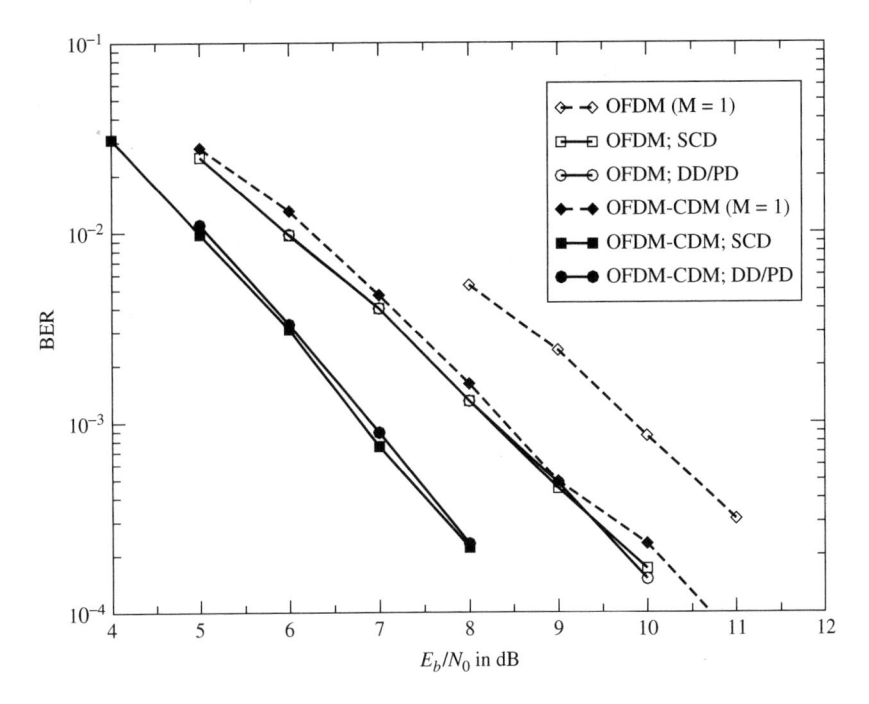

Figure 6-18 BER versus SNR; $M = 2$; $k = 2$; outdoor

system with 1 transmit antenna is given as reference. The corresponding simulation results for an outdoor environment are shown in Figure 6-18. It can be observed that time-variant phase diversity outperforms the other investigated transmit diversity schemes and that phase diversity and delay diversity are superior to sub-carrier diversity in the indoor environment. Moreover, the performance can be improved by up to 2 dB with an additional CDM component.

Finally, some general statements should be made about spatial transmit diversity concepts. The disadvantages of spatial transmit diversity concepts are that multiple transmit chains and antennas are required, increasing system complexity. Moreover, accurate oscillators are required in the transmitters, such that the sub-carrier patterns at the individual transmit antennas fit together and ICI can be avoided. As long as this is done in the base station, e.g., in a broadcasting system or in the downlink of a mobile radio system, the additional complexity is reasonable. Nevertheless, it can also be justified in a mobile transmitter.

The clear advantage of the presented spatial transmit and receive diversity concepts is that significant performance improvements of several dB can be achieved in critical propagation scenarios. Moreover, the presented diversity techniques are standard-compliant and can be applied in already standardized systems such as DAB, DVB-T, HIPERLAN/2, or IEEE 802.11a.

Transmit and receive diversity techniques can easily be combined. Performance improvements with phase diversity in the transmitter and MRC in the receiver are shown for a DVB-T transmission in Figure 6-19 [7]. The chosen DVB-T parameters are 4-QAM,

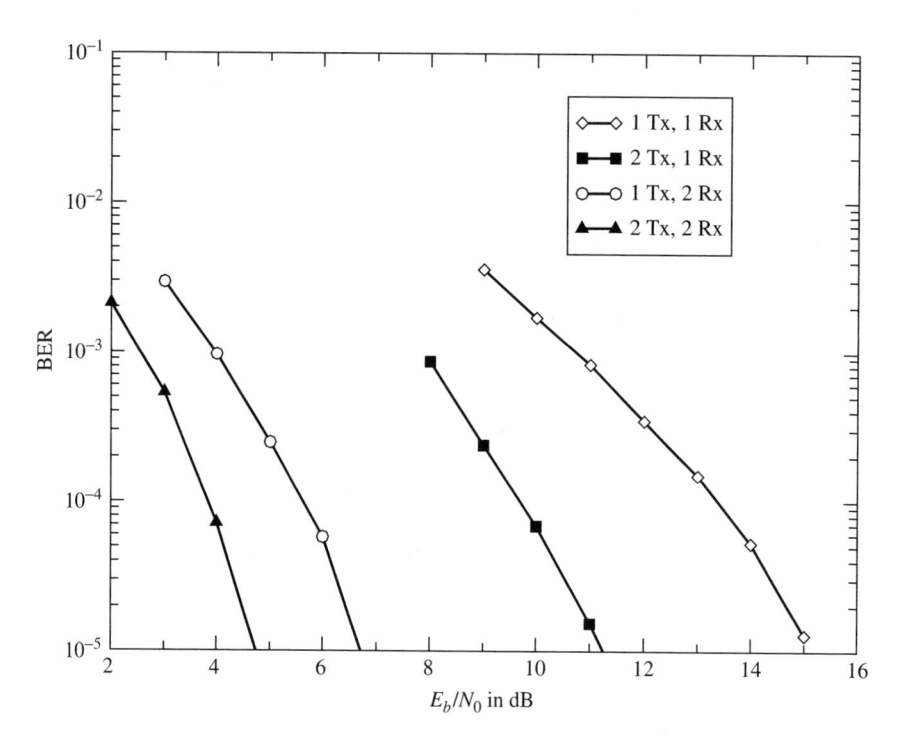

Figure 6-19 Performance of DVB-T with Tx and Rx diversity; 4-QAM; rate 1/2; 2k FFT; indoor; $f_D = 10$ Hz

code rate 1/2, and 2k FFT in an indoor environment with maximum Doppler frequency of 10 Hz. A single antenna system is given as reference.

Further performance improvements can be obtained by combining transmit and receive diversity techniques with beamforming [8]. Beamforming reduces interference within a propagation environment and can efficiently cancel interference. Since the channel of each beam has a small delay spread, the channel appears nearly flat. The diversity techniques presented in Sections 6.3.1 and 6.3.2 can artificially introduce frequency- and time-selectivity and, thus, improve performance.

6.3.4 OFDM and MC-CDMA with Space–Frequency Coding

Transmit antenna diversity in form of space–time block codes exploits time and space diversity and achieves a maximum diversity gain for 2 transmit antennas without rate loss. They have to be applied under the assumption that the channel coefficients remain constant for two subsequent symbol durations in order to guarantee the diversity gain. This is a tough precondition in OFDM systems, where the OFDM symbol duration T_s is N_c times the duration of a serial data symbols T_d. To overcome the necessity to use two successive OFDM symbols for coding, symbols belonging together can be sent on different sub-carriers in multi-carrier systems. We call this approach *space–frequency coding* (SFC).

6.3.4.1 Space–Frequency Block Codes (SFBC)

A feature of OFDM can be exploited, that two adjacent narrowband sub-channels are affected by almost the same channel coefficients. Thus, a space–frequency block code requires only the reception of one OFDM symbol for detection, avoids problems with coherence time restrictions and reduces delay in the detection process.

Figure 6-20 shows an OFDM transmitter with space–frequency block coding. Sequences of N_c interleaved data symbols S_n are transmitted in one OFDM symbol. The data symbols are interleaved by the block Π before space–frequency block coding such that the data symbols combined with space–frequency mapping and, thus, affected by the same fading coefficient are not subsequent data symbols in the original data stream. The interleaver with size I performs frequency interleaving for $I \leqslant N_c$ and time and frequency interleaving for $I > N_c$.

The mapping scheme of the data symbols S_n for SFBCs with 2 transmit antennas and code rate 1 is shown in Table 6-1. The mapping scheme for SFBCs is chosen such that on the first antenna the original data is transmitted without any modification (see Section 6.4.1). Thus, the mapping of the data symbols on the sub-carriers for the first antenna corresponds to the classical inverse discrete Fourier transform with

$$x_{0,v} = \frac{1}{N_c} \sum_{n=0}^{N_c-1} S_n e^{j2\pi nv/N_c}, \qquad (6.15)$$

where n is the sub-carrier index and v is the sample index of the time signal. Only the data symbol mapping on the second antenna has to be modified according to the mapping scheme for space–frequency block coding given in Table 6-1. The data symbols of the second transmit antenna are mapped on the sub-carriers as follows

$$x_{1,v} = \frac{1}{N_c} \sum_{n=0}^{N_c/2-1} (S_{2n}^* e^{j2\pi(2n+1)v/N_c} - S_{2n+1}^* e^{j2\pi 2nv/N_c}). \qquad (6.16)$$

The OFDM block comprises inverse fast Fourier transform (IFFT) and cyclic extension of an OFDM symbol.

A receiver with inverse OFDM operation and space–frequency block decoding is shown in Figure 6-21. The received signals on sub-channels n and $n+1$ after guard interval removal and fast Fourier transform (FFT) can be written as

$$R_n = H_{0,n} S_n - H_{1,n} S_{n+1}^* + N_n$$
$$R_{n+1} = H_{0,n+1} S_{n+1} + H_{1,n+1} S_n^* + N_{n+1}, \qquad (6.17)$$

Table 6-1 Mapping with space–frequency block codes and two transmit antennas

Sub-carrier #	Antenna 1	Antenna 2
n	S_n	$-S_{n+1}^*$
$n+1$	S_{n+1}	S_n^*

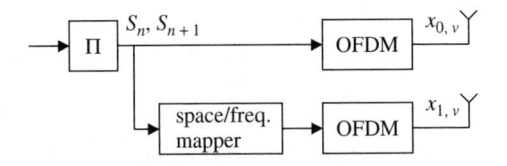

Figure 6-20 Space–frequency block coding in an OFDM transmitter

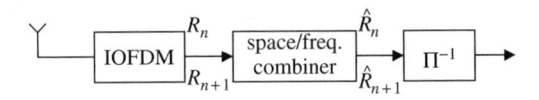

Figure 6-21 Space–frequency block decoding in an OFDM receiver

where $H_{m,n}$ is the flat fading coefficient of sub-channel n assigned to transmit antenna m and N_n is the additive noise on sub-carrier n. OFDM systems are designed such that the fading per sub-channel can be considered as flat, from which it can be concluded that the fading between adjacent sub-carriers can be considered as flat as well and $H_{m,n}$ can be assumed to be equal to $H_{m,n+1}$. Thus, when focusing analysis on an arbitrary pair of adjacent sub-channels n and $n+1$, we can write H_m as the flat fading coefficient assigned to the pair of sub-channels n and $n+1$ and to transmit antenna m. After using the combining scheme

$$\hat{R}_n = H_0^* R_n + H_1 R_{n+1}^*$$
$$\hat{R}_{n+1} = -H_1 R_n^* + H_0^* R_{n+1}, \qquad (6.18)$$

the received signals result in

$$\hat{R}_n = (|H_0|^2 + |H_1|^2) S_n + H_0^* N_n + H_1 N_{n+1}^*$$
$$\hat{R}_{n+1} = (|H_0|^2 + |H_1|^2) S_{n+1} - H_1 N_n^* + H_0^* N_{n+1}. \qquad (6.19)$$

After deinterleaving Π^{-1}, symbol detection, and demapping, the soft decided bit w is obtained, which in the case of channel coding is fed to the channel decoder. Optimum soft decision channel decoding is guaranteed when using log-likelihood ratios (LLRs) in the Viterbi decoder. The normalized LLRs for space–frequency block coded systems result in

$$\Gamma = \frac{2\sqrt{|H_0|^2 + |H_1|^2}}{\sigma^2} w. \qquad (6.20)$$

6.3.4.2 Performance Results

In this section, the performance of a classical OFDM system, a space–frequency block coded OFDM system, an OFDM-CDM system, and combinations of these are compared in a Rayleigh fading channel. It is important to note that the performance of OFDM-CDM is comparable to the performance of a fully loaded MC-CDMA scheme in the downlink and that the performance of OFDM is comparable to the performance of OFDMA or MC-TDMA with perfect interleaving. The transmission bandwidth of the systems is $B = 2$ MHz and the carrier frequency is located at 2 GHz. The number of sub-carriers is 512. The guard interval duration is 20 µs. As channel codes, punctured convolutional codes with memory 6 and variable code rate R between 1/3 and 4/5 are used. QPSK is used for symbol mapping. The depth of the interleaver Π is equal to 24 subsequent OFDM symbols, so that time and frequency interleaving is applied. SFBCs are realized with $M = 2$ transmit antennas. There is no correlation between the antennas in the space–frequency block coded OFDM system. In the case of OFDM-CDM, short Hadamard codes of length $L = 8$ are applied for spreading. The mobile radio channel is modeled as an uncorrelated Rayleigh fading channel, assuming perfect interleaving in the frequency and time directions.

In Figure 6-22, the BER versus the SNR per bit is shown for space–frequency block coding in OFDM and OFDM-CDM systems. The results are presented for OFDM-CDM with different single symbol detection techniques. No FEC coding is used. The corresponding results without SFBC are shown in Figure 2.14.

The BER performance of various OFDM based diversity schemes with and without SFBC is shown in Figure 6-23. The channel code rate is $R = 2/3$. OFDM-CDM applies soft interference cancellation with one iteration at the receiver.

In Figure 6-24, the gain, i.e., SNR reduction, with SFBC OFDM-CDM compared to SFBC OFDM is shown versus the channel code rate for a BER of 10^{-5}. Additionally,

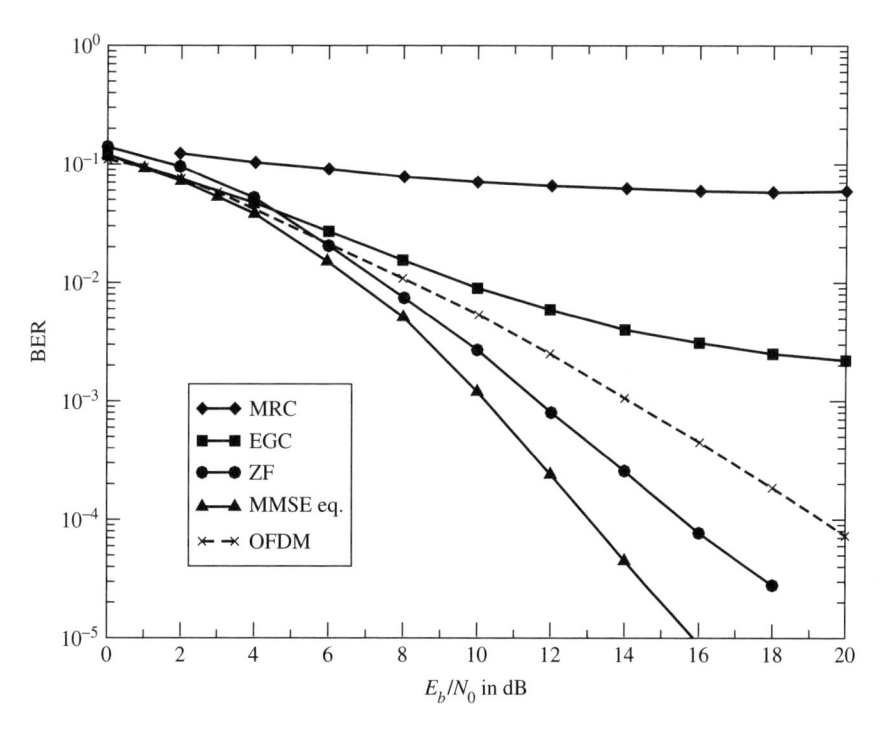

Figure 6-22 BER versus SNR for space–frequency block coding applied to OFDM and OFDM-CDM with different single symbol detection techniques; no FEC coding

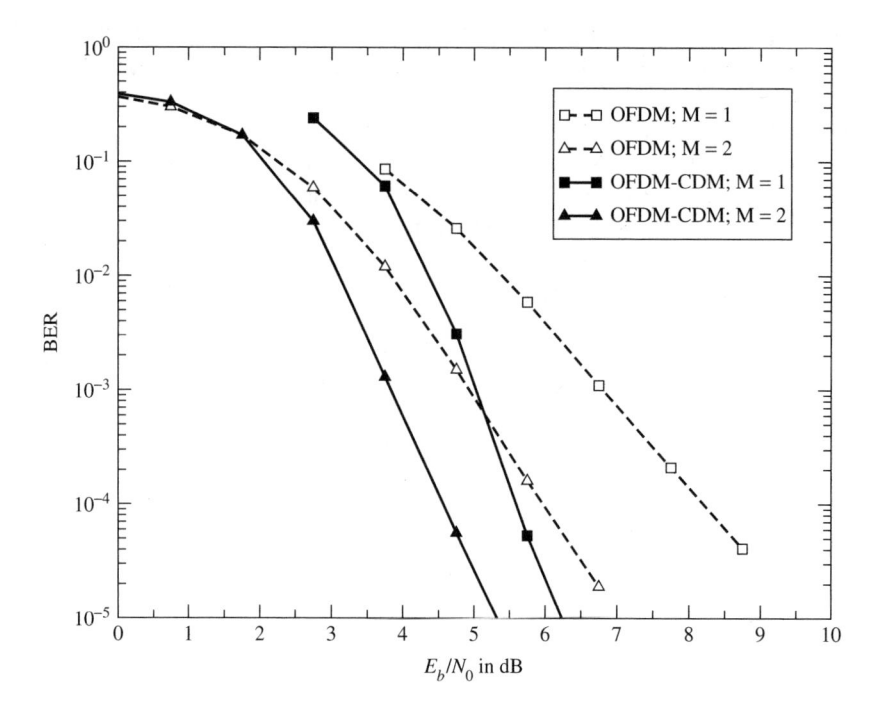

Figure 6-23 BER versus SNR for different OFDM schemes; code rate $R = 2/3$

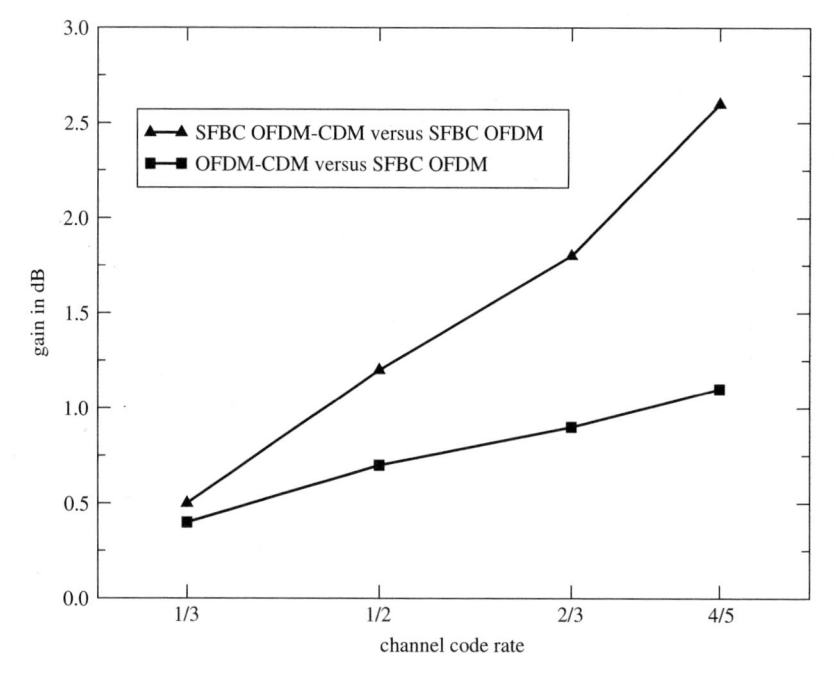

Figure 6-24 Gain with OFDM-CDM compared to OFDM with space–frequency block coding in dB versus channel code rate R; BER $= 10^{-5}$

OFDM-CDM without SFBC is compared to SFBC OFDM. In the case of OFDM-CDM, soft interference cancellation with one iteration is applied. It can be observed that the gains due to CDM increase with increasing code rate. This result shows that the weaker the channel code is, the more diversity can be exploited by CDM.

6.4 Examples of Applications of Diversity Techniques

Two concrete examples of the application of space–time coding for mobile and fixed wireless access (FWA) communications are given below. First we consider the UMTS standard and then look at the multi-carrier-based draft FWA standard below 10 GHz.

6.4.1 UMTS-WCDMA

A modified version of the Alamouti STBC is part of the UMTS-WCDMA standard [10]. Here the mapper B is given by

$$B = \begin{bmatrix} x_0 & -x_1^* \\ x_1 & x_0^* \end{bmatrix}, \tag{6.21}$$

which is used before spreading.

The symbols are transmitted from the first antenna, whereas the conjugates are transmitted in the second antenna. The advantage is the compatibility with systems without STBC if the second antenna is not implemented or simply switched off in the UMTS base station (Node B). At the mobile terminal (TS), a linear combination can be applied in each arm of the rake receiver, as given in Figure 6-25.

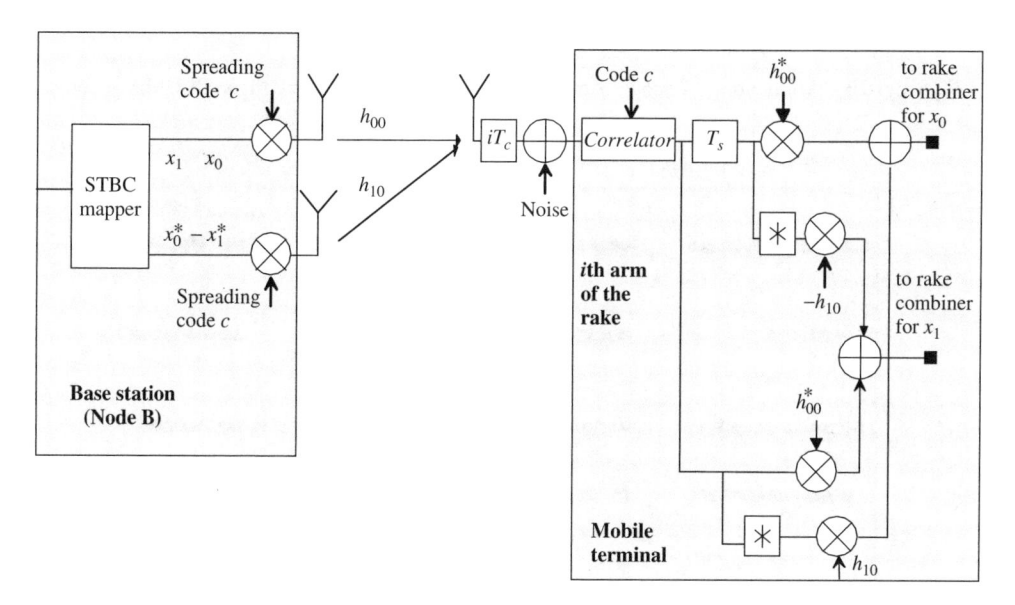

Figure 6-25 Application of STBC for UMTS receivers (only a single Rx antenna) [3]

6.4.2 FWA Multi-Carrier Systems

The Alamouti scheme is used only for the downlink (from BS to TS) to provide a second order of diversity, as described in the draft HIPERMAN specification [9]. There are two transmit antennas at the base station and one (or more) receive antenna(s) at the terminal station. The decoding can be done by MRC. Figure 6-26 shows the STBC in the FWA OFDM or OFDMA mode. Each transmit antenna has its own OFDM chain. Both antennas transmit two different OFDM symbols at the same time, and they share the same local oscillator. Thus, the received signal has exactly the same autocorrelation properties as for a single antenna and time and frequency coarse and fine estimation can be performed in the same way as for a single transmit antenna. The receiver requires a MISO channel estimation, which is allowed by splitting some preambles and pilots between the two transmit antennas (see Figure 6-27).

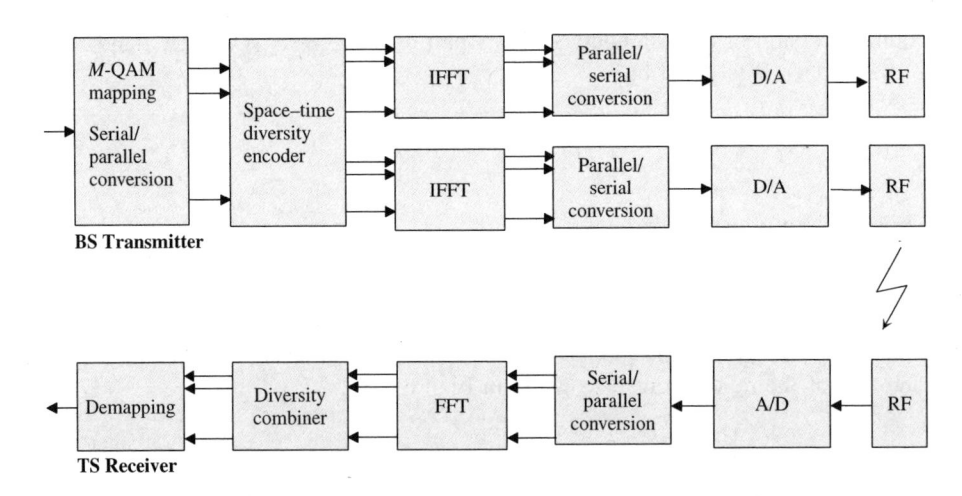

Figure 6-26 Application of space–time block coding for FWA (OFDM or OFDMA mode)

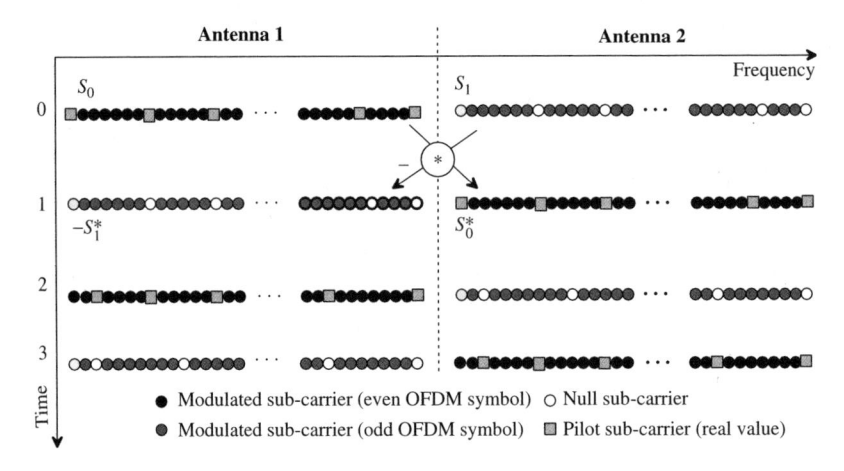

Figure 6-27 Alamouti scheme with OFDM/OFDMA

The basic scheme transmits two complex-valued OFDM symbols S_0 and S_1 over two antennas where at the receiver side one antenna is used. The channel values are h_0 (from Tx antenna 0) and h_1 (from Tx antenna 1). The first antenna transmits S_0 and $-S_1^*$ and the second antenna transmits S_1 and S_0^*. The receiver combines the received signal as follows,

$$\hat{S}_0 = h_0^* r_0 + h_1 r_1^*$$
$$\hat{S}_1 = h_1^* r_0 - h_0 r_1^*. \qquad (6.22)$$

OFDM symbols are taken in pairs. In the transmission frame, variable location pilots are identical for two symbols. At the receiver side, the receiver waits for two OFDM symbols and combines them on a sub-carrier basis according to the above equations.

6.5 Software-Defined Radio

The transmission rate for the future generation of wireless systems may vary from low rate messages up to very high rate data services up to 100 Mbit/s. The communication channel may change in terms of its grade of mobility, the cellular infrastructure, the required symmetrical or asymmetrical transmission capacity, and whether it is indoor or outdoor. Hence, air interfaces with the highest flexibility are required in order to maximize the area spectrum efficiency in a variety of communication environments. Future systems are also expected to support various types of services based on IP or ATM transmission protocols, which require a varying quality of services (QoS).

Recent advances in digital technology enable the faster introduction of new standards that benefit from the most advanced physical (PHY) and data link control (DLC) layers (see Table 6-2). These trends are still growing and new standards or their enhancements are being added continuously to the existing network infrastructures. As we explained in Chapter 5, the integration of all these existing and future standards in a common platform is one of the major goals of the next generation (4G) of wireless systems.

Hence, a fast adaptation/integration of existing systems to emerging new standards would be feasible if the 4G system has a generic architecture, while its receiver and transmitter parameters are both reconfigurable per software.

6.5.1 General

A common understanding of a software-defined radio (SDR) is that of a transceiver, where the *functions* are realized as programs running on suitable processors or reprogrammable components [21]. On the hardware, different transmitter/receiver algorithms, which describe transmission standards, could be executed per corresponding application software. For instance, the software can be specified in such a manner that several standards can be loaded via parameter configurations. This strategy can offer a seamless change/adaptation of standards, if necessary.

The software-defined radio can be characterized by the following features:

— the radio functionality is configured per software and
— different standards can be executed on the hardware according to the parameter lists.

Table 6-2 Examples of current wireless communication standards

Mobile communication systems		Wireless LAN/WLL	
CDMA based	TDMA based	Multi-carrier or CDMA based	Non MC, non CDMA based
IS-95/-B: Digital cellular standard in the USA	*GSM*: Global system for mobile communications	*HIPERLAN/1*: WLAN based on CDMA	*DECT*: Digital enhanced cordless telecommunications
W-CDMA: Wideband CDMA	*PDC*: Personal digital cellular system	*IEEE 802.11b*: WLAN based on CDMA	*HIPERACCESS*: WLL based on single-carrier TDMA
CDMA-2000: Multi-carrier CDMA based on IS-95	*IS-136*: North American TDMA system	*HIPERLAN/2*: WLAN based on OFDM	*IEEE 802.16*: WLL based on single-carrier TDMA
TD-CDMA: Time division synchronous CDMA	*UWC136*: Universal wireless communications based on IS-136	*IEEE. 802.11a*: WLAN based on OFDM	
	GPRS: General packet radio service	Draft *HIPERMAN*: WLL based on OFDM	
	EDGE: Enhanced data rate for global evolution	Draft *IEEE 802.16a*: WLL based on OFDM	

A software-defined radio offers the following features:

— The radio can be used everywhere if all major wireless communication standards are supported. The corresponding standard-specific application software can be downloaded from the existing network itself.
— The software-defined radio can guarantee compatibility between several wireless networks. If UMTS is not supported in a given area, the terminal station can search for another network, e.g., GSM or IS-95.
— Depending on the hardware used, SDR is open to adopt new technologies and standards.

Therefore, SDR plays an important role for the success and penetration of 4G systems.

Software-Defined Radio

A set of examples of the current standards for cellular networks is given in Table 6-2. These standards, following their multi-access schemes, can be characterized as follows:

— Most of the 2G mobile communication systems are based on TDMA, while a CDMA component is adopted in 3G systems.
— In conjunction with TDMA many broadband WLAN and WLL standards support multi-carrier transmission (OFDM).

For standards beyond 3G we may expect that a combination of CDMA with a multi-carrier (OFDM) component is a potential candidate. Hence, a generic air interface based on multi-carrier CDMA using software-defined radio would support many existing and future standards (see Figure 6-28).

6.5.2 Basic Concept

A basic implementation concept of software-defined radio is illustrated in Figure 6-29. The digitization of the received signal can be performed directly on the radio frequency (RF) stage with a direct down-conversion or at some intermediate (IF) stage. In contrast to the conventional multi-hardware radio, channel selection filtering will be done in the

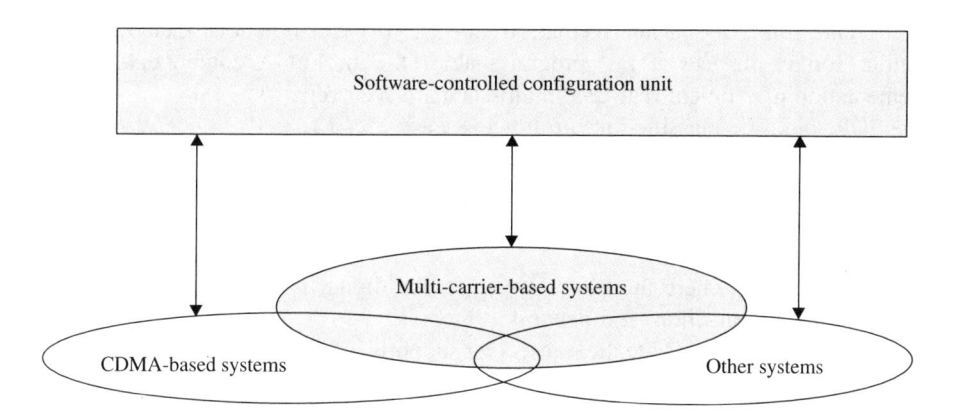

Figure 6-28 Software configured air interface

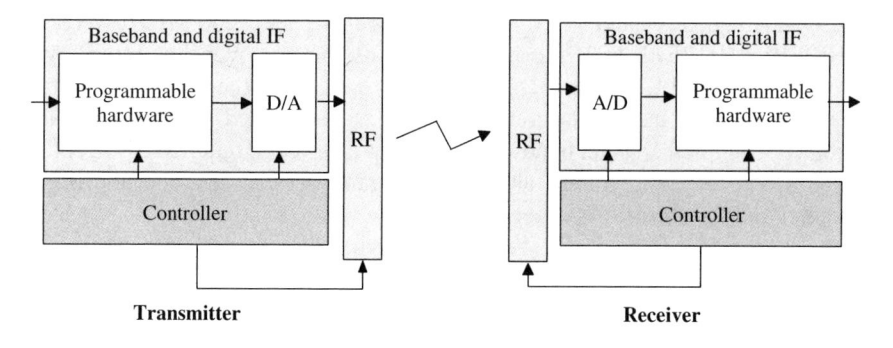

Figure 6-29 Basic concept of SDR implementation

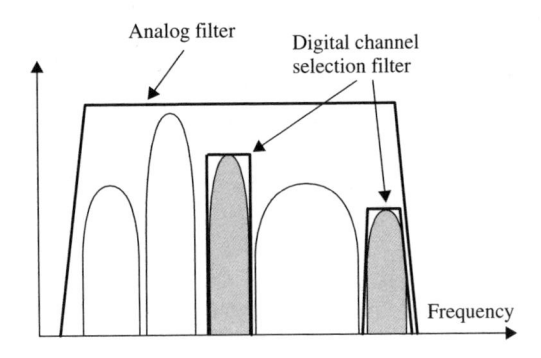

Figure 6-30 Channel selection filer in the digital domain

digital domain (see Figure 6-30). However, it should be noticed that if the A/D converter is placed too close to the antenna, it has to convert a lot of useless signals together with the desired signal. Consequently, the A/D converter would have to use a resolution that is far too high for its task, therefore leading to a high sampling rate that would increase the cost. Digital programmable hardware components such as digital signal processors (DSPs) or field programmable gate arrays (FPGAs) can, beside the baseband signal processing tasks, execute some digital intermediate frequency (IF) unit functions including channel selection. Today, the use of fast programmable DSP or FPGA components allow the implementation of efficient real-time multi-standard receivers.

The SDR might be classified into following categories [21]:

— *Multi-band radio*, where the RF head can be used for a wide frequency range, e.g., from VHF (30–300 MHz) to SHF (30 GHz) to cover all services (e.g., broadcast TV to microwave FWA).
— *Multi-role radio*, where the transceiver, i.e., the digital processor, supports different transmission, connection, and network protocols.
— *Multi-function radio*, where the transceiver supports different multimedia services such as voice, data, and video.

The first category may require quite a complex RF unit to handle all frequency bands. However, if one concentrates the main application, for instance, in mobile communications using the UHF frequency band (from 800 MHz/GSM/IS-95 to 2200 MHz/UMTS to even 5 GHz/HIPERLAN/2/IEEE 802.11a) it would be possible to cover this frequency region with a single wide band RF head [21]. Furthermore, regarding the transmission standards that use this frequency band, all parameters such as transmitted services, allocated frequency region, occupied channel bandwidth, signal power level, required SNR, coding and modulation are known. Knowledge about these parameters can ease the implementation of the second and the third SDR categories.

6.5.3 MC-CDMA-Based Software-Defined Radio

A detailed SDR transceiver concept based on MC-CDMA is illustrated in Figure 6-31. At the transmitter side, the higher layer, i.e., the protocol layer, will support several connections at the user interface (TS), e.g., voice, data, video. At the base station it can offer several network connections, e.g., IP, PSTN, ISDN. The data link controller (DLC)/medium access controller (MAC) layer according to the chosen standard takes care of the scheduling (sharing capacity among users) to guarantee the required quality of service (QoS). Furthermore, in adaptive coding, modulation, spreading, and power leveling the task of the DLC layer is the selection of appropriate parameters such as FEC code rate, modulation density and spreading codes/factor. The protocol data units/packets (PDUs) from the DLC layer are submitted to the baseband processing unit, consisting mainly of FEC encoder, mapper, spreader, and multi-carrier (i.e., OFDM) modulator. After digital I/Q generation (digital IF unit), the signal can be directly up-converted to the RF analog signal, or it may have an analog IF stage. Note that the digital I/Q generation has the advantage that only one converter is needed. In addition, this avoids problems of I and Q sampling mismatch. Finally, the transmitted analog signal is amplified, filtered, and tuned by the local oscillator to the radio frequency and submitted to the Tx antenna. An RF decoupler is used to separate the Tx and Rx signals.

Similarly, the receiver functions, being the inverse of the transmitter functions (but more complex), are performed. In case of an analog IF unit, it is shown in [21] that the filter dimensioning and sampling rate are crucial to support several standards. The sampling rate is related to the selected wideband analog signal, e.g., in case of direct down-conversion [19]. However, the A/D resolution depends on many parameters: i) the ratio between the narrowest and the largest selected channel bandwidths, ii) used modulation, iii) needed dynamic for different power levels, and iv) the receiver degradation tolerance.

As an example, the set of parameters that might be configured by the controller given in Figure 6-31 could be:

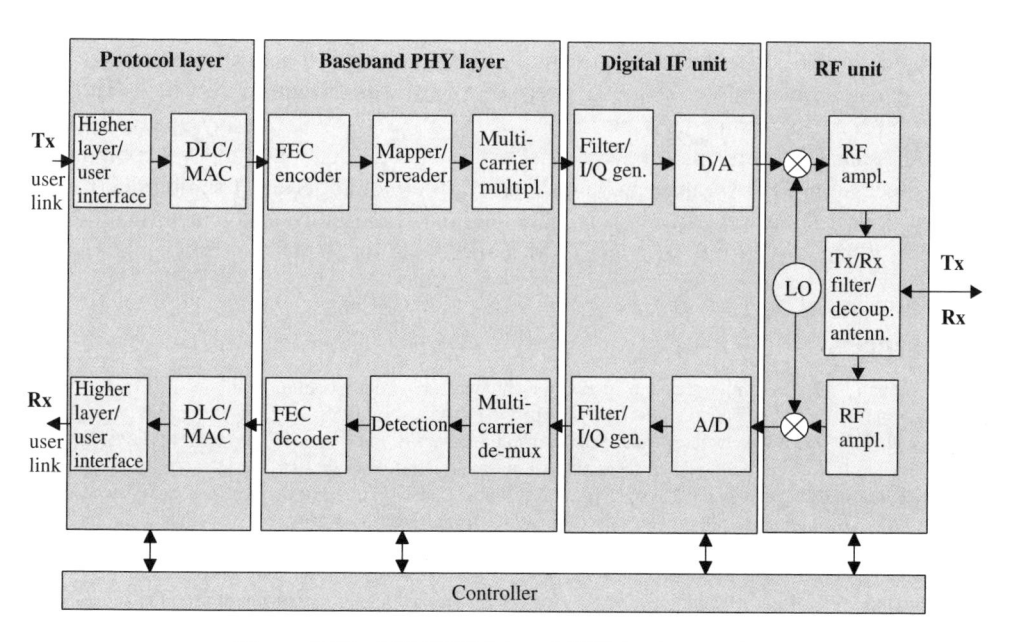

Figure 6-31 MC-CDMA-based SDR implementation

- higher layer connection parameters (e.g., port, services)
- DLC, MAC, multiple access parameters (QoS, framing, pilot/reference, burst formatting and radio link parameters)
- ARQ/FEC (CRC, convolutional, block, Turbo, STC, SFC)
- modulation (M-QAM, M-PSK, MSK) and constellation mapping (Gray, set partitioning, pragmatic approach)
- spreading codes (one- or two-dimensional spreading codes, spreading factors)
- multi-carrier transmission, i.e., OFDM (FFT size, guard time, guard band)
- A/D, sampling rate and resolution
- channel selection
- detection scheme (single- or multiuser detection)
- diversity configuration
- duplex scheme (FDD, TDD).

Hence, SDR offers elegant solutions to accommodate various modulation constellations, coding, and multi-access schemes. Besides its flexibility, it also has the potential of reducing the cost of introducing new technologies supporting sophisticated future signal processing functions.

However, the main limitations of the current technologies employed in SDR are:

- A/D and D/A conversion (dynamic and sampling rate),
- power consummation and power dissipation,
- speed of programmable components, and
- cost.

The future progress in A/D conversion will have an important impact on the further development of SDR architectures. A high A/D sampling rate and resolution, i.e., high signal dynamic, may allow to use a direct down-conversion with a very wideband RF stage [19], i.e., the sampling is performed at the RF stage without any analog IF unit, "zero IF" stage. The amount of power consumption and dissipation of today's components (e.g., processors, FPGAs) may prevent its use in the mobile terminal station due to low battery lifetimes. However, its use in base stations is currently under investigation, for instance, in the UMTS infrastructure (UMTS BS/Node-B).

References

[1] Alamouti S.M., "A simple transmit diversity technique for wireless communications," *IEEE Journal on Selected Areas in Communications*, vol. 16, pp. 1451–1458, Oct. 1998.

[2] Bauch G., *"Turbo-Entzerrung" und Sendeantennen-Diversity mit "Space–Time-Codes" im Mobilfunk*. Düsseldorf: Fortschritt-Berichte VDI, series 10, no. 660, 2000, PhD thesis.

[3] Bauch G. and Hagenauer J., "Multiple antenna systems: Capacity, transmit diversity and turbo processing," in *Proc. ITG Conference on Source and Channel Coding*, Berlin, Germany, pp. 387–398, Jan. 2002.

[4] Chuang J. and Sollenberger N., "Beyond 3G: Wideband wireless data access based on OFDM and dynamic packet assignment," *IEEE Communications Magazine*, vol. 38, pp. 78–87, July 2000.

[5] Cimini L., Daneshrad B. and Sollenberger N.R., "Clustered OFDM with transmitter diversity and coding," in *Proc. IEEE Global Telecommunications Conference (GLOBECOM'96)*, London, UK, pp. 703–707, Nov. 1996.

[6] Dammann A. and Kaiser S., "Standard conformable diversity techniques for OFDM and its application to the DVB-T system," in *Proc. Global Telecommunications Conference (GLOBECOM 2001)*, San Antonio, USA, pp. 3100–3105, Nov. 2001.
[7] Dammann A. and Kaiser S., "Transmit/receive antenna diversity techniques for OFDM systems," *European Transactions on Telecommunications (ETT)*, vol. 13, pp. 531–538, Sept./Oct. 2002.
[8] Damman A., Raulefs R. and Kaiser S., "Beamforming in combination with space-time diversity for broadband OFDM systems," in *Proc. IEEE International Conference on Communications (ICC 2002)*, New York, pp. 165–172, May 2002.
[9] ETSI HIPERMAN (Draft TS 102 177), "High performance metropolitan area network, Part 1: Physical layer," Sophia Antipolis, France, Feb. 2003.
[10] ETSI UMTS (TR-101 112 V 3.2.0), "Universal mobile telecommunications system (UMTS)," Sophia Antipolis, France, April 1998.
[11] Foschini G.J., "Layered space–time architecture for wireless communication in a fading environment when using multi-element antennas," *Bell Labs Technical Journal*, vol. 1, pp. 41–59, 1996.
[12] Kaiser S., "OFDM with code division multiplexing and transmit antenna diversity for mobile communications," in *Proc. IEEE International Symposium on Personal, Indoor and Mobile Radio Communications (PIMRC 2000)*, London, UK, pp. 804–808, Sept. 2000.
[13] Kaiser S., "Spatial transmit diversity techniques for broadband OFDM systems," in *Proc. IEEE Global Telecommunications Conference (GLOBECOM 2000)*, San Francisco, USA, pp. 1824–1828, Nov./Dec. 2000.
[14] Li Y., Chuang J.C., and Sollenberger N.R., "Transmit diversity for OFDM systems and its impact on high-rate data wireless networks," *IEEE Journal on Selected Areas in Communications*, vol. 17, pp. 1233–1243, July 1999.
[15] Lindner J. and Pietsch C., "The spatial dimension in the case of MC-CDMA," *European Transactions on Telecommunications (ETT)*, vol. 13, pp. 431–438, Sept./Oct. 2002.
[16] Seshadri N. and Winters J.H., "Two signaling schemes for improving the error performance of frequency division duplex transmission system using transmitter antenna diversity," *International Journal of Wireless Information Network*, vol. 1, pp. 49–59, 1994.
[17] Tarokh V., Jafarkhani H., and Calderbank A.R., "Space–time block codes from orthogonal designs," *IEEE Transactions on Information Theory*, vol. 45, pp. 1456–1467, June 1999.
[18] Tarokh V., Seshadri N. and Calderbank A.R., "Space–time codes for high data rate wireless communications," *IEEE Transactions on Information Theory*, vol. 44, pp. 744–765, March 1998.
[19] Tsurumi H. and Suzuki Y., "Broadband RF stage architecture for software-defined radio in handheld terminal applications," *IEEE Communications Magazine*, vol. 37, pp. 90–95, Feb. 1999.
[20] Wolniansky P.W., Foschini G.J., Gloden G.D. and Valenzuela R.A., "V-BLAST: An architecture for realizing very high data rates over the rich-scattering wireless channel," in *Proc. International Symposium on Advanced Radio Technologies*, Boulder, USA, Sept. 1998.
[21] Wiesler A. and Jondral F.K., "A software radio for second- and third-generation mobile systems," *IEEE Transactions on Vehicular Technology*, vol. 51, pp. 738–748, July 2002.

Definitions, Abbreviations, and Symbols

Definitions

Adjacent channel interference (ACI): interference emanating from the use of adjacent channels in a given coverage area, e.g., dense cellular system.

Asynchronous: users transmitting signals without time constraints.

Base station (BS): equipment consisting of a base station controller (BSC) and several base station transceivers (BST).

Burst: transmission event consisting of a symbol sequence (preamble and the data symbols).

Cell: geographical area controlled by a base station. A cell can be split into sectors.

Co-channel interference (CCI): interference emanating from the reuse of the same frequency band in a given coverage area, e.g., dense cellular system.

Detection: operation for signal detection in the receiver. In a multiuser environment *single-user* (SD) or *multiuser* (MD) detection can be used. Multiuser detection requires the knowledge of the signal characteristics of all active users.

  Single-user detection techniques for MC-CDMA: MRC, EGC, ZF, MMSE.

  Multiuser detection techniques for MC-CDMA: MLSE, MLSSE, IC, JD.

Doppler spread: changes in the phases of the arriving waves that lead to time-variant multipath propagation.

Downlink (DL): direction from the BS to the TS.

Downlink channel: channel transmitting data from the BS to the TS.

FEC block: block resulting from the channel encoding.

Frame: ensemble of data and pilot/reference symbols sent periodically in a given time interval, e.g., OFDM frame, MAC frame.

Frequency division duplex (FDD): the transmission of uplink (UL) and downlink (DL) signals performed at different carrier frequencies. The distance between the UL and DL carrier frequencies is called duplex distance.

Full-duplex: equipment (e.g., TS) which is capable of transmitting and receiving data at the same time.

Full load: simultaneous transmission of all users in a multiuser environment.

Guard time: cyclic extension of an OFDM symbol to limit the ISI.

Inter-channel interference (ICI): interference between neighboring sub-channels (e.g., OFDM sub-channels) in the frequency domain, e.g., due to Doppler effects.

Interference cancellation (IC): operation of estimating and subtracting interference in case of *multiuser* signal detection.

Inter-symbol interference (ISI): interference between neighboring symbols (e.g., OFDM symbols) in the time domain, e.g., due to multipath propagation.

Multipath propagation: consequence of reflections, scattering, and diffraction of the transmitted electromagnetic wave at natural and man-made objects.

Multiple access interferences (MAI): interference resulting from other users in a given multiple access scheme (e.g., with CDMA).

OFDM frame synchronization: generation of a signal indicating the start of an OFDM frame made up of several OFDM symbols. Closely linked to OFDM symbol synchronization.

OFDM symbol synchronization: FFT window positioning, i.e., the start time of the FFT operation.

Path loss: mean signal power attenuation between transmitter and receiver.

PHY mode: combination of a signal constellation (modulation alphabet) and FEC parameters.

Point to multi-point (PMP): a topological cellular configuration with a base station (BS) and several terminal stations (TSs). The transmission from the BS towards the TS is called downlink and the transmission from the TS towards the BS is called uplink.

Preamble: sequence of channel symbols with a given autocorrelation property assisting modem synchronization and channel estimation.

Puncturing: operation for increasing the code rate by not transmitting (i.e., by deleting) some coded bits.

Rake: bank of correlators, e.g., matched filters, to resolve and combine multipath propagation in a CDMA system.

Ranging: operation of periodic timing advance (or power) adjustment to guarantee the required radio link quality.

Sampling rate control: control of the sampling rate of the A/D converter.

Sector: geometrical area resulting from cell splitting by the use of a sector antenna.

Shadowing: obstruction of the transmitted waves by, e.g., hills, buildings, walls, and trees, which results in more or less strong attenuation of the signal strength, modeled by a log-normal distribution.

Shortening: operation for decreasing the length of a systematic block code that allows an adaptation to different information bit/byte sequence lengths.

Tail bits: zero bits inserted for trellis termination of a convolutional code in order to force the trellis to go to the zero state.

Time division duplex (TDD): the transmission of uplink (UL) and downlink (DL) signals is carried out in the same carrier frequency bandwidth. The UL and the DL signals are separated in the time domain.

Spectral efficiency: efficiency of a transmission scheme given by the maximum possible data rate (in bit/s) in a given bandwidth (in Hz). It is expressed in bit/s/Hz.

Area spectrum efficiency gives the spectral efficiency per geographical coverage area, e.g., cell or sector. It is expressed in bit/s/Hz/cell or sector.

Spreading: operation of enlarging/spreading the spectrum. Several spreading codes can be used for spectrum spreading.

Synchronous: users transmitting following a given time pattern.

Uplink (UL): direction from the TS to the BS.

Uplink channel: channel transmitting data from the TS to the BS.

Abbreviations

ACF	Autocorrelation Function
ACI	Adjacent Channel Interference
A/D	Analog/Digital (converter)
AGC	Automatic Gain Control
ARIB	Association of Radio Industries and Businesses (Japanese association)
ARQ	Automatic Repeat re-Quest
ASIC	Application-Specific Integrated Circuit
A-TDMA	Advanced TDMA (EU-RACE project)
ATM	Asynchronous Transfer Mode
AWGN	Additive White Gaussian Noise
BCH	Bose–Chaudhuri–Hocquenghem (FEC Code)
BER	Bit Error Rate
BLAST	Bell-Labs Layered Space–Time
BPSK	Binary Phase Shift Keying
BRAN	Broadband Radio Access Network
BS	Base Station (= Access Point, AP)
BSC	BS Controller

BST	BS Transceiver
BTC	Block Turbo Code
BU	Bad Urban (radio channel model)
CAZAC	Constant Amplitude Zero Autocorrelation
CC	Convolutional Code
CCF	Cross-Correlation Function
CCI	Co-Channel Interference
CDD	Cyclic Delay Diversity
CDM	Code Division Multiplexing
CDMA	Code Division Multiple Access
cdma2000	Code Division Multiple Access standard 2000 (American 3G standard)
CF	Crest Factor (square root of PAPR)
C/I	Carrier-to-Interference power ratio
C/N	Carrier-to-Noise power ratio
C/(N+I)	Carrier-to-Noise and -Interference power ratio
CODIT	Code Division Testbed (EU-RACE project)
COST	European Cooperation in the Field of Scientific and Technical Research
CPE	Common Phase Error
CRC	Cyclic Redundancy Check
CSI	Channel State Information
CTC	Convolutional Turbo Code
D/A	Digital/Analog (converter)
DAB	Digital Audio Broadcasting
D-AMPS	Digital-Advanced Mobile Phone Service
DC	Direct Current
DD	Delay Diversity
DECT	Digital Enhanced Cordless Telecommunications
DFT	Discrete Fourier Transform
DiL	Direct Link
DL	Downlink
DLC	Data Link Control
D-QPSK	Differential QPSK
DS	Direct Sequence (DS-CDMA)
DSP	Digital Signal Processor
DVB	Digital Video Broadcasting
DVB-RCT	DVB Return Channel Terrestrial
DVB-S	DVB standard for Satellite broadcasting
DVB-T	DVB standard for Terrestrial broadcasting
EDGE	Enhanced Data for Global Evolution
EGC	Equal Gain Combining
EIRP	Effective Isotopic Radiated Power
ETSI	European Telecommunication Standard Institute
EU	European Union
FDD	Frequency Division Duplex

FDM	Frequency Division Multiplexing
FDMA	Frequency Division Multiple Access
FEC	Forward Error Correction
FFH	Fast FH
FFT	Fast Fourier Transform
FH	Frequency Hopping (FH-CDMA)
FIR	Finite Impulse Response
FPGA	Field Programmable Gate Array
FRAMES	Future Radio Wideband Multiple Access System
FWA	Fixed Wireless Access
GMSK	Gaussian Minimum Shift Keying
GPP	Third Generation Partnership Project
GPRS	General Packet Radio Services
GSM	Global System for Mobile communications
H-FDD	Half-duplex Frequency Division Duplex
HIPERLAN	High Performance Local Area Network
HIPERMAN	High Performance Metropolitan Area Network
HL	HIPERLAN
HM	HIPERMAN
HPA	High Power Amplifier
HSDPA	High Speed Downlink Packet Access (UMTS)
HT	Hadamard Transform/Hilly Terrain (radio channel model)
IBO	Input Back Off
IC	Interference Cancellation
ICI	Inter-Channel Interference
IDFT	Inverse Discrete Fourier Transform
IEEE	Institute of Electrical and Electronics Engineers
IF	Intermediate Frequency
IFDMA	Interleaved FDMA
IFFT	Inverse Fast Fourier Transform
IHT	Inverse Hadamard Transform
IMT-2000	International Mobile Telecommunications 2000
IP	Internet Protocol
I/Q	In-phase/Quadrature
IR	Infrared
IS	Interim Standard (e.g., American Standard IS-95)
ISDN	Integrated Service Digital Network
ISI	Inter-Symbol Interference
ISM	Industrial Scientific and Medical (ISM-license free Band)
ISO	International Standards Organization
JD	Joint Detection
JTC	Joint Technical Committee
LAN	Local Area Network
LLF	Log Likelihood Function
LLR	Log Likelihood Ratio
LMDS	Local Multi-point Distribution System

LO	Local Oscillator
LOS	Line Of Sight
MA	Multiple Access
MAC	Medium Access Control
MAI	Multiple Access Interference
MAP	Maximum A Posteriori
MBS	Mobile Broadband System
MC	Multi-Carrier
MC-CDMA	Multi-Carrier CDMA
MC-DS-CDMA	Multi-Carrier DS-CDMA
MCM	Multi-Carrier Modulation
MC-SS	Multi-Carrier Spread Spectrum
MC-TDMA	Multi-Carrier TDMA (OFDM and TDMA)
MD	Multiuser Detection
MF	Match Filter
MIMO	Multiple Input Multiple Output
MISO	Multiple Input Single Output
ML	Maximum Likelihood
MLD	ML Decoder (or Detector)
MLSE	ML Sequence Estimator
MLSSE	ML Symbol-By-Symbol Estimator
MMAC	Multimedia Mobile Access Communication
MMDS	Microwave Multi-point Distribution System
MMSE	Minimum Mean Square Error
MPEG	Moving Picture Expert Group
M-PSK	Phase Shift Keying constellation with M points, e.g., 16-PSK
M-QAM	QAM constellation with M points, e.g. 16-QAM
MRC	Maximum Ratio Combining
MSK	Minimum Shift Keying
MT-CDMA	Multi-Tone CDMA
NLOS	Non Line Of Sight
Node-B	UMTS Base Station
OBO	Output Back Off
OFCDM	Orthogonal Frequency and Code Division Multiplexing
OFDM	Orthogonal Frequency Division Multiplexing
OFDMA	Orthogonal Frequency Division Multiple Access
OSI	Open System Interconnect
PAPR	Peak-to-Average Power Ratio
PD	Phase Diversity
PDC	Personal Digital Cellular (Japanese mobile standard)
PDU	Protocol Data Unit
PER	Packet Error Rate
PHY	PHYsical (layer)
PIC	Parallel Interference Cancellation
PLL	Phase Lock Loop

PMP	Point to Multi-Point
PN	Pseudo Noise
POTS	Plain Old Telephone Services
PPM	Pulse Position Modulation
PRBS	Pseudo Random Binary Sequence
P/S	Parallel-to-Serial (converter)
PSD	Power Spectral Density
PSK	Phase Shift Keying
PSTN	Public Switched Telephone Network
QAM	Quadrature Amplitude Modulation
QEF	Quasi Error Free
QoS	Quality of Service
QPSK	Quaternary Phase Shift Keying
RA	Rural Area (radio channel model)
RACE	Research in Advanced Communications in Europe (EU research projects)
RF	Radio Frequency
RMS	Root Mean Square
RS	Reed–Solomon (FEC code)
Rx	Receiver
SCD	Sub-Carrier Diversity
SD	Single user Detection
SDR	Software-Defined Radio
SF	Spreading Factor
SFBC	Space–Frequency Block Code
SFC	Space–Frequency Coding
SFH	Slow Frequency Hopping
SHF	Super High Frequency
SI	Self Interference
SIC	Soft Interference Cancellation
SIMO	Single Input Multiple Output
SIR	Signal-to-Interference power Ratio
SISO	Soft-In/Soft-Out
SNI	Service Node Interface
SNR	Signal-to-Noise Ratio
SOHO	Small Office/Home Office
SOVA	Soft Output Viterbi Algorithm
S/P	Serial-to-Parallel (converter)
SS	Spread Spectrum
SS-MC-MA	Spread Spectrum Multi-Carrier Multiple Access
SSPA	Solid State Power Amplifier
STC	Space–Time Coding
STBC	Space–Time Block Code
STTC	Space–Time Trellis Code
TC	Turbo Code
TD	Total Degradation

TDD	Time Division Duplex
TDM	Time Division Multiplexing
TDMA	Time Division Multiple Access
TF	Transmission Frame
TIA	Telecommunication Industry Association (American association)
TPC	Turbo Product Code
TPD	Time-variant Phase Diversity
TS	Terminal Station
TU	Typical Urban (radio channel model)
TWTA	Travelling Wave Tube Amplifier
Tx	Transmitter
UHF	Ultra High Frequency
UL	Uplink
UMTS	Universal Mobile Telecommunication System
UNI	User Network Interface
UTRA	UMTS Terrestrial Radio Access
UWB	Ultra Wide Band
VCO	Voltage Controlled Oscillator
VHF	Very High Frequency
VSF	Variable Spreading Factor
WARC	World Administration Radio Conference
W-CDMA	Wideband CDMA
WH	Walsh–Hadamard
WLAN	Wireless Local Area Network
WLL	Wireless Local Loop
WMAN	Wireless Metropolitan Area Network
xDSL	Digital Subscriber Line (e.g., x: A = Asymmetric)
ZF	Zero Forcing (equalization)

Symbols

$a^{(k)}$	source bit of user k
$\mathbf{a}^{(k)}$	source bit vector of user k
a_p	amplitude of path p
$b^{(k)}$	code bit of user k
$\mathbf{b}^{(k)}$	code bit vector of user k
B	bandwidth
B_s	signal bandwidth
c	speed of light
$c_l^{(k)}$	chip l of the spreading code vector $\mathbf{c}^{(k)}$
$\mathbf{c}^{(k)}$	spreading code vector of user k
C	capacity
$\mathbf{C}$	spreading code matrix
$d^{(k)}$	data symbol of user k
$\mathbf{d}^{(k)}$	data symbol vector of user k

D_O	overall diversity
D_f	frequency diversity
D_t	time diversity
dB	decibel
dBm	decibel relative to 1 mW
$E\{.\}$	expectation
E_b	energy per bit
E_c	energy per chip
E_s	energy per symbol
f	frequency
f_c	carrier frequency
f_D	Doppler frequency
$f_{D,filter}$	maximum Doppler frequency permitted in the filter design
f_{Dmax}	maximum Doppler frequency
$f_{D,p}$	Doppler frequency of path p
f_n	nth sub-carrier frequency
F	noise figure in dB
F_s	sub-carrier spacing
$G_{Antenna}$	antenna gain
$G_{l,l}$	lth diagonal element of the equalizer matrix $\mathbf{G}$
$\mathbf{G}$	equalizer matrix
$\mathbf{G}^{[j]}$	equalizer matrix used for IC in the jth iteration
$h(t)$	impulse response of the receive filter or channel impulse response
$h(\tau,t)$	time-variant channel impulse response
$H(f,t)$	time-variant channel transfer function
$H_{l,l}$	lth diagonal element of the channel matrix $\mathbf{H}$
$H_{n,i}$	discrete-time/frequency time-variant channel transfer function
$\mathbf{H}$	channel matrix
I_c	size of the bit interleaver
I_{TC}	size of the Turbo code interleaver
j	$\sqrt{-1}$
J_{it}	number of iterations in the multistage detector
K	number of active users or the number of information symbols of an RS code
K_{Rice}	Rice factor
L	spreading code length or number of R_x antennas
L_a	length of the source bit vector $\mathbf{a}^{(k)}$
L_b	length of the code bit vector $\mathbf{b}^{(k)}$
L_d	length of the data symbol vector $\mathbf{d}^{(k)}$
M	number of data symbols transmitted per user and OFDM symbol or number of Tx antennas
m	number of bits transmitted per modulated symbol
$n(t)$	additive noise signal
$\mathbf{n}$	noise vector
N_c	number of sub-carriers
N_f	pilot symbol distance in frequency direction

N_{grid}	number of pilot symbols per OFDM frame
N_{ISI}	number of interfering symbols
N_l	lth element of the noise vector $\mathbf{n}$
N_p	number of path
N_s	number of OFDM symbols per OFDM frame
N_t	pilot symbol distance in time direction
N_{tap}	number of filter taps
$p(.)$	probability density function
$P\{.\}$	probability
P_b	BER
P_G	processing gain
Q	number of user groups
$\mathbf{r}$	received vector after inverse OFDM
$\mathbf{r}^{(k)}$	received vector of the k-th user after inverse OFDM
R	code rate
R_b	bit rate
rect(x)	rectangle function
R_l	lth element of the received vector $\mathbf{r}$
R_s	symbol rate
$\mathbf{s}$	symbol vector before OFDM
$\mathbf{s}^{(k)}$	symbol vector of user k before OFDM
S_l	lth element of the vector $\mathbf{s}$
sinc(x)	$\sin(x)/x$ function
t	time or number of error correction capability of a RS code
T	symbol duration
T_c	chip duration
T_d	data symbol duration
T_{fr}	OFDM frame duration
T_g	duration of guard interval
T_s	OFDM symbol duration without guard interval
T_s'	OFDM symbol duration with guard interval
T_{samp}	sampling period
$\mathbf{u}$	data symbol vector at the output of the equalizer
U_l	lth element of the equalized vector $\mathbf{u}$
v	velocity
$v^{(k)}$	soft decided value of the data symbol $d^{(k)}$
$\mathbf{v}^{(k)}$	soft decided value of the data symbol vector $\mathbf{d}^{(k)}$
V_{pilot}	loss in SNR due to the pilot symbols
$w^{(k)}$	soft decided value of the code bit $b^{(k)}$
$\mathbf{w}^{(k)}$	soft decided value of the code bit vector $\mathbf{b}^{(k)}$
$x(t)$	transmitted signal
$X(f)$	frequency spectrum of the transmitted signal $x(t)$
$y(t)$	received signal
α	roll-off factor
β	primitive element of the Galois field
$\Delta^2(.,.)$	squared Euclidean distance

Symbols

σ^2	variance of the noise
δ	delay
ϕ	autocorrelation function
φ_p	phase offset of path p
θ	cross-correlation function
τ	delay
τ_p	delay of path p
Γ	log-likelihood ratio
Ω	average power
$(.)^H$	Hermitian transposition of a vector or a matrix
$(.)^T$	transposition of a vector or a matrix
$(.)^{-1}$	inversion
$(.)^*$	complex conjugation
$\lvert.\rvert$	absolute value
$\lVert.\rVert$	norm of a vector
$\otimes$	convolution operation
$\lceil x \rceil$	smallest integer larger than or equal to x
∞	infinity

Index

Adaptive techniques   170
  Adaptive channel coding and
    modulation   171
  Adaptive power control   172
  Nulling of weak sub-carriers
    171
Advantages and drawbacks of OFDM
  30
Advantages and drawbacks of
  MC-CDMA   43–44
Alamouti space-time block code
  (STBC)   238
AM/AM (AM/PM) conversion (HPA)
  179–180
Analog-to-digital (A/D) conversion
  120–121
Antenna diversity   234
Antenna gain   188
Automatic gain control (AGC)
  139

Bad urban (BA) channel model   20
Beyond 3G   6, 198
BLAST architecture   235–236
  Diagonal (D-BLAST)   235
  Vertical (V-BLAST)   235
Blind and semi-blind channel
  estimation   153
Block codes   160–162, 164–166
Block linear equalizer   61
Blotzmann constant   188
Bluetooth   4

cdma2000   2–3
Cellular systems beyond   3G   198
Centralized mode   208
Channel coding and decoding   158
Channel estimation   139–158
  Adaptive design   144
  Autocorrelation function   142
  Boosted pilot symbols   146
  Cross-correlation function   142
  Downlink   154
  MC-SS systems   154
  One-dimensional   143
  Overhead due to pilot symbols
    143, 145
  Performance analysis   147
  Pilot distance   145
  Robust design   144
  Two-dimensional   140
  Uplink   154
Channel fade statistics   18–19
Channel impulse response   16–17
Channel matrix   29, 51
Channel modeling   16–18
Channel selection in digital domain
  258
Channel state information (CSI)   159,
  169, 186
Channel transfer function   16–17, 21
Chips   34, 49
Clock error   128
Code division multiple access (CDMA)
  5, 33–34, 94

Code division multiplexing (CDM) 94, 100
Coding (FEC) for packet transmission 161
CODIT 40
Coherence bandwidth 22
Coherence time 23
Common phase error (CPE) correction 176
Complexity of FFT 120
Concatenated coding 159–162
Constellation 167–169
Convolutional coding (FEC) 159–164
COST channel models 20–21
Crest factor 54
Cyclic delay diversity (CDD) 243
Cyclic extension 27

Decision directed channel estimation 152
Delay diversity 240, 245
Delay power density spectrum 17
Delay spread 17–18
Demapping 169
Detection techniques 57–64
    Multiuser detection 60–64
    Single-user detection 58–60
Digital AMPS (D-AMPS) 3
Digital audio broadcasting (DAB) 7, 31, 196
Digital signal processor (DSP) 258
Digital-to-analog (D/A) conversion 120–121
Digital video broadcasting (DVB-T) 7, 31, 197, 222
Direct current (DC) sub-carrier 120
Direct mode 208
Direct sequences CDMA (DS-CDMA) 33–37
    Receiver 36
    Transmitter 36–37
Discrete Fourier transform (DFT) 26–28, 119
Diversity 22–24, 233
    Diversity in multi-carrier transmission 240–253

Frequency diversity 22
Receive diversity 244–245
Time diversity 23
Transmit diversity 240
Doppler frequency 17–18
Doppler power density spectrum 18
Doppler spread 16, 18
DVB return channel (DVB-RCT) 221
    Channel characteristics 223
    FEC coding and modulation 227
    Frame structure 224
    Link budget 229
    Network topology 221

Equal gain combining (EGC) 59, 66
Equalization 58–60, 169–170
Euclidean distance 61

Fast Fourier transform (FFT) 26–27, 119–120
    Complexity 120
FDD frame structure 214
Field programmable gate array (FPGA) 258
Filter design 144–145
    Adaptive 144
    Non adaptive 144–145
Fixed wireless access (FWA) 44, 210, 254
    Channel characteristics 212
    Network topology 211
Flexibility with MC-CDMA 72
Forward error correction (FEC) 158–167, 208–209, 216–220, 227–29
Fourier codes 52
Fourier transform 26–27, 119–120
Fourth generation (4G) 198, 233
Frequency diversity 22, 233
Frequency division duplex (FDD) 93
Frequency division multiple access (FDMA) 4, 93
Frequency division multiplexing (FDM) 93
Frequency error 127

Frequency hopping CDMA (FH-CDMA) 33–34
Frequency hopping OFDMA 99
Frequency synchronization 136–139
   Coarse frequency synchronization 136–138
      CAZAC/M sequence 137
      Schmidl and Cox, 137–138
   Fine frequency synchronization 138–139

Galois field 160
General packet radio service (GPRS) 2
Generator polynomial 159–160, 165
Global system for mobile communications (GSM) 2–3
Golay code 53
Gold code 53
Guard time 27, 50, 119

Hadamard matrix 52
Hamming codes 164–165
High power amplifier (HPA) models 179
   Traveling wave tube amplifier (TWTA) 179
   Solid state power amplifier (SSPA) 180
High speed downlink packet access (HSDPA) UMTS 198
Hilly terrain (HT) channel model 20
HIPERLAN/2 4, 7, 31, 197, 205–209
   Channel model/characteristics 21, 206
   FEC coding and modulation 208–209
   Link budget 209–210
   MAC frame structure 206–207
HIPERMAN 32, 197, 210–220
   FEC coding and modulation 216–220
   Link budget 221
   MAC frame structure 213–216
Hybrid multiple access schemes 93
   Performance comparison 110

IDFT 26, 119
IEEE 802.11a 4, 7, 31, 197, 205–209
   Channel characteristics 206
   FEC coding and modulation 208–209
   Link budget 209–210
   MAC frame structure 206–207
IEEE 802.11b 4
IEEE 802.16a 32, 197, 210–220
   FEC coding and modulation 216–220
   Link budget 221
   MAC frame structure 213–216
IFFT 26, 119
Implementation issues 115
IMT-2000 3, 40
Indoor/outdoor channel models 20–21
Inter-channel interference (ICI) 19
Interference cancellation (IC) 57, 62–64
Interference estimation 186–187
Interleaved FDMA (IFDMA) 104–105
Inter-symbol interference (ISI) 19
Inverse OFDM (IOFDM) 28, 119
I/Q generation 120–123
   Analog quadrature method 121–122
   Digital FIR filtering method 122–123
IS-95 standard 2, 3, 37–39

Joint detection 57
JTC channel model 21

Line of sight (LOS) 16, 18
Link budget 188–189, 209–210, 229–230
Local multipoint distribution system (LMDS) 44
Local oscillator 122–124
Log likelihood ratio (LLR) 64, 69–72, 129
   MC-CDMA 70–72
      Interference cancellation 72

Log likelihood ratio (LLR) (*continued*)
  Maximum likelihood detection (MLD)   71
  Single-user detection   70–71
  OFDM systems   69–70
Low rate convolutional codes   53–54

Mapping   167–169
Maximum a posteriori (MAP)   60
Maximum likelihood parameter estimation   129–132
Maximum likelihood sequence estimation (MLSE)   60–61
Maximum likelihood symbol-by-symbol estimation (MLSSE)   61
Maximum ratio combining (MRC)   59, 66, 244–245
MC-CDMA   8–10, 41–44, 49–83
  Downlink signal   50–51
  MC-CDMA software defined radio   258–260
  Performance   74–84
  Spreading   51–54
  Uplink signal   51
MC-DS-CDMA   8–10, 41–44, 83–90
  Downlink signal   86
  Performance   87–90
  Spreading   86
  Uplink signal   86
Mean delay   17
Medium access control (MAC)   93, 99–100, 106–107
Microwave multipoint distribution system (MMDS)   44
Minimum mean square error (MMSE) equalization   59
  MMSE Block linear equalizer   61
  MMSE Pre-equalization   67
M-Modification   72–73
Monocycle   108
Moose maximum likelihood frequency estimation   131–132
M-QAM constellation   167–169
$M\&Q$-Modification   74
Multiband radio   258

Multi-carrier CDMA (MC-CDMA)   8–10, 41–44, 49–83
Multi-carrier channel modeling   21–22
Multi-carrier FDMA (MC-FDMA)   94–105
Multi-carrier modulation and demodulation   116–123
Multi-carrier spread spectrum (MC-SS)   8–10, 41–44
Multi-carrier TDMA (MC-TDMA)   105–107
Multi-carrier transmission   24–30
Multi-function radio   258
Multimedia services   195–196
Multipath propagation   15
Multiple access interference (MAI)   35, 37, 70
Multiple input multiple output (MIMO)   234
  MIMO capacity   235
Multiple input single output (MISO)   234
Multi-role radio   258
Multitone CDMA (MT-CDMA)   85
Multiuser detection   57, 60–64

Narrowband interference rejection   185–188
Noise factor   188
Noise power   188
Noise variance   28
Nonlinearities   177–185
  Effects   179–182
    Influence in DS-CDMA   180–181
    Influence in MC-CDMA   181–182
  High power amplifier models   179–180
  Pre-distortion techniques   182–183
NTT-DoCoMo concept for 4G   199–203
  System parameters   200, 204
Null symbol   126
Nyquist pulse shaping   99

Index

OFDMA with code division multiplexing (SS-MC-MA) 100–104
OFDM-CDM 166–167
OFDM-CDMA 41
One-dimensional channel estimation 143
One-dimensional spreading code 55
Orthogonal frequency division multiplexing (OFDM) 25–30, 119–120
  frame 29, 126
  frame duration 28
  spectrum 26
  standards 31–32
  symbol 25, 29
  symbol duration 25
Orthogonal frequency division multiple access (OFDMA) 95–100
  CDM 100–104
  Frequency hopping OFDMA 99
  Pulse shaping 98–99
  Synchronization sensitivity 97–98
  Transceiver 99–100
Output back-off (OBO) 179

Parallel interference cancellation (PIC) 62–63
Path loss 16, 188
Peak-to-average power ratio (PAPR) 54–55, 178
  Downlink PAPR 54–55
  Uplink PAPR 54
Phase diversity 241–243, 245–247
Phase lock loop (PLL) 174
Phase noise 173–177
  Effects 175–177
  Modeling 173–175
    Lorenzian power density spectrum 173
    Measured power density spectrum 174–175
Pilot symbol grid 144
Pre-distortion techniques 182–183
Pre-equalization 65–67
  Downlink 65–67

Power constraint 66
  Uplink 67
Processing gain 33, 49
Product codes 164–165
Pseudo noise (PN) sequence 53
Pulse position modulation (PPM) 107–109
Pulse shaping 98–99, 119, 224
Punctured convolutional codes 159
  Puncturing tables 159
  SNR performance 161

QAM constellation 167–169
Q-Modification 73–74

Radio channel characteristics 15–24
Raised cosine filtering 99, 118–119
Rake receiver 19, 36–37, 39–41
Rayleigh distribution 18
Receive diversity in multi-carrier transmission 244–245
  Delay and phase diversity 245
  Maximum ratio combining 244–245
Receiver sensitivity 188
Rectangular band-limited transmission filter 118
Rectangular time-limited transmission filter 118
Reed Solomon coding 159–162
Requirements on OFDM frequency and clock accuracy 129
RF issues 172–189
Rice distribution 18
RMS delay spread 17–18
Rotated constellations 56–57
Rural area (RA) channel model 20

Sampling clock adjustment 135–136
Sampling rate 121
Services 196
Shadowing 16
Signal constellation 167–169
Single input multiple output (SIMO) 234

Single-user detection 57–60
Soft bit 64
Soft channel decoding 67–72
Soft-in soft-out (SISO) 162
Soft interference cancellation 63–64
Soft-output Viterbi decoding algorithm
 (SOVA) 163, 166
Software-defined radio (SDR)
 255–260
Space division multiple access (SDMA)
 94, 234
Space division multiplexing 94
Space-frequency coding (SFC)
 248–250
 SF block codes (SFBC) 249–250
Space-time coding (STC) 236–240
 FWA 254–255
 ST block codes (STBC) 238–240
 ST trellis codes (STTC) 237–238
 UMTS 253
Spatial diversity 233
Spreading codes 52–54
 PAPR 54, 184–185
Spreading length 34, 49, 55, 201
Spread spectrum 30–45
Spread spectrum multi-carrier multiple
 access (SS-MC-MA) 100–104
Squared Euclidean distance 61
SSPA characteristics 180
Sub-carrier 24–26
 Spacing 25
Sub-carrier diversity 243–244
Sub-channel 24
Suboptimum MMSE equalization 60
Successive interference cancellation
 (SIC) 63
Synchronization 123–139
 Frequency synchronization
  136–139
 Maximum likelihood parameters
  estimation 129–132
 Time synchronization 132–136
Synchronization sensitivity 97–98,
 126–129
System matrix 51

System performance 74–83, 87–90
 MC-CDMA synchronous downlink
  76–80
 MC-CDMA synchronous uplink
  80–83
 MC-DS-CDMA asynchronous uplink
  88–90
 MC-DS-CDMA synchronous uplink
  88

TDD frame structure 207
Time diversity 23, 233
Time division duplex (TDD) 65, 93
Time division multiple access (TDMA)
 4, 93
Time division multiplexing (TDM)
 5, 93
Time domain channel estimation
 151–152
Time synchronization 132–136
 Coarse symbol timing 132–134
  Guard time exploitation 134
  Null symbol detection 133
  Two identical half reference
   symbols 133–134
 Fine symbol timing 134–135
Time-variant phase diversity 243
Transmit diversity in multi-carrier
 transmission 240–244
 Delay diversity 240–241
 Phase diversity 241–243
 Sub-carrier diversity 243–244
 Time-variant phase diversity 243
Total degradation 183
Turbo codes 162–166
 Block Turbo codes 164–166
 Convolutional Turbo codes
  162–164
Two-dimensional channel estimation
 140–143
 Two cascaded one-dimensional
  filters 142–143
 Two-dimensional filter 140–141
 Two-dimensional Wiener filter 141
Two-dimensional spreading 55–56,
 200–201

Index

TWTA characteristics 179–180
Typical urban (TU) channel model 20

Ultra wide band (UWB) systems
   107–110
  PPM UWB signal generation
    107–109
  Transmission scheme 109–110
UMTS standard 2–3, 40–41, 197,
   253
  UMTS with STC 253
UMTS/UTRA channel model 21
Uncorrelated fading channel models
   22

Variable spreading factor (VSF) 42,
   200–201
VHF/UHF 221, 223

Virtual sub-carriers 120
Viterbi decoding 64, 69, 159, 163,
   237
VSF-OFCDM 199–201

Walsh-Hadamard code 52
Wideband CDMA (W-CDMA) 3,
   40–41
Wireless local area networks (WLAN)
   7, 31, 203–210
  Channel characteristics 206
  Network topology 205
Wireless local loop (WLL) 7, 32
Wireless standards 256

Zadoff-Chu code 53
Zero forcing (ZF) 59, 67
ZF Block linear equalizer 61